Studies in Logic
Volume 4

Algebra, Logic, Set Theory
Festschrift für Ulrich Felgner
zum 65. Geburtstag

Volume 1
Proof Theoretical Coherence
Kosta Dosen and Zoran Petric

Volume 2
Model Based Reasoning in Science and Engineering
Lorenzo Magnani, editor

Volume 3
Foundations of the Formal Sciences IV: The History of the Concept of the Formal Sciences
Benedikt Löwe, Volker Peckhaus and Thoralf Räsch, editors

Volume 4
Algebra, Logic, Set Theory. Festschrift für Ulrich Felgner zum 65. Geburtstag
Benedikt Löwe, editor

Studies in Logic Series Editor
Dov Gabbay   dov.gabbay@kcl.ac.uk

Algebra, Logic, Set Theory

Festschrift für Ulrich Felgner zum 65. Geburtstag

edited by

Benedikt Löwe

© Individual author and King's College 2007. All rights reserved.

ISBN 1-904987-28-1
College Publications
Scientific Director: Dov Gabbay
Managing Director: Jane Spurr
Department of Computer Science
Strand, London WC2R 2LS, UK

Original cover design by Richard Fraser
Cover produced by orchid creative www.orchidcreative.co.uk
Printed by Lightning Source, Milton Keynes, UK

All rights reserved. No part of this publication may be reproduced, stored in a retrieval system or transmitted, in any form, or by any means, electronic, mechanical, photocopying, recording or otherwise, without prior permission, in writing, from the publisher.

CONTENTS

Preface ... vii

F. HAUG
Das wissenschaftliche Werk von Ulrich Felgner ... 1

A. BAUDISCH, M. ZIEGLER, A. MARTIN-PIZARRO
Hrushovski's Fusion ... 15

O. BELEGRADEK
The Theory of Square-like Abelian Groups is Decidable ... 33

A. BOROVIK, J. BURDGES, G. CHERLIN
Simple Groups of Finite Morley Rank of Unipotent Type ... 47

J. BRENDLE
Independence for Distributivity Numbers ... 63

R. ECKARDT
The Lower Part of Event Ontology ... 85

A. ECKER
Existentially Closed Locally Finite CA-groups ... 103

C. GRÜNENWALD, F. HAUG
On Stable Groups in Some Soluble Group Classes ... 141

B. MÜLLER-CLOSTERMANN
Formal Methods in Concurrent Systems Engineering:
Survey and Examples ... 157

A. PILLAY
On Externally Definable Sets and a Theorem of Shelah ... 175

K. U. SCHULZ
A Note on the Categoricity of Countable Interval Structures ... 183

J. K. TRUSS
Countable Homogeneous and Partially Homogeneous Ordered
Structures   193

R. VILLEMAIRE, M. HÉBERT
Theories of Abelian Groups and Modules Preserved under
Extensions   239

A. C. WALCZAK-TYPKE
A Model Theoretic Approach to Set Theory without the Axiom
of Choice   255

A. WEBER
Quantifier Elimination on Real Closed Fields and Differential
Equations   291

Preface

This volume is the *Festschrift* in honour of Ulrich Felgner's 65th birthday. It is a product of the work of his students, friends and colleagues, but it is not just a celebration of Felgner's work, but rather a contribution to research in itself. Almost all papers in this volume are original research papers, and all have been thoroughly refereed to high standards.

This book contains fifteen papers, starting with a *laudatio* by Frieder Haug that lists Ulrich Felgner's achievements, his students and his publications. The reader studying the list of the former students will see that nine of them have contributed to this volume: Jörg Brendle, Roger Villemaire, and Andreas Weber have produced new research papers for this volume; Andreas Ecker, Claus Grünenwald, Frieder Haug, and Klaus Schulz have used the opportunity to publish unpublished material; Regine Eckardt provided a paper that was written for a collection on event ontologies; and Bruno Müller-Clostermann wrote a survey paper on formal methods in applied computer science.

In addition to Ulrich Felgner's students, also his collaborators wish to honour him in this volume. The reader will find four original research papers by researchers close to Ulrich Felgner (the papers by Baudisch, Ziegler, and Martin-Pizarro, by Belegradek, by Borovik, Burdges, and Cherlin, and by Pillay). One of Felgner's coauthors, John Truss, offered the written version of a survey lecture series on countable homogeneous ordered structures that he gave in Rome. Truss's own student, Agatha Walczak-Typke, has written her PhD thesis about model theory without the axiom of choice; since Ulrich Felgner has been active in both of these areas, it was natural to ask her to write a survey of this area for the *Festschrift*.

With such a strong collection of interesting papers, covering Algebra, Logic, and Set Theory, this volume that will be useful for researchers in all areas covered by Ulrich Felgner's work for many years to come.

The high scientific standard of this volume could only be maintained with the help of numerous referees who wrote critical and insightful reports and kept to the production schedule. Frieder Haug and Torsten Schatz supported this book project from the moment the idea was first mentioned and provided the contact with Tübingen. Originally, this volume was supposed to be edited by the three of us; and only in the last minute, after galley proofs had been sent out, they modestly decided not to be officially listed as editors. All attempts to convince them otherwise were unsuccessful, but it should be known that their support and encouragement were crucial for the production of this volume.

Special thanks are due to Sara and Joel Uckelman for help with typesetting, Stefan Bold for proofreading, and the *Universitätsclub Tübingen* for financial support. We would like to thank *Walter de Gruyter Verlag* (in particular Heiko Hartmann) and Johannes Dölling, Tatjana Heyde-Zybatow, and Martin Schäfer for the permission to republish Eckardt's paper, and Jane Spurr of *College Publications* for her assistance with the production of the volume.

Amsterdam,
December 2006   B. L.

Das wissenschaftliche Werk von Ulrich Felgner

FRIEDER HAUG

Mathematisches Institut
Eberhard-Karls-Universität Tübingen
Auf der Morgenstelle 10
72076 Tübingen, Germany
E-mail: frieder.haug@uni-tuebingen.de

Dieser Band von Arbeiten aus der Algebra, Mengenlehre und Logik entstand aus Anlass des 65. Geburtstags von Prof. Dr. Ulrich Felgner. Es soll daher hier in aller Kürze sein wissenschaftliches Werk dargestellt und gewürdigt werden.

Das mathematische Werk von Ulrich Felgner ist kein Werk aus einem Guss, das man in einfacher Weise einem Teilgebiet der Mathematik zuordnen könnte. Mit Hilfe des Diagramms auf Seite 2 soll versucht werden, einen Überblick über einige der wichtigsten mathematischen Arbeiten von Ulrich Felgner zu geben.

Im Kernbereich befindet sich das Dreieck mit den "Säulen" **Algebra**, **Mengenlehre** und **Logik** oder –genauer gesagt– **Gruppentheorie**, **Modelltheorie** und **Infinitäre Kombinatorik**, denn dort liegt der Schwerpunkt der Arbeiten.

Zu diesen drei Gebieten selber hat Ulrich Felgner wertvolle Beiträge geleistet. Einige dieser Beiträge sind im Diagramm direkt neben den Säulen aufgeführt.

Für die **Algebra** sind Arbeiten zur Zahlentheorie (hier handelt es sich um Diophantische Gleichungen oder Primzahlabschätzungen) oder über Zentrumsfaktorgruppen (gemeinsam mit Peter Schmid) aufgeführt. In der **Modelltheorie** gibt es unter anderem die wichtigen Beiträge zur Kategorizität. Für die **infinitäre Kombinatorik** seien hier Arbeiten über Aronszajn-Bäume oder zur Kardinalzahlarithmetik erwähnt. Wie man schon an der Größe der Kreise im Diagramm erkennen kann, liegt der Schwerpunkt des Werks von Ulrich Felgner aber auf den Verbindungen dieser Gebiete.

Zwischen **Logik** und **Mengenlehre** sind einige der älteren Arbeiten, die um 1970 entstanden sind, anzusiedeln. Damals war die Cohensche

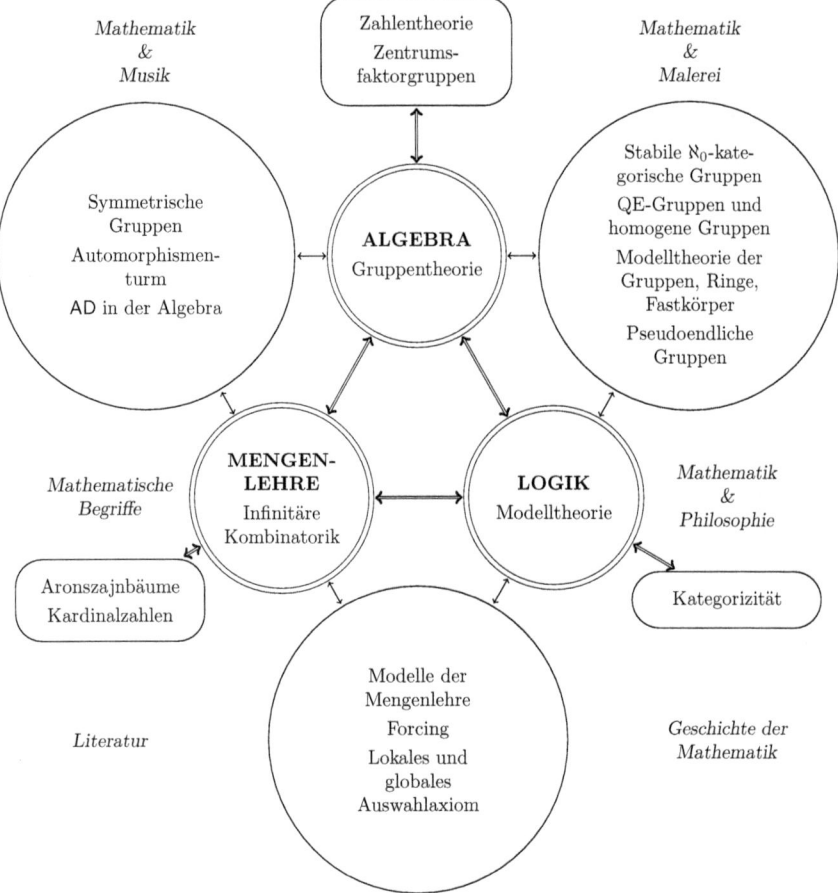

Erzwingungsmethode (*Forcing*), mit der die Unabhängigkeit der Kontinuums-Hypothese (das berühmte erste Hilbertsche Problem) gezeigt wurde, noch ganz neu. Ulrich Felgner hat diese *Forcing*-Technik weiterentwickelt und weitere Unabhängigkeitsresultate bewiesen. Einige dieser Resultate sind in seinem Buch "Modelle der Mengenlehre" zu finden.

Zwischen **Mengenlehre** und **Gruppentheorie** sind viele seiner Arbeiten, z.B. zum Automorphismenturm, über Anwendungen des Axioms der Determiniertheit (zusammen mit Klaus Schulz) oder über unendliche symmetrische Gruppen anzusiedeln. Aus diesem Themenkreis, d.h. Anwendung von infinitärer Kombinatorik in der Gruppentheorie, wurden bis heute viele gute Themen zu Diplom- und Doktorarbeiten vergeben. Sie entstammen der Einsicht von Ulrich Felgner, dass in bestimmten unendlichen Gruppen die zugrunde liegenden Mengen und ihre Kombinatorik den algebraischen Aspekt dominieren, so dass man mit infinitärer Kombinatorik diese Gruppen genauer untersuchen kann.

Zwischen **Algebra** und **Logik** liegen bedeutende Arbeiten von Ulrich Felgner, die seit den siebziger Jahren bis heute entstanden sind. Einige zentrale Sätze dieser Arbeiten werden weiter unten aufgeführt.

Ganz außen, um diesen mathematischen Kern herum, sind im Diagramm einige der zahlreichen weiteren Interessensgebiete von Ulrich Felgner dargestellt. Diese Gebiete haben auf verschiedenartige Weise sein Werk beeinflusst und er hat in diesen wertvolle Beiträge geleistet. Es entstanden Arbeiten über Mathematik und Musik, Mathematik und Malerei, vor allem aber Mathematik und Philosophie oder Mathematik und Geschichte. Auch hier handelt es sich um Verbindungen zwischen unterschiedlichen Gebieten. Ulrich Felgner konnte solche Brücken zwischen verschiedenen Gebieten schlagen, da er ein großes Wissen in den beteiligten Gebieten und vom kulturellen Hintergrund hat. Er hat die grundlegenden Ideen der beteiligten Gebiete vollkommen durchdrungen und konnte sich dadurch parallele Kernideen erschließen, die er schließlich in unverwechselbarer Weise zu Verbindungen entwickeln konnte.

Man erkennt, dass Verbindungen ein einendes Element hinter dem Werk von Ulrich Felgner darstellen. Immer wieder konnte er die Ideen, die hinter den Gebieten stehen, herausarbeiten und zusammenzuführen. Dies gelang in ganz besonderer Weise bei der Verbindung zwischen Modelltheorie und Gruppentheorie. Die Techniken der Stabilitätstheorie und Klassifikationstheorie, die in der Modelltheorie entstanden waren, wurden von ihm in der Gruppentheorie angewandt, und dort wurden erfolgreich Klassifikationsresultate erzielt. Exemplarisch dafür seien hier die folgenden bedeutenden Sätze von Ulrich Felgner aufgeführt:

Satz. Stabile $\aleph_0$-kategorische Gruppen sind endliche Erweiterungen von nilpotenten Gruppen.

Satz. Eine endliche Gruppe G ist genau dann homogen, wenn G direktes Produkt zweier endlicher homogener Gruppen H und K von teilerfremder Ordnung ist, wobei

- H auflösbar ist, und
- K entweder trivial oder von der Form $\mathrm{SL}(2,5)$ oder $\mathrm{PSL}(2,p)$ mit $p=5$ oder $p=7$ ist.

Die homogenen endlichen auflösbaren Gruppen wurden dabei auch vollständig klassifiziert. Es ist eine längere Liste, zu der z.B. bestimmte endliche Erweiterungen von homozyklischen p-Gruppen oder Q^* (eine bestimmte auflösbare Gruppe der Ordnung 64) gehören.

Der erste dieser Sätze wurde 1977 bewiesen. Ein zweiter Beweis entstand unabhängig und zeitgleich durch Walter Baur, Greg Cherlin und Angus Macintyre. Der zweite Satz ist über einen längeren Zeitraum in Zusammenarbeit mit Greg Cherlin entstanden.

Ulrich Felgner hat aber nicht nur hervorragende Forschungsresultate vorzuweisen, er ist auch ein brillanter akademischer Lehrer. Dies zeigt sich einerseits in seinen ausgezeichneten Vorlesungen und Vorträgen, in denen er die tiefer liegenden Grundlagen der Mathematik und ihre Geschichte vermittelt hat. Es zeigt sich andererseits aber auch in der hohen Zahl von über 50 Diplom- und Staatsexamensarbeiten und 12 Doktorarbeiten, die von ihm betreut wurden. Unter seiner Betreuung wurden die Studierenden durch Offenheit zum Gespräch und durch andauernde Motivation und Anregung zu mathematischer Eigenständigkeit geführt, so dass sie in optimaler Weise profitiert haben.

In diesem Sinne möchten seine Schülerinnen und Schüler mit diesem Festband sich bei Ulrich Felgner ganz herzlich für seine engagierte Arbeit bedanken. Wir wünschen ihm alles Gute für die Zukunft.

Ein besonderer Dank gilt an dieser Stelle Herrn Benedikt Löwe, der alleine die gesamte immense Arbeit getragen hat, die zur Erstellung des Bandes erforderlich war. Ohne sein Engagement hätte dieser Festband nicht entstehen können.

Ulrich Felgners Doktoranden.

- Cankaya Alkor: *Untersuchungen über die Metamathematik der Ackermannschen Mengenlehre* (1977).

- Jörg Brendle: *Some contributions to combinatorial set theory and its applications* (1991).

- Andreas Ecker: *Zur Modelltheorie der CA-Gruppen* (1999).

- Claus Grünenwald: *Beiträge zur Stabilitätstheorie auflösbarer Gruppen* (1994).

- Frieder Haug: *Existenziell abgeschlossene LFC-Gruppen* (1987).

- Peter Kissel: *Beiträge zur Modelltheorie der linearen und der symplektischen Gruppen* (1991).

- Karl Mütz: *Computus chirometralis, Lehrbuch für Kalenderrechnung um 1330* (2003).

- Torsten Ingo Schatz: *Kürzeste konfinale Ketten im Untergruppenverband unendlicher Permutationsgruppen* (2002).

- Peter Schmitt: *Categorical Lattices* (1975).

- Klaus Schulz: *Beiträge zur Modelltheorie der Fastkörper* (1986).

- Klaus Sperlich: *Die Anzahl der Bahnen einer unendlichen Permutationsgruppe* (1998).

- Roger Villemaire: $\aleph_0$-*Categoricity over a Predicate* (1988).

- Horst Zeitler: *Modeltheoretic investigations on the amalgamation property* (1976).

Ulrich Felgners Habilitanden.

- Frieder Haug: *Existentially closed locally finite extensions* (1994).

- Jörg Brendle: *Combinatorial aspects of the meager and null ideals and of other ideals on the reals* (1995).

Ulrich Felgners Diplomanden.

- Christine Altseimer: *Bender-Gruppen von endlichem Morley-Rang* (1995).

- Jörg Brendle: *Kardinalzahlinvarianten unendlicher Gruppen* (1988)

- Claudia Charitius: *Gruppen, die in verallgemeinerten Baer-Specker-Gruppen involviert sind* (2006).

- Regine Eckardt: Faktor-$\aleph_0$-Kategorizität (1990).

- Stefan Elser: *Mengentheoretische Probleme in der Topologie* (2006).

- Jürgen Endreß: *Beweisvarianten für das kubische und biquadratische Reziprozitäts-Gesetz* (1983).

- Gebhard Engelhart: *Suslin-Bäume und Jensens kombinatorisches Prinzip* $\diamondsuit$ (1994).

- Christhard Flothow: *Elementare Eigenschaften von Kardinalzahl-Paaren mit endlichem Abstand* (1975).

- Peter Gasche: *Über die Anzahl nicht-konjugierter Untergruppen* (1994).

- Claus Grünenwald: *Modelltheorie von Gruppen und ihren Automorphismengruppen* (1989).

- Albrecht Haug: *Kombinatorische Prinzipien bei der Konstruktion endo-starrer Gruppen* (1995).

- Magdalena Kohler: *Maximale Untergruppen beschränkter symmetrischer Gruppen* (1996).

- Barbara Kreh: *Ein Entscheidungsverfahren für teilbare abelsche Gruppen mittels Quantorenelimination* (1990).

- Ralf Lieder: *Die Automorphismengruppen abzählbarer, rekursiv-saturierter Modelle der Peano-Arithmetik* (1996).

- Rainer Lüdecke: $\bigvee$-*freie Gruppen von singulärer Kardinalität* (2006).

- Petra Maier: *Nichtfundierte Mengenlehre als linguistisches Beschreibungsmittel* (1990).

- Hendrik Maryns: *Kompaktheit und Vollständigkeit in verallgemeinerten Logiken* (2004).

- Claus Meyer-Cording: *Nonstandard Topologie* (1980).

- Margit Meßmer: *Quasi-total-transzendente Theorien* (1988).

- Bruno Müller: *Die Modelltheorie der reell-abgeschlossenen Körper* (1975).

- Hedwig Schmucker: *Pseudo-endliche Gruppen* (1988).

- Thomas Schock: *Sylow-Theorie für Gruppen mit endlichem Morley-Rang* (1996).

- Karsten Schrempp: *Rekursive Unentscheidbarkeit des Körpers der gebrochen rationalen Funktionen über einem endlichen Grundkörper* (1989).

- Klaus Schumacher: *Boolesche Algebren in der 'Alternativen Mengenlehre'* (1990).

- Bernhard Six: *Modelle $\aleph_1$-kategorischer Theorien* (1973).

- Klaus Sperlich: *Die Anzahl der Bahnen von Untergruppen von* $\mathrm{Sym}(\omega)$ *auf der Potenzmenge der natürlichen Zahlen* (1994).

- Heike Steinwand: *Meßbarkeit von Kardinalzahlen unter* AD (1999).

- Günter Straub: *Konfinalität von Kardinalzahlen in generischen Modellen* (1975).

- Andreas Weber: *Paare von Modellen* (1990).

- Oliver Wienand: *Potenzen singulärer Kardinalzahlen und die Shelah'sche pcf-Theorie* (2004).

- Frank Zühlke: *Rabinowitschs Beiträge zum De Bruijn'schen Problem für beschränkte symmetrische Gruppen* (1993).

Ulrich Felgners Staatsexamenskandidaten.

- Gabriele Buck: *Elementar äquivalente Strukturen* (1990).

- Richard Bühler: *Die rekursive Unentscheidbarkeit des rationalen Zahlkörpers* $\mathbb{Q}$ (1982).

- Volker Dunst: *Arithmetik in endlichen Körpern* (1988).

- Johannes Guther: *Aleph-Hypothesen und die zugehörigen Kardinalzahl-Exponentiationen* (1993).

- Frieder Haug: *Erhaltungssätze für elementare Äquivalenz* (1983).

- Rainer Isbert: *Anwendungen der Theorie der kubischen Zahlringe auf die Bachet-Mordell-Gleichung* $x^3 = y^2 + k$ (1987).

- Jürgen Kaletta: *Die mengentheoretischen Voraussetzungen im Beweis, daß* $(\mathbb{Q},+)$ *ein direkter Summand von* $(\mathbb{R},+)$ *ist* (1985).

- Ulrich Mayer: *Zum Konsistenzproblem der Peano-Arithmetik* (1990).

- Adelheid Pfaff: *Unentscheidbarkeit in der Körpertheorie* (1980).

- Manfred Pfeffer: *Die Feinstruktur der Hierarchie der konstruktiblen Mengen* (1998).

- Konrad Pomm: *Definierbare Teilmengen endlicher Körper* (1996).

- Torsten Ingo Schatz: *Die Bedeutung großer Kardinalzahlen für die Determiniertheit von Mengen* (1998).

- Manfred Schiller: *Über die positive Theorie freier Gruppen und freier Produkte von Gruppen* (1978).

- Ute Katrin Schmid: *Partitionskardinalzahlen und unendliche Spiele* (1999).

- Klaus Schulz: *Algebraische Konsequenzen des Axioms der Determiniertheit* (1983).

- Margot Spielvogel: *Die Diophantische Gleichung $x^3 = y^2 + k$ von Bachet-Mordell* (1982).

- Heike Steinwand: *Konsequenzen aus dem Axiom der Determiniertheit* (1998).

Ulrich Felgners Publikationen.

Bücher.

1. Ulrich Felgner, Models of ZF-Set Theory. Lecture Notes in Mathematics 223, Springer-Verlag, Berlin 1971.

2. Ulrich Felgner (*ed.*), Mengenlehre, Wissenschaftliche Buchgesellschaft, Darmstadt 1979.

3. Egbert Brieskorn, Shrishti D. Chatterji, Moritz Epple, Ulrich Felgner, Horst Herrlich, Miroslav Hušek, Vladimir Kanovei, Peter Koepke, Gerhard Preuß, Walter Purkert, Erhard Scholz (*eds.*), Felix Hausdorff, Gesammelte Werke, Band II: "Grundzüge der Mengenlehre", Springer Verlag, Berlin 2002.

4. Egbert Brieskorn, Ulrich Felgner, Walter Purkert (*eds.*), Felix Hausdorff, Gesammelte Werke, Band I: Mengenlehre, Springer Verlag, Berlin 2007.

Abhandlungen und Aufsätze.

1. Ulrich Felgner, Untersuchungen über das Zornsche Lemma, Compositio Mathematica 18 (1967), pp. 170-180.

2. Ulrich Felgner, Die Inklusionsrelation zwischen Universa und ein abgeschwächtes Fundierungsaxiom, Archiv der Mathematik 20 (1969), pp. 561-566.

3. Ulrich Felgner, Die Existenz wohlgeordneter, konfinaler Ketten und das Auswahlaxiom, Mathematische Zeitschrift 111 (1969), pp. 221-232.
   Berichtigung dazu: Mathematische Zeitschrift 115 (1970), p. 392.

4. Ulrich Felgner, Comparison of the axioms of local and universal choice, Fundamenta Mathematicae 71 (1971), pp. 43-62.

5. Ulrich Felgner, Über das Ordnungstheorem, Zeitschrift für mathematische Logik und Grundlagen der Mathematik 17 (1971), pp. 257-272.

6. Ulrich Felgner, Thomas J. Jech, Variants of the axiom of choice in set theory with atoms, Fundamenta Mathematicae 79 (1973), pp. 79-85.

7. Ulrich Felgner, Catégoricité en $\aleph_1$ dans la Théorie des Anneaux non-Commutatifs, Fonds National de la Recherche Scientifique, Groupes de Contact: Sciences Mathématiques, Brüssel 1973, pp. 17-22.

8. Ulrich Felgner, Abzählbarkeit und Wohlordenbarkeit, Commentarii Mathematici Helvetici 49 (1974), pp. 114-124.

9. Ulrich Felgner, On $\aleph_0$-categorical extra-special p-groups, Logique et Analyse 71-72 (1975), pp. 407-428.
   Gesondert erschienen in: P. Henrard (ed.), Six days of model theory, Proceedings Conference Louvain-la-Neuve 1975, Editions Castella, Albeuve/Schweiz, 1977, pp. 175-196.

10. Ulrich Felgner, $\aleph_1$-Kategorische Theorien nicht-kommutativer Ringe, Fundamenta Mathematicae 82 (1975), pp. 331-346.

11. Ulrich Felgner, Stability and $\aleph_0$-categoricity of non-abelian groups, in: R. Gandy, M. Hyland (eds.), Logic Colloquium '76, North-Holland, Amsterdam 1977, pp. 301-324.

12. Ulrich Felgner, Choice functions on sets and classes, in: G. H. Müller (ed.), Sets and Classes — on the work of Paul Bernays, Studies in Logic and the Foundatiuons of Mathematics 84, North-Holland, Amsterdam 1976, pp. 217-255.

13. Ulrich Felgner, Das Problem von Suslin für geordnete algebraische Strukturen, in: W. Marek, M. Srebrny, A. Zarach (eds.), Set Theory and Hierarchy Theory — a memorial tribute to Andrzej Mostowski, Proceedings Conference Bierutowice 1975, Lecture Notes in Mathematics 537, Springer Verlag, Berlin 1976, pp. 83-107.

14. Ulrich Felgner, Einige gruppentheoretische Äquivalente zum Auswahlaxiom, Acta Mathematica Academiae Scientiae Hungaricae 28 (1976), pp. 13-18.

15. Ulrich Felgner, Timothy Flannagan, Wellordered subclasses of proper classes, in: G. H. Müller, D. Scott (eds.), Higher Set Theory, Proceedings Conference Oberwolfach 1977, Lecture Notes in Mathematics 669, Springer Verlag Berlin 1978, pp. 1-14.

16. Ulrich Felgner, $\aleph_0$-Categorical Stable Groups, Mathematische Zeitschrift 160 (1978), pp. 27-49.

17. Ulrich Felgner, The Model Theory of FC-Groups, in: A. I. Arruda, R. Chuaqui, N. da Costa (eds.), Mathematical Logic in Latin America, Proceedings IV Latin-American Symposium on Mathematical Logic, Santiago (Chile) 1978, North-Holland, Amsterdam 1980, pp. 163-190.

18. Ulrich Felgner, Bericht über die Cantorsche Kontinuums-Hypothese, *in:* U. Felgner (*ed.*), Mengenlehre, Wissenschaftliche Buchgesellschaft Darmstadt 1979, pp. 166-205.

19. F. Rudolf Beyl, Ulrich Felgner, Peter Schmid, On Groups Occurring as Center Factor Groups, Journal of Algebra 61 (1979), pp. 161-177.

20. Ulrich Felgner, Horn theories of abelian groups, *in:* L. Pacholski, J. Wierzejewski, A. Wilkie (*eds.*), Model Theory of algebra and arithmetic, Proceedings Conference Karpacz 1979, Lecture Notes in Mathematics 834, Springer-Verlag, Berlin 1980, pp. 163-173.

21. Ulrich Felgner, Kategorizität, Jahresbericht der Deutschen Mathematiker-Vereinigung 82 (1980), pp. 12-32.

22. Gregory Cherlin, Ulrich Felgner, Quantifier Eliminable Groups, *in:* D. van Dalen, D. Lascar, J. Smiley (*eds.*), Logic Colloquium '80, North-Holland, Amsterdam 1982, pp. 69-81.

23. Ulrich Felgner, The Classification of all Quantifier-Eliminable FC-Groups, *in:* E. Agazzi, M. Mondadori, S. Tugnoli Pattaro (*eds.*), Atti del Congresso Logica e Filosofia della Scienza, oggi (S. Gimignano, 7-11 dicembre 1983), Clueb, Bologna, 1986, pp. 27-33

24. Ulrich Felgner, On Bachet's Diophantine Equation $x^3 = y^2 + k$, Monatshefte für Mathematik 98 (1984), pp. 185-191.

25. Ulrich Felgner, Klaus Schulz, Algebraische Konsequenzen des Determiniertheits-Axioms, Archiv der Mathematik 42 (1984), pp. 557-563.

26. Ulrich Felgner, Über Wittgensteins Bemerkungen zur Geometrie, *in:* W. Leinfellner (*ed.*), Logik, Wissenschaftstheorie und Erkenntnistheorie, Akten des 11. Internationalen Wittgenstein-Symposiums 1986, Kirchberg am Wechsel, Schriftenreihe Wittgenstein-Gesellschaft 13, Hölder-Pichler-Tempsky, Wien 1987, pp. 91-96.

27. Ulrich Felgner, Pseudo-finite Near-Field, *in:* G. Betsch (*ed.*), Near-rings and Near-fields, Proceedings Conference Tübingen 1985, North-Holland, Amsterdam 1987, pp. 15-29.

28. Ulrich Felgner, Estimates for the sequence of primes, Elemente der Mathematik 46 (1990), pp. 17-25.

29. Ulrich Felgner, Pseudo-endliche Gruppen, Jahrbuch der Kurt Gödel-Gesellschaft 1990, Wien 1991, pp. 94-108.

auch in: Proceedings der 8. Oster-Konferenz über Modelltheorie in Wendisch-Rietz, 1990, Seminarbericht der Humboldt-Universität Berlin 110, pp. 82-96.

30. Gregory Cherlin, Ulrich Felgner, Homogeneous Solvable Groups, Journal London Mathematical Society (2) 44 (1991), pp. 102-120.

31. Ulrich Felgner, Frieder Haug, The homomorphic images of infinite symmetric groups, Forum Mathematicum 5 (1993), pp. 505-520.

32. Ulrich Felgner, Warum mich das alles nicht überzeugt – Erwiderung auf R. Taschner "Mathematik, Logik, Wirklichkeit", Ethik und Sozialwissenschaften 9 (1998), pp. 438-439.

33. Ulrich Felgner, John K. Truss, The independence of the Prime-Ideal-Theorem from the Order-Extension-Principle, Journal of Symbolic Logic 64 (1999), pp. 199-215.

34. Gregory Cherlin, Ulrich Felgner, Homogeneous Finite Groups, Journal London Mathematical Society (2) 62 (2000), pp. 784-794.

35. Ulrich Felgner, Regressive Funktionen, Aronszajn-Bäume und Automorphismentürme, Bulletin of the Mathematical Society Simon Stevin 2001 suppl., pp. 81-91

36. Ulrich Felgner, Zur Geschichte des Mengenbegriffs, *in:* B. Buldt, E. Köhler, M. Stöltzner, P. Weibel, C. Klein, W. Depauli-Schimanovich-Göttig (*eds.*), Kurt Gödel: Wahrheit & Beweisbarkeit, Band 2: Kompendium zum Werk, ÖBV & HPT Verlags-GmbH, Wien 2002, pp. 169-185.

37. Norbert Brunner, Ulrich Felgner, Gödels Universum der konstruktiblen Mengen *in:* B. Buldt, E. Köhler, M. Stöltzner, P. Weibel, C. Klein, W. Depauli-Schimanovich-Göttig (*eds.*), Kurt Gödel: Wahrheit & Beweisbarkeit, Band 2: Kompendium zum Werk, ÖBV & HPT Verlags-GmbH, Wien 2002, pp. 189-198.

38. Ulrich Felgner, Ein Brief Gödels zum Fundierungsaxiom, *in:* B. Buldt, E. Köhler, M. Stöltzner, P. Weibel, C. Klein, W. Depauli-Schimanovich-Göttig (*eds.*), Kurt Gödel: Wahrheit & Beweisbarkeit, Band 2: Kompendium zum Werk, ÖBV & HPT Verlags-GmbH, Wien 2002, pp. 205-213.

39. Ulrich Felgner, Der Begriff der Funktion, *in:* E. Brieskorn, S. D. Chatterji, M. Epple, U. Felgner, H. Herrlich, M. Hušek, V. Kanovei,

P. Koepke, G. Preuß, W. Purkert, E. Scholz (*eds.*), Felix Hausdorff, Gesammelte Werke, Band II: "Grundzüge der Mengenlehre", Springer Verlag, Berlin 2002, pp. 621-633.

40. Ulrich Felgner, Der Begriff der Kardinalzahl, *in:* E. Brieskorn, S. D. Chatterji, M. Epple, U. Felgner, H. Herrlich, M. Hušek, V. Kanovei, P. Koepke, G. Preuß, W. Purkert, E. Scholz (*eds.*), Felix Hausdorff, Gesammelte Werke, Band II: "Grundzüge der Mengenlehre", Springer Verlag, Berlin 2002, pp. 634-644.

41. Ulrich Felgner, Die Hausdorffsche Theorie der $\eta\alpha$-Mengen und ihre Wirkungsgeschichte, *in:* E. Brieskorn, S. D. Chatterji, M. Epple, U. Felgner, H. Herrlich, M. Hušek, V. Kanovei, P. Koepke, G. Preuß, W. Purkert, E. Scholz (*eds.*), Felix Hausdorff, Gesammelte Werke, Band II: "Grundzüge der Mengenlehre", Springer Verlag, Berlin 2002, pp. 645-674.

42. Ulrich Felgner, Torsten Schatz, The cofinality of normal subgroups and homomorphic images of infinite symmetric groups, Communications in Algebra 33 (2005), pp. 2601-2606.

43. Ulrich Felgner, Über den Ursprung des Wurzelzeichens, Mathematische Semesterberichte 52 (2005), pp. 1-7.

44. Ulrich Felgner, Zurück zur Natur? – Erwiderung auf B. Kanitscheider "Naturalismus und logisch-mathematische Grundlagenprobleme", Erwägen, Wissen Ethik (EWE) 3 (2006), pp. 34-36.

45. Ulrich Felgner, Dichtung und Wahrheit – zur Geschichte der Antinomie vom Lügner, *in:* G. Löffladt (*ed.*), Mathematik, Logik und Philosophie, Verlag Harry Deutsch, Frankfurt/Main 2007, 21 Seiten.

46. Ulrich Felgner, Kommentar zu Hausdorffs Abhandlung "Über zwei Sätze von G. Fichtenholz und L. Kantorovitch", *in:* C. F. Bödigheimer, F. Hirzebruch, H. Herrlich, M. Hušek, V. Kanovei, P. Koepke, G. Preuß, W. Purkert, E. Scholz (*eds.*), Felix Hausdorff, Gesammelte Werke, Band III: Deskriptive Mengenlehre und Topologie, Springer Verlag, Berlin 2007.

Sonstiges.
1. Ulrich Felgner, Das Tafelklavier der Familie Uhland, Nachrichten aus dem Stadtmuseum, Heft 5/6 (1989/90), Tübingen 1990, pp. 25-28.

2. Ulrich Felgner, Eine einheitliche Theorie für alles?, *Beitrag zum* Symposium "Auseinandersetzung mit dem Buch 'Der Anfang aller Dinge' von Hans Küng", http://www.weltethos.org, 2006.

Hrushovski's Fusion

ANDREAS BAUDISCH[1]
MARTIN ZIEGLER[2]
AMADOR MARTIN-PIZARRO[1]

[1] Institut für Mathematik
Humboldt-Universität zu Berlin
Rudower Chaussee 25
10099 Berlin, Germany

[2] Mathematisches Institut
Albert-Ludwigs-Universität Freiburg
Abteilung für Mathematische Logik
Mathematisches Institut
Eckerstraße 1
79104 Freiburg, Germany
E-mail: {baudisch,pizarro}@mathematik.hu-berlin.de, ziegler@uni-freiburg.de

ABSTRACT. We present a detailed and simplified exposition of Hrushovki's fusion of two strongly minimal theories.

1 Introduction

A definable set whose definable subsets (in some saturated model) are either finite or cofinite is called *strongly minimal*. Examples of strongly minimal structures are the trivial one, infinite dimensional vector spaces, or algebraically closed fields. It was observed that the geometrical behaviour of these archetypical examples could be generalized to the pregeometry of algebraic closure on strongly minimal sets, exhibiting a first example of *regular types*. This led B. Zilber to conjecture that all strongly minimal sets could be classified according to these three basic ones. This conjectured was refuted by E. Hrushovski [$H_3$93], who adapted Fraïssé's construction in a ingenious way in order to *collapse* to finite rank a candidate for the counterexample. The amalgamation procedure can be described in the following way: the goal is to construct a countable universal model starting from a given collection of finitely generated structures. In this model there is a unique type of rank ω. The decisive part (or *collapse*) is to modify this construction in order to algebraize types of finite rank. In order to do so, a collection of representatives (or *codes*) of these types needs to be chosen and one assigns a maximal length of an independence sequence of realisations to each code. The structure so obtained after amalgamation has now finite rank. Note that the prescribed maximal length must reflect any interaction between different codes, since some realizations of one code may yield realizations for another.

Using the same procedure, E. Hrushovski also merged two strongly minimal theories over a trivial geometry into a new one, their *fusion*. This answered negatively a question of G. Cherlin on the existence of a maximal strongly minimal theory. More precisely, consider two countable strongly minimal theories T_1 and T_2 with the definable multiplicity property (in short, DMP) whose respective languages L_1 and L_2 are disjoint. Recall that a ω-stable theory has the DMP if Morley rank and degree are definable on the parameters of any given definable set.

In this survey we give a detailed and slightly simplified exposition of the following theorem proved by E. Hrushovski.

Theorem 1 ([H$_3$92]). The theory $T_1 \cup T_2$ admits a strongly minimal completion T^μ. Its models satisfy the following: Let tr_i denote the transcendence degree in T_i. Given a finite subset A of M, then

$$|A| \leq \mathrm{tr}_1(A) + \mathrm{tr}_2(A).$$

Our presentation grew out of a seminar held at the Humboldt-Universität Berlin in 2003. Several articles on this topic (among others [H$_2$99] and [H$_0$H$_1$06]) have been published, however we believe that this survey will be beneficial for the mathematical community in order to become more acquainted with Hrushovski's fusion method. The authors used the simplified approach in three subsequent articles: In [B$_1$MZ06] to reprove a theorem of Poizat and Baldwin–Holland ([P99], [B$_0$H$_2$00]) about the existence of fields of Morley rank 2 with a distinguished subset in any characteristic, in [B$_1$MZ∞b] to construct fields of Morley rank 2 with a distinguished additive subgroup, and in [B$_1$MZ∞a] to prove the fusion over a vector space over a finite field, which had been proposed by E. Hrushovski.

The simplified technique was also helpful in [B$_1$H$_1$MW∞], where a bad field was constructed: a field of Morley rank 2 with a distinguished multiplicative subgroup. This solved a long standing open problem.

Finally our techniques were used in [Z∞] to prove the following generalization of Theorem 1.

Theorem 2. Let T_1 and T_2 be two countable complete theories in disjoint languages of finite Morley rank and of the same Morley degree. Assume that in T_1 and T_2 Morley rank and Morley degree are definable. Then $T_1 \cup T_2$ has a "nice" complete extension of any rank which is a common multiple of the ranks of T_1 and T_2.

2 Codes

All throughout the following sections (and until specified otherwise) T denotes a countable strongly minimal theory with the DMP.

First, let us fix some notation: $\operatorname{tr}(a/B)$ is the transcendence degree of a over B[1] and $\operatorname{MR}(p)$ denotes Morley rank of the type p. Note that

$$\operatorname{tr}(a/B) = \operatorname{MR}(\operatorname{tp}(a/B)).$$

We write

$$\phi(x) \sim^k \psi(x)$$

or $\phi(x) \sim_x^k \psi(x)$ if the symmetric difference of ϕ and ψ has smaller Morley rank than k.

A formula $\chi(x, b)$ is *simple* if it has Morley degree 1 and the components of a generic realization are pairwise different and not in $\operatorname{acl}(b)$. If a is an n–tuple and s is some subset of $\{1, \ldots, n\}$, then a_s is $\{a_i \mid i \in s\}$.

A *code* c is a parameter-free formula

$$\phi_c(x, y),$$

where $|x| = n_c$ and y lies in some sort of T^{eq}, with the following properties.

(i) $\phi_c(x, b)$ is either empty[2] or simple. Furthermore, $\phi_c(x, b)$ implies that the components of x are pairwise different.

(ii) All non-empty $\phi_c(x, b)$ have Morley rank k_c and Morley degree 1.

(iii) For each subset s of $\{1, \ldots, n_c\}$ there exists an integer $k_{c,s}$ such that for every realization a of $\phi_c(x, b)$

$$\operatorname{tr}(a/ba_s) \leq k_{c,s}.$$

Equality holds for generic a.

(iv) If both $\phi_c(x, b)$ and $\phi_c(x, b')$ are non-empty and $\phi_c(x, b) \sim^{k_c} \phi_c(x, b')$, then $b = b'$.

For $\phi_c(x, b)$ to have Morley rank k_c in (ii) is equivalent to $k_{c,\varnothing} = k_c$. The simplicity of $\phi_c(x, b)$ in (i) is equivalent to Morley degree 1 and $k_{c,\{i\}} = k_c - 1$ for all i.

---
[1] The maximal length of a B-independent subtuple of a.
[2] We assume that $\phi_c(x, b)$ is non-empty for some b.

Corollary 3. Let $p \in S(b)$ be the unique type of Morley rank k_c containing $\phi_c(x, b)$. Then b is the canonical basis of p.

Proof. This follows easily from (iv). q.e.d.

Lemma 4. Let $\chi(x, d)$ be a simple formula. Then there is some code c and some $b_0 \in \mathrm{dcl}^{\mathrm{eq}}(d)$ such that $\chi(x, d) \sim^{k_c} \phi_c(x, b_0)$.

We say that c *encodes* $\chi(x, d)$.

Proof. Set $k_c = \mathrm{MR}(\chi(x,d))$ and $n_c = |x|$. Let p be the global type of rank k_c containing $\chi(x, d)$, with canonical basis b_0. We find a formula $\phi(x, b_0)$ in p of rank k and degree 1. Choose a generic realization a_0 of $\phi(x, b_0)$. For each $s \subset \{1, \ldots, n_c\}$ set $k_{c,s} = \mathrm{MR}(a_0/b_0 a_{0s})$. By strengthening $\phi(x, b_0)$ appropriately we may assume that $\phi_c(a, b_0)$ implies that the components of a are pairwise different, and that $\mathrm{tr}(a/b_0 a_s) \leq k_{c,s}$ and $\mathrm{tr}(a_s/b_0) \leq (k_c - k_{c,s})$ for all realizations a of $\phi(x, b_0)$.

Consider now the following property $\mathsf{E}(b, b')$:

- $\phi(x, b)$ implies that the components of x are pairwise disjoint.
- $\phi(x, b)$ has Morley rank k_c and degree 1.
- $\mathrm{tr}(a/ba_s) \leq k_{c,s}$ and $\mathrm{tr}(a_s/b) \leq (k_s - k_{c,s})$ for all realizations a of $\phi(x, b)$.
- $\phi(x, b) \sim^{k_c} \phi(x, b')$ implies that $b = b'$.

E holds for all b, b' realizing the type of b_0. Moreover, E is equivalent to an infinite disjunction of formulae $\epsilon(y, y')$. Therefore, there is some $\theta(y) \in \mathrm{tp}(b_0)$ such that $\models \theta(y) \wedge \theta(y') \to \mathsf{E}(y, y')$. Set
$$\phi_c(x, y) = \phi(x, y) \wedge \theta(y).$$

Let a be a generic realization of $\phi_c(x, b)$. Then $\mathrm{tr}(a/ba_s) = k_{c,s}$ follows from $\mathrm{tr}(a/ba_s) \leq k_{c,s}$, $\mathrm{tr}(a_s/b) \leq (k_c - k_{c,s})$ and $\mathrm{tr}(a/b) = k_c$. By simplicity of $\chi(x, d)$ we have $k_{c,\{i\}} < k_c$ for all i, which in turn implies that all non-empty $\phi_c(x, b)$ are simple. q.e.d.

Let c be a code, $\phi_c(x, b)$ non-empty and $p \in S(b)$ the type of rank k_c determined by $\phi(x, b)$. Hence, b lies in the definable closure of a sufficiently large segment of a Morley sequence of p. Let m_c be some upper bound for the length of such a segment.

Lemma 5. For every code c and every integer $\mu \geq m_c - 1$ there exists some formula $\Psi_c(x_0, \ldots, x_\mu, y)$ without parameters satisfying the following:

(v) Given a Morley sequence $e_0, \ldots, e_\mu$ of $\phi_c(x, b)$, then $\models \Psi_c(e_0, \ldots, e_\mu, b)$.

(vi) For all $e_0, \ldots, e_\mu, b$ realizing Ψ_c the e_i's are pairwise disjoint realizations of $\phi_c(x, b)$.

(vii) Let $e_0, \ldots, e_\mu, b$ realize Ψ_c. Then b lies in the definable closure of any m_c many e_i's.

We say that "$x_0, \ldots, x_\mu$ is a pseudo Morley sequence of c over y".

Proof. The statement "(e_i) is a Morley sequence of $\phi_c(x, b)$" can be described by a partial type $\mathsf{M}(e_0, \ldots, b)$. Likewise, properties (vi) and (vii) can be described by an infinite disjunction $\mathsf{D}(e_0, \ldots, b)$. Since non-empty $\phi_c(x, b)$ are simple, it follows that $\models \mathsf{M} \to \mathsf{D}$. Hence we may choose a sufficiently strong formula Ψ_c in M with the desired properties. q.e.d.

Choose now for every code (and every μ)[3] a formula Ψ_c as above.

Let c be a code and σ some permutation of $\{1, \ldots, n_c\}$. Then c^σ defined by
$$\phi_{c^\sigma}(x^\sigma, y) = \phi_c(x, y)$$
is also a code. Similarly,
$$\Psi_{c^\sigma}(\bar{x}^\sigma, y) = \Psi_c(\bar{x}, y)$$
defines a pseudo Morley sequence of c^σ.

We consider two codes c and c' to be *equivalent* if $n_c = n_{c'}$, $m_c = m_{c'}$, and

- For every b there is some b' such that
$$\phi_c(x, b) \equiv \phi_{c'}(x, b') \text{ and } \Psi_c(\bar{x}, b) \equiv \Psi_{c'}(\bar{x}, b')$$
in T.

- Similarly permuting c and c'.

Theorem 6. There is a collection of codes C such that:

(viii) Every simple formula can be encoded by exactly one $c \in C$.

---
[3] In the proof of Theorem 6 this choice may be modified.

(ix) For every $c \in C$ and every permutation σ, we have that c^σ is equivalent to a code in C.[4]

In [H$_3$92] it was stated that one could find such a set C closed under permutations, which is stronger than (ix) in Theorem 6. This is not true.

Proof. Work inside some countable ω–saturated model M of T and list all simple formulae χ_i, $i = 1, 2, \ldots$, with parameters in M. We need only show that every χ_i may be encoded by some $c \in C$. We build up C as the union of an increasing sequence $\varnothing = C_0 \subset C_1 \subset \cdots$ of finite sets. Suppose by induction on i that C_{i-1} has already been constructed and satisfies (ix). If χ_i may be encoded by some element in C_{i-1}, then set $C_i = C_{i-1}$. Otherwise, choose some code c and some b_0 with $\phi_c(x, b_0) \sim^{k_c} \chi_i$. Replace ϕ_c by

$$\phi_c(x, y) \wedge \text{``}\phi_c(x, y) \text{ cannot be encoded by any element of } C_{i-1}\text{.''}$$

We obtain a new code which still encodes χ_i. Therefore, we may assume that no permutation of c encodes a formula which may be also encoded by some element of C_{i-1}. Let G be now the group of all permutations $\sigma \in \mathrm{Sym}(n_c)$ with

$$\phi_c(x, b_0) \sim^{k_c} \phi_{c^\sigma}(x, b_0')$$

for some realization b_0' of p, the type of b_0. It follows that b_0' is uniquely determined and hence given by a $\varnothing$-definable function of b_0. Write $b_0' = b_0^\sigma$.

After strengthening $\phi_c(x, y)$ with an appropriate subset[5] of p, we may assume that for all b with non-empty $\phi_c(x, b)$ and all σ there is a b^σ with $\phi_c(x, b) \sim^{k_c} \phi_{c^\sigma}(x, b^\sigma)$ iff $\sigma \in G$.

It is easy to see that

$$\phi_d(x, y) = \bigwedge_{\sigma \in G} \phi_{c^\sigma}(x, y^\sigma)$$

defines a code which still encodes χ_i. Likewise,

$$\Psi_d(\bar{x}, y) = \bigwedge_{\sigma \in G} \Psi_{c^\sigma}(\bar{x}, y^\sigma)$$

defines pseudo Morley sequences of d.

---

[4] In fact, we find a collection C such that every c^σ is equivalent to some permutation of c which lies in C.

[5] Choose $\rho'(y) \in \mathrm{tp}(b_0)$ such that $\models \neg \rho'(b_0^\sigma)$ for all $\sigma \notin G$. The aforementioned subset of p is

$$\rho(y) = \bigwedge_{\sigma \in G} \rho'(b^\sigma) \wedge \bigwedge_{\sigma \notin G} \neg \rho'(b^\sigma).$$

Moreover, $\phi_d(x,y) \equiv \phi_{d^\sigma}(x, y^\sigma)$ and $\Psi_d(\bar{x}, y) \equiv \Psi_{d^\sigma}(\bar{x}, y^\sigma)$ for all $\sigma \in G$. Hence d and d^σ are equivalent. Finally, choose representatives $\rho_1, \ldots, \rho_r$ of the right cosets of G in $\mathrm{Sym}(n_c)$ and set $C_i = C_{i-1} \cup \{d^{\rho_1}, \ldots, d^{\rho_r}\}$. q.e.d.

3 The δ–function

From now on, T_1 and T_2 are two strongly minimal theories[6], formulated in two disjoint languages L_1 and L_2. We assume the following **QE-Hypothesis**:

> T_1 and T_2 have quantifier elimination, and L_1 and L_2 are pure relational languages.

by considering Morleyizations of T_1 and T_2. We may also assume that ϕ_c and Ψ_c are quantifier-free for the T_1–Codes and the T_2–Codes. Types $\mathrm{tp}_i(a/B)$ in each theory T_i are always quantifier-free types. These assumptions will be dropped in §7.

Let $\mathcal{K}$ be the class of all models of $T_1^\forall \cup T_2^\forall$. We also allow $\varnothing$ to be in $\mathcal{K}$. If $\mathbb{C}_i$ is some monster model of T_i, we may see elements of $\mathcal{K}$ simultaneously as subsets of $\mathbb{C}_1$ and $\mathbb{C}_2$. Given a finite $A \in \mathcal{K}$, define

$$\delta(A) = \mathrm{tr}_1(A) + \mathrm{tr}_2(A) - |A|.$$

The following hold:

$$\delta(\varnothing) = 0, \tag{1}$$
$$\delta(\{a\}) \leq 1, \tag{2}$$
$$\delta(A \cup B) + \delta(A \cap B) \leq \delta(A) + \delta(B). \tag{3}$$

If $A \setminus B$ is finite, we set

$$\delta(A/B) = \mathrm{tr}_1(A/B) + \mathrm{tr}_2(A/B) - |A \setminus B|.$$

For B finite, it follows that $\delta(A/B) = \delta(A \cup B) - \delta(B)$.

We say that B is *strong* in A if $B \subset A$ and $\delta(A'/B) \geq 0$ for all finite $A' \subset A$. Denote this by

$$B \leq A.$$

---

[6] Countability and DMP of T_i will not be used in this section.

An element $a \in A$ is *algebraic* over $B \subset A$ if a is algebraic over B either in the sense of T_1 or in the sense of T_2. A is transcendental over B if no $a \in A \setminus B$ is algebraic over B. A set $B \subsetneq A$ is *minimal* if $B \le A' \le A$ for no A' properly contained between B and A.

Lemma 7. A proper strong extension $B \le A$ is minimal if and only if $\delta(A/A') < 0$ for all A' properly contained between B and A.

Proof. One direction is clear, since $A' \le A$ implies that $\delta(A/A') \ge 0$. On the other hand, if $\delta(A/A') \ge 0$ for some A', we may choose A' such that $\delta(A/A')$ is maximal. Hence $A' \le A$, which implies that A is not minimal over B. q.e.d.

Note that $A \setminus B$ is finite for minimal extensions.

Lemma 8. Let $B \le A$ be a minimal extension of elements in $\mathcal{K}$. Then one of the following hold:

(I) $\delta(A/B) = 0$ and $A = B \cup \{a\}$ for some element $a \in A \setminus B$ algebraic over B. *(algebraic minimal extension)*

(II) $\delta(A/B) = 0$ and A is transcendental over B. *(prealgebraic minimal extension)*

(III) $\delta(A/B) = 1$ and $A = B \cup \{a\}$ for some a in A transcendental over B. *(transcendental minimal extension)*

Note that $|A \setminus B| \ge 2$ in the prealgebraic case.

Proof. If $A \setminus B$ contains some algebraic element a, then $\delta(a/B) = 0$. Hence $B \cup \{a\} = A$.

Otherwise, two cases apply: if $\delta(A/B) = 0$, then the extension is prealgebraic minimal. Otherwise, $\delta(A'/B) \ge 1$ for each $B \subsetneq A' \subset A$. Given any $a \in A \setminus B$, we have that $B \cup \{a\} \le A$ and hence $B \cup \{a\} = A$.

q.e.d.

Define $\mathcal{K}^0 \subset \mathcal{K}$ as the subclass

$$\mathcal{K}^0 = \{M \in \mathcal{K} \mid \varnothing \le M\}.$$

It is easy to see that $\mathcal{K}^0$ may be described by a collection of universal $L_1 \cup L_2$–sentences. The following lemmas follow easily from (1), (2) and (3).

Lemma 9. Let M in $\mathcal{K}^0$ and A a finite subset of M. Set
$$d(A) = \min_{A \subset A' \subset M} \delta(A').$$
Then d is the dimension function of a pregeometry, i.e., d satisfies (1), (2), (3) and
$$d(A) \geq 0 \qquad (4)$$
$$A \subset B \Rightarrow d(A) \leq d(B) \qquad (5)$$

Lemma 10. Let $M \in \mathcal{K}^0$ and A a finite subset of M. Take a minimal superset A' of A with $\delta(A') = d(A)$. Then A' is the smallest strong subset $cl(A)$ of M containing A, called the *closure* of A.

4 Prealgebraic codes

From now on, T_1 and T_2 are two countable strongly minimal theories with the DMP as in Theorem 1. The **QE-Hypothesis** from §3 holds all throughout this and the next sections (§§5,6).

Fix for each T_i a collection C_i of codes as in Theorem 6. A *prealgebraic code* $c = (c_1, c_2)$ consists of a code $c_1 \in C_1$ and a code $c_2 \in C_2$ with the following properties:

- $n_c := n_{c_1} = n_{c_2} = k_{c_1} + k_{c_2}$,
- For each proper non-empty subset s of $\{1, \ldots, n_c\}$,
$$k_{c_1,s} + k_{c_2,s} - (n_c - |s|) < 0.$$

Set $m_c = \max(m_{c_1}, m_{c_2})$. Note that simplicity of $\phi_{c_i}(x, b)$ implies that
$$n_c \geq 2.$$
For each permutation σ, the code
$$c^\sigma = (c_1^\sigma, c_2^\sigma)$$
is also prealgebraic.

Some explanatory remarks: T_1^{eq} and T_2^{eq} share only their home sort. An element $b \in \text{dcl}^{eq}(B)$ is a pair $b = (b_1, b_2)$ with $b_i \in \text{dcl}^{eq}{}_i(B)$ for $i = 1, 2$. Likewise for $\text{acl}^{eq}(B)$. A *generic realization* of $\phi_c(x, b)$ (over B) is a generic realization of $\phi_{c_i}(x, b_i)$ (over B) in T_i for $i = 1, 2$. A *Morley sequence* of $\phi_c(x,b)$ is a Morley sequence both of $\phi_{c_1}(x,b_1)$ and $\phi_{c_2}(x,b_2)$. A *pseudo Morley sequence* of c over b is a realization of both $\Psi_{c_1}(\bar{x},b_1)$ and $\Psi_{c_2}(\bar{x},b_2)$. We say that M is *independent* from A over B if M is independent from A over B both in T_1 and T_2.

Lemma 11. Let $B \leq B \cup \{a_1,\ldots,a_n\}$ be a prealgebraic minimal extension and $a = (a_1,\ldots,a_n)$. Then there is some prealgebraic code c and $b \in \mathrm{acl}^{\mathrm{eq}}(B)$ such that a is a generic realization of $\phi_c(a,b)$.

Proof. Fix $i \in \{1,2\}$ and choose $d_i \in \mathrm{acl}^{\mathrm{eq}}{}_i(B)$ such that $\mathrm{tp}_i(a/Bd_i)$ is stationary, and $\chi_i(x,d_i) \in \mathrm{tp}_i(a/Bd_i)$ with Morley rank $\mathrm{MR}_i(a/Bd_i)$ and degree 1. Since A/B is transcendental, the formula $\chi_i(x,d_i)$ is simple. Choose some T_i–code $c_i \in C_i$ and some $b_i \in \mathrm{dcl}^{\mathrm{eq}}{}_i(d_i)$ with $\chi_i(x,d_i) \sim^{k_{c_i}} \phi_{c_i}(x,b_i)$. It follows from $\delta(A/B) = 0$ that $k_{c_1} + k_{c_2} = n$. Moreover, $k_{c_1,s} + k_{c_2,s} - (n - |s|) < 0$ holds by Lemma 7. q.e.d.

The following lemma is proved similarly.

Lemma 12. Let $B \in \mathcal{K}$, c a prealgebraic code and $b \in \mathrm{acl}^{\mathrm{eq}}(B)$. Take a generic realization $a = (a_1,\ldots,a_{n_c})$ of $\phi_c(x,b)$ over B. Then $B \cup \{a_1,\ldots,a_{n_c}\}$ is a prealgebraic minimal extension of B. q.e.d.

Note that the isomorphism type of a over B is uniquely determined.

Lemma 13. Let $B \subset A$ in $\mathcal{K}$, c a prealgebraic code, b in $\mathrm{acl}^{\mathrm{eq}}(B)$ and $a \in A$ a realization of $\phi_c(x,b)$ which does not lie completely in B. Then

(i) $\delta(a/B) \leq 0$.

(ii) If $\delta(a/B) = 0$, then a is a generic realization of $\phi_c(x,b)$ over B.

Proof. Let $s = \{i \mid a_i \in B\}$. Since a is not completely contained in B, then s is a proper subset of $\{1,\ldots,n_c\}$. Therefore

$$\delta(a/B) = \mathrm{tr}_1(a/B) + \mathrm{tr}_2(a/B) - (n - |s|) \leq k_{c_1,s} + k_{c_2,s} - (n - |s|).$$

If $s \neq \varnothing$, then the right-hand side is negative. If $s = \varnothing$, we have that

$$\delta(a/B) = \mathrm{tr}_1(a/B) + \mathrm{tr}_2(a/B) - n \leq k_{c_1} + k_{c_2} - n = 0.$$

So, $\delta(a/B) = 0$ implies that $\mathrm{tr}_i(a/B) = k_{c_i}$. q.e.d.

Lemma 14. Let $M \leq N$ be a strong extension of structures in $\mathcal{K}$ and $e_0,\ldots,e_\mu$ a pseudo Morley sequence of c in N over b. Then one of the following hold:

- $b \in \mathrm{dcl}^{\mathrm{eq}}(M)$.
- At least $\mu - n_c m_c + 1$ many e_i's lie in $N \setminus M$.

Proof. Permute the e_i's so that $e_0, \ldots, e_{r_0-1}$ are in M and $e_{r_1}, \ldots, e_{\mu(c)}$ lie in $N \setminus M$. Hence $0 \leq r_0 \leq r_1 \leq \mu(c) + 1$. Possibly the e_i's do not form a pseudo Morley sequence of c after permutation, however they are still disjoint realizations of $\phi_c(x, b)$. Assume $b \notin \mathrm{dcl}^{\mathrm{eq}}(M)$. Then (vii) implies $r_0 < m_c$. We need only to show that $r_1 \leq m_c n_c$. Suppose that $m_c \leq r_1$.

Define $\delta(i) = \delta(e_i/Me_0 \cdots e_{i-1})$. For $i < r_1$ the following upper bound holds[7] $\delta(i) \leq (n_c - 1)$. If $m_c \leq i < r_1$, then $\delta(i) < 0$ since $b \in \mathrm{dcl}^{\mathrm{eq}}(Me_0 \cdots e_{i-1})$ by Lemma 13. Therefore

$$0 \leq \delta(e_0 \cdots e_{r_1-1}/M) = \sum_{i<r_1} \delta(i) = \sum_{i<m_c} \delta(i) + \sum_{m_c \leq i < r_1} \delta(i)$$
$$\leq m_c(n_c - 1) - (r_1 - m_c).$$

The above inequality proves the claim. q.e.d.

5 The class $\mathcal{K}^\mu$

Let μ^* be a function that assigns to each prealgebraic code c some natural number $\mu^*(c)$. We suppose that

- $\mu^*(c) \geq m_c - 1$
- For all triples l, m, n with $m > 0$ there are only finitely many c's with $\mu^*(c) = l$, $m_c = m$ and $n_c = n$ for each $m > 0$ and n. (Such μ^* exist since there are only countably many codes.)
- $\mu^*(c) = \mu^*(d)$ if c is equivalent to some permutation of d.[8]

Define
$$\mu(c) = m_c n_c + \mu^*(c).$$
Note that $\mu(c) \geq m_c$.

From now on, a *pseudo Morley sequence* denotes a pseudo Morley sequence of length $\mu(c) + 1$ for a prealgebraic code c. Given such a pseudo Morley sequence (e_i), so is every (e_i^σ) for every permutation σ by (ix).

---

[7] Note that $\delta(A/B) \leq |A/B|$ in general.
[8] Note that each permutation is equivalent to at most one prealgebraic code.

The class $\mathcal{K}^\mu$ consists of the elements $M \in \mathcal{K}^0$ which do not contain any pseudo Morley sequence.

Lemma 15. Let B be a finite strong subset of $M \in \mathcal{K}^\mu$ and A/B a prealgebraic minimal extension. Then there are only finitely many B–isomorphic copies of A in M.

Proof. Let $A = B \cup \{a\}$ for some tuple a and choose $d \in \mathrm{acl}^{\mathrm{eq}}(B)$ with $\mathrm{tp}_i(a/Bd_i)$ stationary. We need only show that $\mathrm{tp}_1(a/Bd_1) \cup \mathrm{tp}_2(a/Bd_2)$ has only finitely many realizations in M. Choose a prealgebraic code c by Lemma 11 and $b \in \mathrm{acl}^{\mathrm{eq}}(B)$ with $\models \phi_c(a,b)$. We show that $\phi_c(x,b)$ cannot be infinitely often realized in M. Otherwise, we obtain at least $(\mu(c)+1)$ many such realizations e_i with $e_i \notin B \cup \{e_0, \ldots, e_{i-1}\}$. It follows from Lemma 13 that the e_i's form a Morley sequence of $\phi_c(x,b)$ over B and hence a pseudo Morley sequence of c over b by (v), which contradicts that M is in $\mathcal{K}^\mu$. q.e.d.

Corollary 16. Let $B \leq M \in \mathcal{K}^\mu$, $B \subset A$ finite with $\delta(A/B) = 0$. Then there are only finitely many A' such that: $B \leq A' \subset M$ and A' is B–isomorphic to A.

Note that we automatically have $A' \leq M$.

Proof. Decompose the extension A/B into a finite sequence of minimal ones. q.e.d.

Corollary 17. Let B be a finite subset of $M \in \mathcal{K}^\mu$. Then the d–closure of B,
$$\mathrm{cl}_\mathrm{d}(B) = \{x \in M \mid \mathrm{d}(Bx) = \mathrm{d}(B)\},$$
is countable.

Proof. Recall that $\mathrm{cl}_\mathrm{d}(B)$ is the union of all finite $A' \subset M$ with $\mathrm{cl}(B) \subset A'$ and $\delta(A'/\mathrm{cl}(B)) = 0$. q.e.d.

Lemma 18. If $M \in \mathcal{K}^\mu$, $M \leq N$ and $|N \setminus M| = 1$, then N is also in $\mathcal{K}^\mu$.

Proof. Let (e_i) be a pseudo Morley sequence of c over b in N. There is at most one e_i not in M. Now $b \in \mathrm{dcl}^{\mathrm{eq}}(M)$ since $\mu(c) \geq m_c$. It follows that every e_i is either in M or in $N \setminus M$ by Lemma 13. The latter cannot hold, since $n_c \geq 2$. Hence (e_i) is completely contained in M. Contradiction. q.e.d.

Theorem 19. $\mathcal{K}^\mu$ (and hence the class of all finite structures in $\mathcal{K}^\mu$) has the amalgamation property with respect to strong embeddings.

Proof. Let $B \leq M$ and $B \leq A$ be structures in $\mathcal{K}^\mu$. We need to find a strong extension $M' \in \mathcal{K}^\mu$ of M and some $B \leq A' \leq M'$ isomorphic to A over B. We may assume that both A/B and M/B are minimal. We will show that either a "free amalgam" M' of M and A over B is in $\mathcal{K}^\mu$ or that M and A are B–isomorphic.

Case 1: A/B is an algebraic minimal extension. Suppose that $A = B \cup \{a\}$ for some a algebraic over B in T_1 and transcendental over B in T_2. Two possible (non-exclusive) cases may arise.

Subcase 1.1: $\text{tp}_1(a/B)$ is realized in M by some a'. Then a'/B is transcendental in T_2. Hence $B \cup \{a'\}$ is B–isomorphic to A and strong in M. By minimality, $M = B \cup \{a'\}$.

Subcase 1.2: There is some $a' \notin M$ realizing $\text{tp}_1(a/B)$. Define $M' = M \cup \{a\}$ by letting a have the type of a' over M in the sense of T_1 and be transcendental over M in the sense of T_2. Then M' is a *free amalgam* of M and A over B, i.e., $M \cap A = B$ and M is independent from A over B. It is easy to see that $M \leq M'$ and $A \leq M'$ for such amalgams. Now, $M' \in \mathcal{K}^\mu$ by Lemma 18.

Case 2: A/B is transcendental. Then there is a free amalgam M' of M and A as above. Suppose that M' is not in $\mathcal{K}^\mu$. Then M' contains a pseudo Morley sequence (e_i) of c over b. Apply Lemma 14 to the extension M'/M to obtain one of the following cases.

Subcase 2.1: $b \in \text{dcl}^{\text{eq}}(M)$. Since M is in $\mathcal{K}^\mu$, not all members of the pseudo Morley sequence lie in M. Let $e_i \notin M$. By Lemma 13 e_i is a generic realization of $\phi_c(x, b)$ over M. Independence of M and e_i over B yields that b in $\text{acl}^{\text{eq}}(B)$ by Corollary 3. Since $A \in \mathcal{K}^\mu$, there is some e_j not completely contained in A. Again, e_j is a generic realization of $\phi_c(x, b)$ over B. It follows that $M = B \cup \{e_j\}$ and $A = B \cup \{e_i\}$ are isomorphic over B.

Subcase 2.2: More than $\mu^*(c)$ many e_i's lie in $M' \setminus M$. Since $\mu^*(c) + 1 \geq m_c$, we have that $b \in \text{dcl}^{\text{eq}}(A)$. Proceed now as in subcase 2.1. q.e.d.

A structure $M \in \mathcal{K}^\mu$ is *rich* if for every finite $B \leq M$ and every finite $B \leq A \in \mathcal{K}^\mu$ there is some B-isomorphic copy of A in M. We will show in the next section that rich structures are models of $T_1 \cup T_2$.

Corollary 20. There is a unique (up to isomorphism) countable rich structure K^μ. Any two rich structures are $(L_1 \cup L_2)_{\infty,\omega}$–equivalent. q.e.d.

6 The theory T^μ

Lemma 21. Let $M \in \mathcal{K}^\mu$, $b \in \mathrm{acl}^{\mathrm{eq}}(M)$, $a \models \phi_c(x,b)$ generic over M and M' the prealgebraic minimal extension $M \cup \{a_1, \cdots a_{n_c}\}$. If M' is not in $\mathcal{K}^\mu$, then one of the following hold.

(a) M' contains a pseudo Morley sequence of c over b, all whose elements but possibly one are contained in M

(b) M' contains a pseudo Morley sequence for some code c' with more than $\mu^*(c')$ many elements in $M' \setminus M$. Moreover, $m_{c'} > 0$.

Proof. Let (e'_i) be a pseudo Morley sequence of c' over b' in M'. If (b) does not hold, it follows that $b' \in \mathrm{dcl}^{\mathrm{eq}}(M)$ by Lemma 14. There must be some e'_i not completely contained in M, which is a M–generic realization of $\phi_{c'}(x, b')$ by Lemma 13. Minimality of M'/M yields that e'_i is some permutation of a. After permutation of the pseudo Morley sequences, we may assume that $e'_i = a$. Hence $\phi_{c'}(x, b') \sim^{k_c} \phi_c(x, b)$, so $c = c'$ and $b = b'$. q.e.d.

Corollary 22.

1. Let c be a prealgebraic code. The statement "M contains no pseudo Morley sequence for c" can be expressed by a universal $L_1 \cup L_2$–sentence.

2. Let c be as above, $M \in \mathcal{K}^\mu$ a model of $T_1 \cup T_2$. The statement "For no $b \in \mathrm{dcl}^{\mathrm{eq}}(M)$ and generic realization a of $\phi_c(x,b)$ is $M \cup \{a_1, \ldots, a_{n_c}\}$ in $\mathcal{K}^\mu$" can be expressed by an inductive $L_1 \cup L_2$–sentence.

Proof. Ad 1. Let $\Psi^i(\bar{x})$ be quantifier-free and T_i–equivalent to $\exists y\, \Psi_{c_i}(\bar{x}, y)$. Hence, the desired sentence is

$$\neg \exists \bar{x}\, (\Psi^1(\bar{x}) \wedge \Psi^2(\bar{x})).$$

Ad 2. Let $i \in \{1,2\}$ and M be some elementary substructure of $\mathbb{C}_i$. Take $m \in M$ and $\phi(x, m)$ be some L_i–formula of rank k and degree 1. Pick some M–generic realization $a \in \mathbb{C}_i$ of $\phi(x, m)$. Then every quantifier-free property $\psi(a, m)$ of a, m is equivalent to some quantifier-free property $\psi^*(m)$ of m: Set

$$\psi^*(y) = \mathrm{MR}_x\big(\phi(x,y) \wedge \psi(x,y)\big) \doteq k.$$

The above shows that for all $M \in \mathcal{K}$ and for all M–generic realization a of $\phi_c(x,b)$ every $L_1 \cup L_2$–sentence over $M \cup \{a_1 \ldots, a_{n_c}\}$ can be transformed into one $L_1 \cup L_2$–sentence over M, b.

The claim follows now by Lemma 21, since only finitely many codes c' need to be considered in case (b), namely those with $m_{c'} > 0$ and

$$(\mu^*(c') + 1)n_{c'} \leq |M' \setminus M| = n_c.$$

q.e.d.

Models M of the $L_1 \cup L_2$-theory T^μ will be described by the following properties. Lemmas 11 and 12 and the above show that the axioms can be first-order described. The following are the axioms of T^μ:

(a) $M \in \mathcal{K}^\mu$

(b) $T_1 \cup T_2$

(c) No prealgebraic minimal extension of M lies in $\mathcal{K}^\mu$.

It is easy to see that M is a model of (b) if and only if M is infinite and has no algebraic minimal extensions. Hence, M is a model of (b) and (c) if and only if M is infinite and has no minimal (or proper) extensions $M' \in \mathcal{K}^\mu$ with $\delta(M'/M) = 0$.

Theorem 23. An $L_1 \cup L_2$ structure is rich if and only if it is an ω-saturated model of T^μ.

Proof. Let $M \models T^\mu$ be ω-saturated. To show that M is rich, we need only consider a finite strong subset B of M and a minimal strong extension A of B in $\mathcal{K}^\mu$. We aim to show that M contains a B-isomorphic copy of A.

Case (I/II): A/B is algebraic or prealgebraic. We can amalgam M and A in $\mathcal{K}^\mu$, however M has no proper algebraic or prealgebraic extensions in $\mathcal{K}^\mu$. Therefore there must be a B–copy of A in M.

Case (III): $A = B \cup \{a\}$ is transcendental. We want some $a' \in M$ transcendental over B with $B \cup \{a'\} \leq M$. By saturation of M (note that a' satisfies a given partial type over B), it suffices to find a' in some elementary extension M' of M. If M' is uncountable, there is some $a' \in M' \setminus \mathrm{cl}_d(B)$ by Corollary 17. Equivalently, $B \cup \{a'\} \leq M'$.

Let M be now a rich structure.

Axiom (b): Let a be some element in $\mathrm{acl}_1(M)$ transcendental over M in T_2. There is some finite subset B of M witnessing 1–algebraicity of a. We may assume that $B \leq M$. By Lemma 18, $B \leq B \cup \{a\} \in \mathcal{K}^\mu$, so there exists some copy of a over B in M. It follows that M is acl_1–closed. Since M is infinite[9] it is a model of T_1. Likewise for T_2.

Axiom (c): Let a be a generic realization of $\phi_c(x,b)$ over M such that $M \cup \{a\}$ is in $\mathcal{K}^\mu$. Choose some finite strong subset C of M with $b \in \mathrm{dcl}^{\mathrm{eq}}(C)$. Then $C \leq C \cup \{a\}$, so there is a copy a' of a in M over C with $C' = C \cup \{a'\} \leq M$ by richness of M. Iterate to obtain a sufficiently large Morley sequence $a', a'', \ldots$ of $\phi_c(x,b)$ in M. This contradicts that $M \in \mathcal{K}^\mu$.

Choose now some ω–saturated $M' \equiv M$. The first part of the proof yields that M' is rich. So $M' \equiv_{\infty,\omega} M$. So M is also ω–saturated. q.e.d.

7 Proof of the Main Theorem

In this section we drop the QE-Hypothesis of §3. Hence in our class $\mathcal{K}$ we replace isomorphic embeddings by *bi-elementary maps*, i.e., maps which are both T_1 and T_2 elementary.

Corollary 24. T^μ is complete. Two tuples a and a' in models M and M' have the same type if and only if there is some bi-elementary bijection

$$f : \mathrm{cl}(a) \to \mathrm{cl}(a')$$

with $f(a) = a'$.

Proof. The structure $\mathcal{K}^\mu$ is a model of T^μ, so T^μ is consistent. Let M be any model of T^μ. By Theorem 23 there is some rich $M' \equiv M$. Since $M' \equiv_{\infty,\omega} \mathcal{K}^\mu$, we have that T^μ is complete.

Let $M \prec N$ and $M' \prec N'$ be two ω–saturated elementary extensions. It is easy to see[10] that $M \leq N$ and $M' \leq N'$, i.e., closure does not change. An isomorphism $f : \mathrm{cl}(a) \to \mathrm{cl}(a')$ belongs to some back-and-forth system of partial isomorphisms between finite strong subsets of M' and N'. Hence f is elementary.

For the other direction, suppose that a and a' have the same type. Then there is some isomorphic embedding $f : \mathrm{cl}(a) \to M'$ mapping a to a'. Write

---

[9]This follows also from Lemma 18.

[10]If $M \not\leq N$, there is some $a \in N$ with $\delta(a/M) < 0$. Find some finite $B \leq M$ with $\delta(a/B) < 0$. a realizes some $L_1 \cup L_2$-formula witnessing this fact. However $\phi(x,b)$ is not realized in M, so $M \not\prec N$.

$A' = f(\mathrm{cl}(a))$. Then $\mathrm{d}(a) = \delta(\mathrm{cl}(a)) = \delta(A')$. Therefore $\mathrm{d}(a') \leq \mathrm{d}(a)$ and by symmetry $\mathrm{d}(a') = \mathrm{d}(a)$. Note that A' has no proper subset A'' containing a' with $\delta(A'') = \mathrm{d}(a')$ since $\mathrm{cl}(a)$ does not. Hence $A' = \mathrm{cl}(a')$. q.e.d.

Theorem 25. T^μ is strongly minimal and d is the dimension function of the natural pregeometry on models of T^μ. In particular
$$\mathrm{MR}(\bar{a}/B) = \mathrm{d}(\bar{a}/B).$$

Proof. All types $\mathrm{tp}(a/B)$ with $\mathrm{d}(a/B) = 0$ are algebraic by Corollary 16. It follows from Corollary 24 that there is only one type with $\mathrm{d}(a/B) = 1$.[11] Therefore T is strongly minimal. The rest follows easily since d describes the algebraic closure. q.e.d.

The above proves Theorem 1.

8 Remarks

It is easy to prove that T^μ has the following properties:

- T^μ has the DMP.

- For each $i = 1, 2$ every L_i-formula $\phi(x, b)$ preserves its Morley rank and degree from T_i in T^μ.

- Let M be a model of T^μ which is an elementary substructure of N according to both T_1 and T_2. Then M is an elementary substructure of N [H$_2$99].

We prove the last property. Let M and N be models of T^μ with $M \upharpoonright L_i \prec N \upharpoonright L_i$ for $i = 1, 2$. We show that $M \leq N$. It follows then from Corollary 24 that $N \prec M$.

Recall that M has no extension M' in N with $\delta(M'/M) = 0$. If M is not strong, there is some $a \in N$ with $\delta(a/M) = -1$ which is algebraic over M in both T_1 and T_2. Contradiction.

References.

[B$_0$H$_2$00]   John Baldwin and Kitty L. Holland. Constructing ω-stable structures: rank 2 fields. *J. Symbolic Logic*, 65:371–391, 2000.

---

[11]This type has a unique extension to $\mathrm{cl}(B)$ hence $\mathrm{cl}(B) \cup \{a\}$ is strong in the model we work in.

[$B_1H_1MW\infty$] Andreas Baudisch, Martin Hils, Amador Martin-Pizarro, and Frank Wagner. Die böse Farbe. *Submitted*.

[B_1MZ06] Andreas Baudisch, Amador Martin-Pizarro, and Martin Ziegler. On fields and colors. *Algebra i Logika*, 45:92–105, 2006.

[$B_1MZ\infty a$] Andreas Baudisch, Amador Martin-Pizarro, and Martin Ziegler. Fusion over a vector space. *To appear in Journal of Mathematical Logic*.

[$B_1MZ\infty b$] Andreas Baudisch, Amador Martin-Pizarro, and Martin Ziegler. Red fields. *To appear in Journal of Symbolic Logic*.

[H_0H_106] Assaf Hasson and Martin Hils. Fusion over sublanguages. *J. Symbolic Logic*, 71:361–398, 2006.

[H_299] Kitty L. Holland. Model completeness of the new strongly minimal sets. *J. Symbolic Logic*, 64:946–962, 1999.

[H_392] Ehud Hrushovski. Strongly minimal expansions of algebraically closed fields. *Israel J. Math.*, 79:129–151, 1992.

[H_393] Ehud Hrushovski. A new strongly minimal set. *Ann. Pure Appl. Logic*, 62:147–166, 1993.

[P99] Bruno Poizat. Le carré de l'egalité. *J. Symbolic Logic*, 64(3):1338–1355, 1999.

[$Z\infty$] Martin Ziegler. Fusion of structures of finite morley rank. *To appear in Isaac Newton Institute Publications*.

Received: March 24, 2006;
In revised version: August 12, 2006;
Accepted by the editors: December 4, 2006.

The Theory of Square-like Abelian Groups is Decidable

OLEG BELEGRADEK

Department of Mathematics
Istanbul Bilgi University
80370 Dolapdere–Istanbul, Turkey
E-mail: olegb@bilgi.edu.tr

ABSTRACT. A group is called square-like if it is universally equivalent to its direct square. It is known that the class of all square-like groups admits an explicit first order axiomatization but its theory is undecidable. We prove that the theory of square-like *abelian* groups is decidable. This answers a question posed by D. Spellman. Also, we show that for any $n \geq 2$ the theory of square-like n-step nilpotent groups is hereditarily undecidable.

Introduction

A group G is called *discriminating* [$B_0MR_0$00] if every group separated by G is discriminated by G. Here G is said to separate (discriminate) a group H if for any non-identity element (finite set of non-identity elements) of H there is a homomorphism from H to G which does not map the element (any element of the set) to the identity. A group G is discriminating iff G discriminates G^2 [$B_0MR_0$00]. In particular, if G embeds G^2 then G is discriminating.

A group G is called *square-like* [FGMS02] if the groups G^2 and G are universally equivalent. Any discriminating group is square-like [FGMS01]. The notions of discriminating and square-like group were studied in [$B_0MR_0$00, $B_1$04, FGMS01, FGMS02, FGS02, FGS04a, FGS04b, FGS05].

The class of square-like groups is first order axiomatizable [FGMS02], and the theory of the class is computably enumerable; an explicit first order axiom system was suggested in [$B_1$03, $B_1$04], and also presented in [FGS04b]. In [FGMS02] square-like abelian groups were characterized in terms of Szmielew invariants.

The subclass of discriminating groups is not first order axiomatizable [FGMS02]. Every square-like group is elementarily equivalent to a discriminating group [$B_1$04, FGS04a]; so the class of square-like groups is the axiomatic closure of the class of discriminating groups.

The theory of square-like groups is undecidable [$B_1$04, FGS04a]. The argument in [FGS04a] is based on the obvious observation that any group embeds in a discriminating group, and so the universal theory of square-like groups coincide with the universal theory of all groups. The latter is undecidable because there exist finitely presented groups with unsolvable word problem. In [$B_1$04] a discriminating group that interprets the ring of integers is constructed; any theory that has the group as a model (and, in particular, the theory of square-like groups) is undecidable.

The main result of the present paper is that the theory of square-like abelian groups is decidable. This answers a question posed by Dennis Spellman [S05]. As a byproduct, we found characterizations of discriminating and square-like Szmielew groups. Also, we show that for any $n \geq 2$ the theory of square-like n-step nilpotent groups is hereditarily undecidable.

The author is grateful to the referee for suggesting an approach to a proof of the latter result.

1 Preliminaries

Here we collect some known definitions and facts we will use in the proofs.

Fact 1. [B_0MR$_0$00, Proposition 1] A group G is discriminating iff G discriminates G^2. In particular, G is discriminating if G embeds G^2.

Fact 2. [B_0MR$_0$00, Proposition 2] The direct product (restricted or not) of any family of discriminating groups is a discriminating group.

Fact 3. [B_0MR$_0$00, Proposition 3] Any torsion-free abelian group is discriminating.

Fact 4. [FGMS01, Lemma 2.1] Any discriminating group is square-like.

Fact 5. [FGMS02, Theorem 3] The class of square-like groups is first order axiomatizable.

Fact 6. [$B_1$04, Proposition 3.5] Any End(G)-invariant subgroup of a discriminating group G is trivial or infinite.

Let A be an abelian group. For a positive integer n we denote

$$nA = \{na : a \in A\}, \quad A[n] = \{a \in A : na = 0\},$$

and write $\delta(A)$ for the largest divisible subgroup of A. We write $nA[k]$ for $(nA)[k]$. The subgroups nA, $A[n]$, $nA[k]$, and $\delta(A)$ are End(A)-invariant. We write $A^{(\kappa)}$ for the direct sum of κ copies of A.

We write $\mathbb{Q}$ for the additive group of all rational numbers, and $\mathbb{Z}_{(p)}$ for the additive group of rational numbers with denominator not divisible by a prime p. We write $\mathbb{Z}(n)$ for the cyclic group of order n, and $\mathbb{Z}(p^\infty)$ for the Prüfer p-group.

A *Szmielew group* is defined to be an abelian group of the form

$$\bigoplus_{p \text{ prime}} [\bigoplus_{n>0} \mathbb{Z}(p^n)^{(\kappa_{p,n-1})} \oplus \mathbb{Z}(p^\infty)^{(\lambda_p)} \oplus \mathbb{Z}_{(p)}^{(\mu_p)}] \oplus \mathbb{Q}^{(\nu)} \quad (\star)$$

where $\kappa_{p,n-1}, \lambda_p, \mu_p, \nu$ are cardinals $\leq \omega$.

For a prime p, we call a Szmielew group of the form

$$\bigoplus_{n>0} \mathbb{Z}(p^n)^{(\kappa_{p,n-1})} \oplus \mathbb{Z}(p^\infty)^{(\lambda_p)} \oplus \mathbb{Z}_{(p)}^{(\mu_p)} \oplus \mathbb{Q}^{(\nu)}$$

a *p-Szmielew* group.

Fact 7. [H93, Lemma A.2.3] Every abelian group is elementarily equivalent to a Szmielew group.

Let p be a prime, and $n, k < \omega$. Let $\Phi_k(p, n)$ and $\Phi^k(p, n)$ be the sentences that say about an abelian group B that

$$\dim_p(p^n B[p]/p^{n+1} B[p]) = k \quad \text{and} \quad \dim_p(p^n B[p]/p^{n+1} B[p]) > k,$$

$\Theta_k(p, n)$ and $\Theta^k(p, n)$ be the sentences that say that

$$\dim_p(p^n B[p]) = k \quad \text{and} \quad \dim_p(p^n B[p]) > k,$$

$\Gamma_k(p, n)$ and $\Gamma^k(p, n)$ be the sentences that say that

$$\dim_p(p^n B/p^{n+1} B) = k \quad \text{and} \quad \dim_p(p^n B/p^{n+1} B) > k,$$

$\Delta_k(p, n)$ and $\Delta^k(p, n)$ be the sentences that say that

$$|p^n B| = k \quad \text{and} \quad |p^n B| > k.$$

The sentences defined above are called the Szmielew invariant sentences. Note that $|B| = k$ and $|B| > k$ can be expressed as $\Delta_k(p, 0)$ and $\Delta^k(p, 0)$, for any prime p.

Fact 8. [H93, Section A.2] If A is the Szmielew group $(\star)$ then

- $A \models \Phi_k(p, n)$ iff $\kappa_{p,n} = k$,
- $A \models \Phi^k(p, n)$ iff $\kappa_{p,n} > k$,
- $A \models \Theta_k(p, n)$ iff $\lambda_p + \kappa_{p,n} + \kappa_{p,n+1} + \cdots = k$,

- $A \models \Theta^k(p,n)$ iff $\lambda_p + \kappa_{p,n} + \kappa_{p,n+1} + \cdots > k$,

- $A \models \Gamma_k(p,n)$ iff $\mu_p + \kappa_{p,n} + \kappa_{p,n+1} + \cdots = k$,

- $A \models \Gamma^k(p,n)$ iff $\mu_p + \kappa_{p,n} + \kappa_{p,n+1} + \cdots > k$.

Fact 9. [H93, Theorem A.2.7] Every sentence of the first order language of abelian groups is equivalent, modulo the theory of abelian groups, to a positive Boolean combination of Szmielew invariant sentences.

Fact 10. [H93, Theorem A.2.7] Two abelian groups are elementarily equivalent iff they satisfy the same Szmielew invariant sentences.

Abusing terminology, we call a sentence of the language of abelian groups *consistent* if it is true in some abelian group. By Fact 7, a sentence is consistent iff it holds in some Szmielew group.

Fact 11. [H93, Theorem A.2.8] There is an algorithm that, given a finite conjunction of Szmielew invariant sentences, decides whether it holds in some Szmielew group.

Facts 9 and 11 are main ingredients of a proof of the Szmielew theorem on decidability of the theory of abelian groups; actually, they immediately imply the result. Indeed, given a sentence ϕ, by Fact 9 and computable enumerability of the theory of abelian groups, we can effectively find a positive Boolean combination θ of Szmielew invariant sentences that is equivalent to $\neg\phi$, modulo the theory. A sentence ϕ is not in the theory iff θ is consistent; the latter can be effectively checked, by Fact 11.

We will use a similar method in our proof of decidability of the theory of square-like abelian groups.

2 Discriminating and square-like Szmielew groups

Let A be the Szmielew group $(\star)$. For a prime p, let $I_p = \{n : \kappa_{p,n-1} > 0\}$. In case when the set I_p is finite and nonempty, ℓ_p denotes its maximal element; clearly, $\kappa_{p,\ell_p-1} > 0$.

Proposition 12. The following are equivalent:

(1) A is discriminating;

(2) for any prime p one of the following holds:

    (i) $\lambda_p = \omega$,

    (ii) $\lambda_p = 0$, and if I_p is finite and nonempty then $\kappa_{p,\ell_p-1} = \omega$.

Proof. (1)⇒(2). Suppose (1). Let p be a prime. The subgroup $\delta(A) \cap A[p]$ is $\mathrm{End}(A)$-invariant, and hence is trivial or infinite, by Fact 6. Then λ_p is 0 or ω. Suppose $\lambda_p = 0$, and I_p is finite and nonempty. Then the $\mathrm{End}(A)$-invariant subgroup $p^{\ell_p-1}A[p]$ is nontrivial and hence infinite, again by Fact 6. Then $\kappa_{p,\ell_p-1} = \omega$.

(2)⇒(1). Suppose (2). Then for any prime p the group

$$\bigoplus_{n>0} \mathbb{Z}(p^n)^{(\kappa_{p,n-1})} \oplus \mathbb{Z}(p^\infty)^{(\lambda_p)}$$

embeds it square. So $A = B \oplus C$, where B embeds B^2, and C is torsion-free. By Facts 1, 3, and 2, A is discriminating.   q.e.d.

Proposition 13. *The following are equivalent:*

(1) *A is square-like;*

(2) *for any prime p one of the following holds:*

   (i) $\lambda_p = \omega$,

   (ii) $\lambda_p = 0$, and if I_p is finite and nonempty then $\kappa_{p,\ell_p-1} = \omega$,

   (iii) $0 < \lambda_p < \omega$, and I_p is infinite.

Proof. (1) ⇒ (2). Suppose (2) fails. Then, for some prime p, (i), (ii), (iii) all fail. There are two possibilities:

(a) $\lambda_p = 0$, the set I_p is finite, nonempty, and $\kappa_{p,\ell_p-1} < \omega$,

(b) $0 < \lambda_p < \omega$, and the set I_p is finite.

Suppose (a). Let $\kappa = \kappa_{p,\ell_p-1}$. We have

$$|p^{\ell_p-1}A[p]| = p^\kappa, \qquad |p^{\ell_p-1}A^2[p]| = p^{2\kappa}.$$

Suppose (b). Put $\ell = \ell_p$ if $I_p \neq \emptyset$, and $\ell = 0$ otherwise. We have

$$|p^\ell A[p]| = p^{\lambda_p}, \qquad |p^\ell A^2[p]| = p^{2\lambda_p}.$$

For any positive integers s and t there is an existential sentence that says about an abelian group B that $|sB[p]| \geq t$. Therefore in both cases (a) and (b) the groups A and A^2 are not universally equivalent, and so (1) fails.

(2) ⇒ (1). Suppose (2). Let A' be the Szmielew group obtained from A by replacing

$$\bigoplus_{n>0} \mathbb{Z}(p^n)^{(\kappa_{p,n-1})} \oplus \mathbb{Z}(p^\infty)^{(\lambda_p)}$$

with
$$\bigoplus_{n>0} \mathbb{Z}(p^n)^{(\kappa_{p,n-1})},$$

for all p satisfying (iii). Then A' is discriminating, by Proposition 12. Hence A' is square-like, by Fact 4. It is easy to check that A and A' satisfy the same Szmielew invariant sentences; therefore, by Fact 10, $A \equiv A'$. Then, by Fact 5, the group A is square-like, too. q.e.d.

Corollary 14. *Any square-like abelian group is elementarily equivalent to a discriminating Szmielew group.*

Proof. Let B be a square-like abelian group. By Fact 7, B is elementarily equivalent to a Szmielew group A. By Fact 5, A is square-like. The argument at the end of the proof of Proposition 13 shows that A is elementarily equivalent to a discriminating Szmielew group A'. q.e.d.

3 Main result

Theorem 15. *The theory of square-like abelian groups is decidable.*

Proof. We need to find an algorithm which, given a sentence ϕ of the language of abelian groups, decides whether ϕ is true in some square-like abelian group, or, equivalently by Corollary 14, in some discriminating Szmielew group. By Fact 9, ϕ is equivalent, modulo the theory of abelian groups, to a positive Boolean combination θ of Szmielew invariant sentences. Since the theory of abelian groups is computably enumerable, θ can be found effectively. We may assume that θ is $\bigvee_i \theta_i$, where each θ_i is a conjunction of finitely many Szmielew invariant sentences. So it suffices to prove

Claim 16. *There exists an algorithm that, given a consistent conjunction ψ of finitely many Szmielew invariant sentences, decides whether ψ holds in some discriminating Szmielew group.*

For a prime p, we call a conjunction of formulas of the forms
$$\Phi_k(p,n),\ \Theta_k(p,n),\ \Gamma_k(p,n), \Delta_k(p,n),$$
$$\Phi^k(p,n),\ \Theta^k(p,n),\ \Gamma^k(p,n),\ \Delta^k(p,n)$$
a p-conjunction. To prove the Claim, we show that

(A) *there exists an algorithm that, given a prime p and a consistent p-conjunction ψ, decides whether ψ holds in some discriminating p-Szmielew group*, and

(B) the Claim follows from (A).

First we show (B): assuming (A), we prove the Claim.

Let ψ be a conjunction of Szmielew invariant sentences, which holds in a Szmielew group A. We have $\psi = \bigwedge_p \psi_p$, where p runs over a finite set of primes, and ψ_p is a p-conjunction. There are three possibilities:

(a) ψ has no conjuncts of the form $\Delta_k(p,n)$;

(b) ψ has some conjuncts $\Delta_k(p,n)$ and $\Delta_\ell(q,m)$ with $p \neq q$;

(c) ψ has a conjunct $\Delta_k(p,n)$, but has no conjuncts $\Delta_\ell(q,m)$ with $p \neq q$.

The following three lemmas prove (B).

Lemma 17. Assume (a). The following are equivalent:

(i) ψ holds in some discriminating Szmielew group,

(ii) for all p the sentence ψ_p holds in some discriminating p-Szmielew group.

Proof. Suppose (i). We have $A = \bigoplus_p A(p)$, where $A(p)$ is a p-Szmielew group. Let p be a prime. Then $A(p) \oplus \mathbb{Q}$ is a discriminating p-Szmielew group, by Proposition 12. Also, $A(p) \oplus \mathbb{Q} \models \psi_p$ because of (a). So (ii) holds.

Suppose (ii). For every prime p choose a discriminating p-Szmielew group $A(p)$ in which ψ_p holds. By Proposition 12, the Szmielew group $A = \bigoplus_p A(p)$ is discriminating. For every p we have $A \models \psi_p$, because $A(p) \models \psi_p$ and ψ satisfies (a). Therefore $A \models \psi$. So (i) holds. q.e.d.

Lemma 18. Let B be a discriminating abelian group.

(1) If $\Delta_k(p,n)$ or $\neg \Delta^k(p,n)$ holds in B then $p^n B = 0$.

(2) Assume (b). If $B \models \psi$ then $B = 0$.

Proof. The subgroup $p^n B$ is $\mathrm{End}(B)$-invariant and finite of order at most k. By Fact 6, (1) follows. In order to prove (2), observe that by (1), $p^n B = q^m B = 0$, and hence $B = 0$. q.e.d.

Thus, for any ψ with (b), in order to decide whether there is a discriminating Szmielew group that satisfies ψ, we need to decide whether ψ holds in the trivial group, which can be done effectively.

Lemma 19. Assume (c). Then ψ holds in some discriminating Szmielew group if and only if

(i) For any $q \neq p$ and $\ell > 0$, in ψ there are no conjuncts of the forms

$$\Phi^\ell(q,m),\ \Theta^\ell(q,m),\ \Gamma^\ell(q,m),\ \Phi_\ell(q,m),\ \Theta_\ell(q,m),\ \Gamma_\ell(q,m);$$

(ii) For any $q \neq p$, in ψ there are no conjuncts of the forms

$$\Phi^0(q,m),\ \Theta^0(q,m),\ \Gamma^0(q,m);$$

(iii) the p-conjunction

$$\psi_p \wedge \bigwedge \{\Delta^s(p,0) : s \in S\}$$

holds in some discriminating p-Szmielew group, where S is the set of all s such that $\Delta^s(q,m)$ is a conjunct of ψ, for some $q \neq p$ and some m.

Proof. First suppose that ψ holds in a discriminating Szmielew group A. By (c) and Lemma 18 (1), $p^n A = 0$, and so A is a p-Szmielew group. Therefore (i) and (ii) hold. Let $s \in S$. Then for some m and $q \neq p$ we have $A \models \Delta^s(q,m)$, that is, $|q^m A| > s$. As $p^n A = 0$, we have $q^m A = A$; thus $|A| > s$. Then $A \models \Delta^s(p,0)$. So (iii) holds.

Now suppose (i)–(iii) hold. By (iii) there is a discriminating p-Szmielew group A in which ψ_p and $\{\Delta^s(p,0) : s \in S\}$ are true. We show that $A \models \psi$. Since $\Delta_k(p,n)$ is a conjunct of ψ, we have $p^n A = 0$, by Lemma 18 (1). As A is a p-Szmielew group, all the sentences $\Phi_0(q,m),\ \Theta_0(q,m),\ \Gamma_0(q,m)$ with $q \neq p$ hold in A. Due to (i) and (ii), it remains to show that if $\Delta^s(q,m)$ is a conjunct of ψ, where $q \neq p$, then it holds in A. Suppose not. Then $q^m A = 0$, by Lemma 18 (1). Therefore $A = 0$, contrary to $A \models \Delta^s(p,0)$.
q.e.d.

Now we prove (A). From now on, *let p be a fixed prime, and ψ be a p-conjunction which holds in some Szmielew group A.* We will show how to decide whether ψ holds in some discriminating p-Szmielew group.

There are four possibilities:

(a) ψ has a conjunct $\Delta_k(p,n)$ with $k \neq 1$;

(b) ψ has a conjunct $\Theta_k(p,n)$ with $k > 0$;

(c) ψ has no conjuncts of the forms $\Delta_k(p,n)$ and $\Theta_k(p,n)$;

(d) ψ has a conjunct $\Delta_1(p,n)$ or $\Theta_0(p,n)$, but (a) and (b) fail.

Lemma 20. If (a) then ψ fails in every discriminating abelian group.

Proof. Suppose ψ holds in an abelian group B. Then $|p^n B| = k \neq 1$, and so $p^n B$ is a nontrivial finite $\mathrm{End}(B)$-invariant subgroup. Therefore B is not discriminating, by Fact 6. q.e.d.

Lemma 21. If (b) then ψ fails in every discriminating Szmielew group.

Proof. Suppose $A \models \psi$, and A is a discriminating Szmielew group. Then
$$\omega > k = \lambda_p + \kappa_{p,n} + \kappa_{p,n+1} + \dots.$$
Hence $\lambda_p < \omega$ and so, by Proposition 12, $\lambda_p = 0$. Then
$$0 < \kappa_{p,n} + \kappa_{p,n+1} + \dots < \omega,$$
and so I_p is finite. Then we have $n < \ell_p$, and $\kappa_{p,\ell_p-1} < \omega$. In this case A is not discriminating, by Proposition 12. A contradiction. q.e.d.

Lemma 22. If (c) then ψ holds in some discriminating p-Szmielew group.

Proof. We have $A = \bigoplus_q A(q)$, where $A(q)$ is a q-Szmielew group. Put
$$A'(p) := A(p) \oplus \mathbb{Z}(p^\infty)^{(\omega)}.$$
By Proposition 12, $A'(p)$ is a discriminating p-Szmielew group. Moreover, $A'(p) \models \psi$. Indeed, for any sentence θ of one of the forms
$$\Phi_k(p, n),\ \Phi^k(p, n),\ \Theta^k(p, n),\ \Gamma_k(p, n),\ \Gamma^k(p, n),\ \Delta^k(p, n)$$
if $A \models \theta$ then $A'(p) \models \theta$. q.e.d.

It remains to consider case (d). We will need

Lemma 23. For any $n \geq k$ the sentence $\Gamma_\ell(p, k)$ is effectively equivalent in abelian groups to a positive Boolean combination of sentences of the forms $\Gamma_i(p, n)$ and $\Phi_j(p, s)$, where $k \leq s < n$ and $0 \leq i, j \leq \ell$.

Proof. It suffices to show that in abelian groups $\Gamma_\ell(p, k)$ is equivalent to
$$\Gamma'_\ell(p, k) := \bigvee_{i=0}^{\ell} (\Gamma_{\ell-i}(p, k+1) \wedge \Phi_i(p, k)).$$

A Szmielew group A satisfies $\Gamma_\ell(p,k)$ if and only if

$$\mu_p + \kappa_{p,k} + \kappa_{p,k+1} + \cdots = \ell;$$

the latter holds if and only if, for some $i \in \{0,1,\ldots,\ell\}$,

$$\mu_p + \kappa_{p,k+1} + \kappa_{p,k+2} + \cdots = \ell - i \quad \text{and} \quad \kappa_{p,k} = i,$$

which means that $\Gamma'_\ell(p,k)$ holds in A. q.e.d.

Let $n < \omega$ be given. Replace in ψ every conjunct $\Gamma_\ell(p,k)$, where $k < n$, with an equivalent positive Boolean combination of sentences of the forms $\Gamma_i(p,n)$ and $\Phi_j(p,s)$. The resulting formula is equivalent to a disjunction of p-conjunctions in each of which there is no conjunct $\Gamma_\ell(p,k)$ with $k < n$. Therefore it remains to prove the following statement, which allows to decide whether ψ holds in some discriminating p-Szmielew group, in case (d).

Lemma 24. Suppose that ψ has

(a) a conjunct $\Delta_1(p,n)$ or $\Theta_0(p,n)$;

(b) no conjuncts $\Delta_k(p,m)$ with $k \neq 1$ and $\Theta_k(p,m)$ with $k > 0$;

(c) no conjuncts $\Gamma_\ell(p,s)$ with $s < n$.

Then the following are equivalent:

(1) ψ fails in any discriminating p-Szmielew group;

(2) there exist m with $m < n$ and $i > 0$ such that

    (i) $\Phi_i(p,m)$ is a conjunct of ψ,

    (ii) for every k with $m < k < n$ there is j such that $\Phi_j(p,k)$ is a conjunct of ψ.

Proof. First we show that (b) implies that ψ holds in some p-Szmielew group. If $\Delta_1(p,n)$ is in ψ then $p^n A = 0$; therefore A is a direct sum of cyclic p-groups and hence a p-Szmielew group. Suppose $\Delta_1(p,n)$ is not in ψ. Let $A = \bigoplus_q A(q)$, where each $A(q)$ is a q-Szmielew group. Since ψ is a p-conjunction without conjuncts of the form $\Delta_k(p,n)$, the p-Szmielew group $A(p) \oplus \mathbb{Q}$ satisfies ψ.

So we may assume that A is a p-Szmielew group. By (a),

$$\lambda_p = \kappa_{p,n} = \kappa_{p,n+1} \cdots = 0.$$

Indeed, if $\Delta_1(p,n)$ is in ψ then $p^n A = 0$; if $\Theta_0(p,n)$ is in ψ then

$$0 = \lambda_p + \kappa_{p,n} + \kappa_{p,n+1} + \ldots.$$

In particular, the set I_p is finite.

Suppose (2). Due to (i), we have $\kappa_{p,m} = i > 0$, and therefore $m < \ell_p \le n$. Let $m < k < n$. By (ii) ψ has a conjunct $\Phi_j(p,k)$; then $\kappa_{p,k} = j$. So $\kappa_{p,k} < \omega$ for all k with $m \le k < n$. In particular, $\kappa_{p,\ell_p-1} < \omega$. By Proposition 12, in this case A cannot be discriminating, and (1) follows.

Assuming that (2) is not true, we show that (1) is not true, too.

If $I_p = \varnothing$ then A itself is discriminating, by Proposition 12.

Suppose $I_p \ne \varnothing$. First we show that there is $k < n$ such that $\kappa_{p,r} = 0$ for $r > k$, and for every j the sentence $\Phi_j(p,k)$ is not a conjunct of ψ. Let $m = \ell_p - 1$ and $i = \kappa_{p,m}$. Then $m < n$ and $i > 0$. If (i) fails, put $k := m$. If (i) holds then (ii) fails, and therefore there is k with $m < k < n$ such that for every j the sentence $\Phi_j(p,k)$ is not a conjunct of ψ.

By Proposition 12, the p-Szmielew group $A \oplus \mathbb{Z}(p^{k+1})^{(\omega)}$ is discriminating. Moreover,

$$A \oplus \mathbb{Z}(p^{k+1})^{(\omega)} \models \psi.$$

Indeed, by (c) and the choice of k, a conjunct θ of ψ can have only the forms

$$\Phi_j(p,r),\ \Theta_0(p,n),\ \Gamma_j(p,s),\ \Delta_1(p,n),$$

where $r \ne k$ and $s \ge n$, or the forms

$$\Phi^j(p,t),\ \Theta^j(p,t),\ \Gamma^j(p,t),\ \Delta^j(p,t).$$

Therefore $A \models \theta$ implies $A \oplus \mathbb{Z}(p^{k+1})^{(\omega)} \models \theta$, for all such θ. Here we use that $s \ge n > k$ when consider θ of the forms $\Theta_0(p,n)$ and $\Gamma_j(p,s)$. q.e.d.

The proof of Theorem 15 is completed. q.e.d. (Theorem 15)

4 Undecidability results for square-like groups

Proposition 25. *The universal theory of square-like nilpotent groups is undecidable. For any n, the universal theory of square-like n-step nilpotent groups is decidable.*

Proof. The universal theory of square-like nilpotent groups coincides with the universal theory of nilpotent groups because any group G embeds in the group G^ω, which is discriminating. As any finitely generated nilpotent group is residually finite, the universal theory of nilpotent groups coincides with the universal theory of finite nilpotent groups. The latter is undecidable [K83].

Similarly, the universal theory of square-like n-step nilpotent groups coincides with the universal theory of finite n-step nilpotent groups. The latter is decidable because, obviously, the universal theory of n-step nilpotent groups is computably enumerable, and the universal theory of finite n-step nilpotent groups is co-computably-enumerable. q.e.d.

We show that for any $n \geq 2$ the theory of square-like n-step nilpotent groups is hereditarily undecidable, by constructing various discriminating n-step nilpotent groups with hereditarily undecidable theory.

A theory is called *hereditarily undecidable* if all of its subtheories are undecidable. Examples of hereditarily undecidable theories are the theories of the rings $\mathbb{Z}$ and $\mathbb{F}_p[x]$, for any prime p [R$_1$51]. If a structure M with hereditarily undecidable theory is interpretable in a structure N (with parameters), then the theory of N is hereditarily undecidable [E65].

For a ring with unit R and $n \geq 2$, let $\mathrm{UT}_n(R)$ denote the group of all upper unitriangular $n \times n$ matrices over R; the group is $(n-1)$-step nilpotent. If $n \geq 3$, the group $\mathrm{UT}_n(R)$ interprets the ring R [B$_1$99, Corollary 1.6.2].

For a nonempty set I, let M^I denote the Cartesian power of a structure M.

Lemma 26. *For any $n \geq 3$, the group $\mathrm{UT}_n(R)^I$ interprets the ring R.*

Proof. The ring R^I interprets R. Indeed, for $i \in I$ let e_i be the element of R^I such that $e_i(i) = 1$, and $e_i(j) = 0$ for $j \neq i$; then the ideal of R^I generated by the central idempotent e_i is definable in R^I and isomorphic to R. Since $\mathrm{UT}_n(R)^I \simeq \mathrm{UT}_n(R^I)$, and $\mathrm{UT}_n(R^I)$ interprets R^I, the result follows. q.e.d.

The lemma and the facts on hereditary undecidability above imply

Proposition 27. *Let I be infinite, $n \geq 3$, and R be $\mathbb{Z}$ or $\mathbb{F}_p[x]$. Then any class of groups containing the group $\mathrm{UT}_n(R)^I$ has undecidable theory.*

If I is infinite, this group embeds its square, and hence is discriminating and, in particular, square-like. It is torsion-free when $R = \mathbb{Z}$, and of exponent p^{n-1} when $R = \mathbb{F}_p[x]$. These observations allow to prove undecidability of the theories of various classes of square-like groups.

References.

[B₀MR₀00] G. Baumslag, A. G. Myasnikov and V. N. Remeslennikov, Discriminating and co-discriminating groups, *J. Group Theory* **3** (2000), 467–479.

[B₁99] O. Belegradek, Model theory of unitriangular groups, in: *Model theory and applications*, Amer. Math. Soc. Transl. (2) vol. **195**, AMS, Providence, RI, 1999, 1–116.

[B₁03] O. Belegradek, Review of [FGMS02], *Math. Reviews*, MR1914831 (2003d: 20003).

[B₁04] O. Belegradek, Discriminating and square-like groups, *J. Group Theory* **7** (2004), 521–532.

[E65] Yu. L. Ershov, Undecidability of certain fields, *Soviet Math. Dokl.* **6** (1965), 349–352.

[FGMS01] B. Fine, A. M. Gaglione, A. G. Myasnikov and D. Spellman, Discriminating groups, *J. Group Theory* **4** (2001), 463–474.

[FGMS02] B. Fine, A. M. Gaglione, A. G. Myasnikov and D. Spellman, Groups whose universal theory is axiomatizable by quasi-identities, *J. Group Theory* **5** (2002), 365–381.

[FGS02] B. Fine, A. M. Gaglione, D. Spellman, Every abelian group universally equivalent to a discriminating group is elementarily equivalent to a discriminating group, in: S. Cleary, R. Gilman, A. G. Myasnikov, V. Shpilrain (eds.), *Combinatorial and geometric group theory*, Contemp. Math. **296** (Amer. Math. Soc., Providence, RI, 2002), 129–137.

[FGS04a] B. Fine, A. M. Gaglione, D. Spellman, The axiomatic closure of the class of discriminating groups, *Arch. Math.* **83** (2004), 106–112.

[FGS04b] B. Fine, A. M. Gaglione, D. Spellman, Discriminating and square-like groups. I. Axiomatics, in: A. G. Myasnikov, V. Shpilrain (eds.), *Group Theory, statistics and cryptography*, Contemp. Math. **360** (Amer. Math. Soc., Providence, RI, 2004), 35–46.

[FGS05] B. Fine, A. M. Gaglione, D. Spellman, Discriminating and square-like groups. II. Examples, *Houston J. Math.* **31** (2005), 649–674.

[H93] W. Hodges, *Model theory*, Cambridge University Press, 1993.

[K83] O. G. Kharlampovich, Universal theory of the class of finite nilpotent groups is undecidable, *Math. Notes* **33** (1983), 254–263.

[R₁51] R. M. Robinson, Undecidable rings, *Trans. Amer. Math. Soc.* **70** (1951), 137–159.

[S05] D. Spellman, *Private communication*, March 14, 2005.

Received: February 10, 2006;
In revised version: March 25, 2006;
Accepted by the editors: April 18, 2006.

Simple Groups of Finite Morley Rank of Unipotent Type

ALEXANDRE BOROVIK[1]
JEFFREY BURDGES[1]
GREGORY CHERLIN[2]

[1] School of Mathematics
The University of Manchester
Oxford Road
Manchester M13 9PL, United Kingdom

[2] Department of Mathematics
Hill Center-Busch Campus
Rutgers, The State University of New Jersey
110 Frelinghuysen Rd
Piscataway, NJ 08854-8019, United States of America

E-mail: borovik@manchester.ac.uk, burdges@math.rutgers.edu

ABSTRACT. A simple group of finite Morley rank of unipotent type contains no involutions. We explain why this is true, and why we think it is interesting. The story involves model theory, finite group theory, and some aspects of the theory of algebraic groups.

1 Introduction

About thirty years ago it was conjectured by Cherlin and Zilber that every simple group of finite Morley rank is algebraic, more precisely a Chevalley group over an algebraically closed field. In the latter case the Morley rank agrees with the algebraic dimension of the Zariski closure.

But when we assume that a group has finite Morley rank, we are not assuming that the rank function has any topological content. Similar issues arise in the classification of the finite simple groups. In most cases the problem is to identify a given finite simple group with a group of Lie type. In the absence of more geometric tools, one aims to reconstruct the underlying combinatorial geometry (the associated building) via identification of a suitable (B,N)-pair.

The second author was supported by an NSF Postdoctoral Fellowship. The third author was supported by NSF Grant **DMS-0100794**. The third author also acknowledges with gratitude a Humboldt Fellowship sponsored by Felgner in 1978-79 at Tübingen, in a year that was stimulating both group theoretically and model theoretically, not to mention the fact that Amélie began to walk and talk, and has not stopped since.

There is now a considerable body of work on simple groups of finite Morley rank which incorporates many techniques originating in the classification of the finite simple groups. These techniques become quite effective as soon as one has some involutions in the group (the more, the better). Of course in finite simple group theory one begins with the Feit-Thompson Odd Order Theorem, after which involutions are guaranteed (and with the assistance of some character theory the supply of involutions increases rapidly). We do not have an analog of the Odd Order Theorem, and indeed there is a very real possibility that torsion-free simple groups of finite Morley rank exist. In fact, this remains the most plausible scenario for a counterexample to the Algebraicity Conjecture (rather than, say, some minor variation on simple algebraic groups, or an analog of some sporadic finite group).

Work on the Algebraicity Conjecture borrows heavily from the theory of finite simple groups, while incorporating features (like connectivity) belonging to the domain of algebraic groups. Indeed, the notion of Morley rank has no useful analog in the finite theory. There are deep theorems suggesting that the dimensions of definable sets should be reflected by their cardinalities in sufficiently large finite groups (more precisely, by $\log_q$ of the cardinality, with q the order of the base field), but there is no sensible way to bring this idea to bear on the finite case.

However, there is another branch of finite group theory which does make use of something analogous to dimension: black box group theory, in which a large finite group is hidden in a box and elements may be picked randomly and independently, and various computations made with them. In this context "high probability" corresponds well with "full dimension" (or, as one says, "genericity"). We have a little more in the finite Morley rank context: we can speak of relative genericity with respect to definable subgroups, something which is difficult in the black box case, where in addition to identifying the subgroup one must also somehow equip it with a suitable probability measure; this is sometimes possible, and in one specific case there has been a flow of useful technical ideas back and forth between the black box case and the finite Morley rank case, beginning with work of Altseimer in both areas, cf. [$A_1B_0$01]. Using these techniques, one can prove the following [B_0B_1C05].

Theorem 1. Let G be a connected group of finite Morley rank containing an involution. Then G contains an infinite 2-subgroup.

One may say this more pungently. Borrowing the term "Sylow 2-subgroup" from finite and locally finite group theory, the theorem states that if the Sylow 2-subgroup of a connected group of finite Morley rank is nontrivial, then it is infinite. Thus, while we cast no light on the Odd Order

Theorem per se, we do get a dichotomy: once one has an involution, one has something substantial to work with. In particular, we prefer to work with "Sylow° 2-subgroups", defined as maximal *connected* 2-groups. In that language, our theorem says that when the Sylow 2-subgroups are nontrivial, the Sylow° 2-subgroups are also nontrivial.

A group in which the Sylow° 2-subgroups are trivial is said to be of *degenerate type*. We now know in view of the above that these groups contain no involutions. One reason to prefer the 2-Sylow°s to the 2-Sylows is the following structure theorem [B_0P90, B_0N94].

Theorem 2. Let S be a Sylow° 2-subgroup in a group G of finite Morley rank. Then
$$S = U * T$$
with U definable, connected, nilpotent, of bounded exponent and T 2-divisible abelian.

The "$*$" here represents a central product, and the intersection $U \cap T$ is also finite in this case.

Accordingly, once one moves beyond the degenerate type case, one has either U or T (and possibly both) nontrivial, giving a substantial foothold for further analysis.

Now U and T are in some abstract sense the *unipotent* and *semisimple* parts of S. We make this formal as follows.

Definition 3. Let p be a prime.

(1) A *p-unipotent* group is a definable connected p-group of bounded exponent.

(2) A *p-torus* is a divisible abelian p-group.

Note that while we do not incorporate nilpotence into the definition of p-unipotence here, it would be reasonable to do so. For $p = 2$, the nilpotence can be proved in any case. For p odd it remains open.

In the present paper we will deal with another notion of unipotence, the broadest one we can imagine.

Definition 4. A group G of finite Morley rank is of *unipotent type* if it contains no p-torus, for any prime p.

Our discussion here will be centered on the following result, which was used in [B_0B_1C05].

Theorem 5. If G is a simple group of finite Morley rank of unipotent type, then G contains no involutions.

There are two ways to prove this. The long way round is first to invoke the absence of 2-tori to prove that G is either algebraic in characteristic two or of degenerate type, and then to conclude by invoking the structure of simple algebraic groups. This proof takes something like 200 pages and involves material which is not yet fully published [$A_0B_0C\infty$]. Another proof can be extracted from the long one by paying a little attention to what the relevant portions of the proof are, and this saves approximately $199\frac{1}{2}$ pages. We will give that proof here (in [B_0B_1C05] we saved the full 200 by referring to [$A_0B_0C\infty$]).

There are two reasons to take the foregoing theorem seriously. In [B_0B_1C05] our concern was the analysis of connected groups satisfying a *generic equation* of the form

$$x^n = 1 \text{ (generically)}.$$

In particular we showed there that for n a power of 2, such an equation must hold everywhere if it holds generically, and more generally we reduced the analysis of such equations to the case in which n is odd. Thus we cast some light on a long-standing test problem put forward by Poizat, in a way which has only become possible recently as the work on the Algebraicity Conjecture has been brought to bear to reveal structural features of groups of finite Morley rank which are not tied to a purely inductive context.

But apart from this specific application, there is also a methodological point here, and the latter is really our motivation for taking the matter up again. The reason that the long version of the proof is so very long, is that it is based on a full classification result in much the spirit of finite simple group theory. That result runs as follows.

Theorem 6. *Let G be a simple group of finite Morley rank containing a nontrivial 2-unipotent subgroup. Then G is an algebraic group over an algebraically closed field of characteristic two.*

One may distinguish three phases in the development of the proof of this classification theorem, in which auxiliary hypotheses were gradually stripped out. In the first phase one supposed that the group G involved no "bad field" and no degenerate type groups (the latter is reasonable if one takes an inductive approach to the Algebraicity Conjecture; but then one is committed to dealing with the Odd Order Theorem at some point). We will not detail the "bad fields" hypothesis; it was excised relatively early on (though it has not entirely vanished from the landscape, nor should it). Bypassing the Odd Order Theorem comes as more of a surprise, and the experience of finite group theory would not be encouraging in this respect. But as Altınel's habilitation showed [$A_0 01$], there is one very special feature in our situation:

Theorem 7. A unipotent 2-group acting on a degenerate type group must act trivially.

This tends to "uncouple" the odd order problems from the rest of the theory when there are unipotent 2-groups available.

Still, problems arise as one implements this idea. Some of them are addressed in [A_0C03, A_0C04, A_0C05a, A_0C05b], and a full account is in preparation as [$A_0 B_0 C\infty$]. One of the features which has emerged over time is the importance of p-tori for the analysis. This came into view only gradually and was eventually extracted in an explicit way in [C05], just in time to be incorporated into the final draft of [A_0C05b].

In retrospect, we find it very striking that if one assumes p-tori out of existence, everything collapses quickly back to the degenerate case. At the other extreme, the paper [A_0C05b] is largely concerned with the analysis, or exploitation, of the p-tori which arise in a configuration where a group which ought to be SL_2 turns out in fact to have copies of SL_2 (and, in particular, their tori) inside it. In earlier treatments under more restrictive hypotheses, the existing theory of Carter subgroups in solvable groups gave sufficient control, and there was no need to focus one's attention on p-tori as such.

When this broad classification project got fairly under way, under the initiative of the first author, the type of analysis that took place was very much bound up with various special assumptions. Lately there has been a discernible flow back toward structural results which apply quite broadly to groups of finite Morley rank. Several such results have been detected at first within very concrete configurations associated with specific classification problems.

At the same time, some of the early results already fit into this framework, notably the 2-Sylow theory itself (which is a necessary prerequisite for this approach to the Algebraicity Conjecture), as well as the lemma on which Altınel's habilitation is based. Both of these results are proved inductively, but in a self-contained way. Subsequently there have been very general results on "characteristic 0 unipotence" theory feeding back into the theory of Carter subgroups and other areas [$B_1\infty$, FJ05]. The theory of "good tori" emerged in the series [A_0C03, A_0C04, A_0C05a, A_0C05b], notably in the last of these, and was finally detached from specific contexts in [C05]. The Carter theory has also developed further at a very general level [FJ05, J06]. There are strong parallels between the Carter theory and the theory of p-tori, and an ideal theory would combine the best features of both. We will come back to this at the end.

2 Theorem 5

While the hypothesis of simplicity in Theorem 5 simplifies the statement, it makes more sense to prove the result in a more global form, as follows.

Theorem 8. Let G be a connected group of finite Morley rank of unipotent type. Then $G/O_2^\circ(G)$ contains no involutions.

Here we use the notation "$O_2^\circ(G)$" to denote the largest *connected* normal definable 2-subgroup of G, without attempting to define $O_2(G)$ separately, which can be problematic, in general. Of course, our theorem implies that $O_2^\circ(G) = O_2(G)$ for any reasonable definition of the latter!

In the proof of this result, we make free use of Theorem 1, a substantial result in its own right, and so we aim only at showing that $G/O_2^\circ(G)$ is of degenerate type. We may pass to a quotient and suppose that

$$O_2^\circ(G) = 1$$

and then our claim is that G is itself of degenerate type.

Before entering into the proof of Theorem 8, which will be given in the following section, there is more to be said both about both its hypothesis and its conclusion.

2.1 Tori

Our assumption on G is that there is no nontrivial p-torus for any prime p. In practice one prefers to work with definable subgroups as far as possible, and this leads to two variations on the notion of p-torus which are of considerable utility.

Definition 9.

(1) A *decent torus* is a definable divisible abelian group T which is the definable hull of its torsion subgroup T_{tor} (that is, the smallest definable subgroup containing T_{tor}).

(2) A *good torus* is a definable divisible abelian group T such that every definable subgroup T_0 of T is the definable hull of its torsion subgroup $(T_0)_{\mathrm{tor}}$.

Good tori have remarkable rigidity properties. For example, any uniformly definable family of subgroups of a good torus is finite [A_0C04]. Furthermore, the multiplicative group of a field of finite Morley rank and nonzero characteristic is a good torus. This follows from the main result of [W01], and is made explicit in [A_0C04]. So the following clears the air considerably.

Lemma 10. Let G be a group of finite Morley rank. Then the following are equivalent.

(1) G contains no nontrivial decent torus.

(2) There is no prime p for which G contains a nontrivial p-torus.

(3) No definable section of G is a good torus.

These conditions are inherited under passage to definable sections, or to elementary extensions.

Proof. We show first that condition (2) passes to elementary extensions. Suppose that in an elementary extension G^* of G we have some nontrivial p-torus T_0, and consider $A = d(T_0)$. Then A is definable and p-divisible, and contains p-torsion; the existence of such a group passes to the elementary substructure G and contradicts (2). So (2) is preserved by passage to elementary extensions.

As the third condition is inherited by definable sections, it will be sufficient to check the stated equivalences.

The equivalence of the first two is clear.

To see that the third condition implies the first, it suffices to show that any nontrivial decent torus T has a nontrivial good torus as a definable quotient. Indeed, let T_0 be any maximal proper connected definable subgroup of T, and pass to $\bar{T} = T/T_0$. Then $\bar{T}$ is again a decent torus, and now any proper definable subgroup of $\bar{T}$ is finite. So $\bar{T}$ is in fact a good torus.

Finally let us check the implication ($2 \implies 3$). Suppose that (2) holds and (3) fails, and that H is a definable section of G (that is $H = K/N$ with K, N definable and $N \triangleleft K$), which is a good torus. Then H contains a nontrivial p-torus $\bar{T}$ for some p. Each element $\bar{t}$ of $\bar{T}$ lifts to a p-element t of G (as H is a *definable* section) and hence the abelian groups $Z(C(t))$ as t varies over G contain p-subgroups of unbounded order. It follows easily that some elementary extension of G contains a nontrivial p-torus, and thus as we have seen G also contains a nontrivial p-torus, contradicting (2). So (2) implies (3). q.e.d.

2.2 The structure of G

The conclusion of Theorem 8 also deserves further elucidation. That theorem says that the 2-elements in a group of unipotent type really must behave in a unipotent way. We can say a little more: these elements cannot interact in a serious way with the rest of the group.

Proposition 11. Let G be a connected group of finite Morley rank containing no decent torus and let $U = O_2^\circ(G)$. Suppose that G/U is of degenerate type. Then $G = U \cdot C_G(U)$.

As we will see, this is a combination of Zilber's Field Theorem with Wagner's results in the case of characteristic two, which takes on the following form.

Proposition 12. Let H be a connected solvable $p^\perp$-group of finite Morley rank acting faithfully on a nilpotent p-group V of bounded exponent. Then H is a good torus.

Here a $p^\perp$-group is a group which contains no element of order p. Elements of infinite order are permitted.

Proof. We work in the group $G = V \rtimes H$, which is again a solvable group, and we use the theory of the Fitting subgroup.

Observe that $F(G) = V(F(G) \cap H)$ and that $F(G) \cap H$ centralizes V since it is a $p^\perp$-subgroup of $F(G)$. As the action of H is faithful, the intersection $F(G) \cap H$ is trivial. Thus $F(G) = V$, and $H \cong G/F(G)$ is abelian divisible by the structure theory for connected solvable groups of finite Morley rank (an analog of Lie-Kolchin).

Take a G-invariant normal series $V = V_0 > V_1 > \cdots > V_n = (0)$ with successive quotients finite or G-minimal (that is, having no proper definable G-invariant subgroups). The stabilizer of this chain in H (that is, the subgroup acting trivially on each factor) is trivial since the chain consists of definable p-groups, and H is a $p^\perp$-group; this is again an analog of a standard fact from finite group theory which goes over to our context.

Now consider the combined action of H on all of the quotients $A_i = V_i/V_{i+1}$. In other words, if $\bar{H}_i$ is the image of H in $\text{Aut}(A_i)$, we have a definable injection of H into $\prod_i \bar{H}_i$.

Now by Zilber's Field Theorem, the groups $\bar{H}_i$ are subgroups of multiplicative groups of fields of finite characteristic (two). As H is connected, these groups are good tori by Wagner's theorem, and a connected subgroup of a product of good tori is again a good torus, as is easily checked [A$_0$C04]. So H is a good torus.   q.e.d.

Proof of Proposition 11. Let G be a counterexample of minimal rank. Let $A = Z^\circ(U)$. Then $\bar{G} = G/C_G(A)$ is a group of degenerate type acting faithfully on A. Since $\bar{G}$ has degenerate type, its Borel subgroups are $2^\perp$-groups, either by Theorem 1 or by the considerably more classical solvable case of the same result. So by Proposition 12, the Borel subgroups of $\bar{G}$ are good tori, hence trivial. So the connected group $\bar{G}$ is trivial, or in other words

G centralizes A.

On the other hand, by the minimality of G, the group $G/A = (U/A) \cdot C_{G/A}(U/A)$. Let $H/A = C_{G/A}(U/A)$. Then $G = UH$ and $[H,U] \leq A$.

For any definable $2^\perp$-subgroup X of H, we have $[X,U] = 1$ since X acts trivially on both factors U/A and A of the chain $1 \leq A \leq U$: see [A₀B₀C99, Cor. 2.45] (or consider $[X,u]$ for $u \in U$).

It follows that $H/C_H(U)$ is a 2-group, since for each $a \in H$ the definable hull of a is the sum of a 2-group and a $2^\perp$-group, and thus $H \leq UC_H(U)$, and $G = UC_G(U)$. q.e.d. (Proposition 11)

3 Proof of Theorem 8

Our claim is the following: for G connected of unipotent type, $O_2°(G)$ is a Sylow° 2-subgroup. Let us introduce the notation $U_2(G)$ for the subgroup of G generated by all its unipotent 2-subgroups. By an early result of Zilber, analogous to a well-known lemma in the theory of algebraic groups, the group $U_2(G)$ is definable. Another way of phrasing our claim is that $O_2°(G) = U_2(G)$. Notice that in a simple algebraic group G in characteristic two, $U_2(G)$ will be the whole group.

We now consider the structure of a minimal counterexample to the claim.

Lemma 13. *Let G be a group of finite Morley rank and unipotent type, and suppose that $O_2°(G)$ is not a Sylow° 2-subgroup of G. Suppose further that G is of minimal rank among all such groups. Then the following hold.*

(1) *$Z(G)$ is finite and $G/Z(G)$ is simple.*

(2) *For $U \leq G$ a nontrivial definable 2-subgroup, not contained in $Z(G)$, setting $H = N°(U)$ and $V = O_2°(H)$, we have $H = VC_H°(V)$ and H/V is of degenerate type.*

Proof. Ad (1). It follows from the minimality hypothesis that $O_2°(G) = 1$ and that $U_2(G) = G$. As G is connected, any finite normal subgroup is central. We must show that G contains no nontrivial proper definable connected normal subgroup.

Supposing the contrary, then for H a nontrivial proper definable connected subgroup of G, setting $\bar G = G/H$, our minimality hypothesis implies that $\bar G/O_2°(\bar G)$ is of degenerate type. Since we also have $\bar G = U_2(\bar G)$, we find that $\bar G = O_2°(\bar G)$ is a 2-group. Let S be a Sylow° 2-subgroup of G. Then $G = HS$ [PW00], and by Theorem 7 we have $S \leq C(H)$. Thus $S \leq O_2°(G)$, so $S = 1$ and $G = H$, a contradiction.

This proves the first point.

Ad (2). With U, H, and V as specified, observe that $H = N°(U) < G$. Thus by minimality H/V is of degenerate type, and by Proposition 11 we have $H = VC_H(V)$.
q.e.d.

Now let us fix our notation in accordance with the preceding lemma. We may take G to be a group of finite Morley rank and unipotent type, with $O_2°(G) < U_2(G)$, and of minimal rank among such groups, and we may factor out $Z(G)$. We then have the following conditions.

> $G = U_2(G)$ is simple, and for $U \leq G$ a nontrivial definable 2-subgroup, setting $H = N°(U)$ and $V = O_2°(H)$, we have $H = VC_H°(V)$, with H/V of degenerate type. $\quad (*)$

We observe that this condition also implies that the Sylow° 2-subgroups of G are unipotent.

Now, using the notion of strong embedding, borrowed directly from finite group theory, we can arrive quickly at a proof of Theorem 8.

Definition 14. Let G be any group, M a subgroup. Then M is *strongly embedded* in G if the following conditions are satisfied:

(1) For $g \in G$ $M^g \cap M$ contains an involution if and only if $g \in M$.

(2) $M < G$.

In groups of finite Morley rank a weaker variant of this notion is at least as useful, because it is more easily verified and has similar consequences.

Definition 15. Let G be any group, and M a subgroup. Then M is *weakly embedded* in G if the following conditions are satisfied:

(1) For $g \in G$ $M^g \cap M$ contains an infinite 2-subgroup if and only if $g \in M$.

(2) $M < G$.

There are more convenient criteria for strong and weak embedding in groups of finite Morley rank, valid more generally whenever there is a 2-Sylow theory available. For strong embedding, once G is known to contain an involution it suffices to check that M contains a Sylow 2-subgroup S of G, and the centralizer of each involution in S. For weak embedding, once it is known that G has a nontrivial and unipotent Sylow° 2-subgroup, it suffices to check that M contains a Sylow° 2-subgroup S of G, and the normalizer of every nontrivial definable connected subgroup of S.

One of the useful consequences of strong embedding is the following, whose proof in our context is just as in the finite case.

Lemma 16. *Let M be a strongly embedded subgroup in the group G of finite Morley rank. Then the involutions of M are conjugate in M.*

Lemma 17. *Let G be a group of finite Morley rank satisfying the conditions $(*)$ above. Let S be a Sylow° 2-subgroup of G. Then $N(S)$ is strongly embedded in G.*

Proof. We show first that

$$N(S) \text{ is weakly embedded in } G.$$

For this, it suffices to show that for $U \leq S$ nontrivial, connected, and definable, we have $N(U) \leq N(S)$.

Supposing the contrary, take $U \leq S$ maximal connected definable such that $H = N(U)$ is not contained in $N(S)$. Evidently $U < S$. Let $V = O_2°(N(U))$. As V is a Sylow° 2-subgroup of H, it follows that $U < V$, and by maximality of U we have $N(V) \leq N(S)$. Now $N(U) \leq N(V) \leq N(S)$, a contradiction. This proves that $N(S)$ is weakly embedded in G.

Now it suffices to prove that $C(i) \leq N(S)$ for any involution $i \in N(S)$ (note that $N(S)$ contains a Sylow 2-subgroup of G). Let $U = C_S°(i)$, a nontrivial connected definable 2-group (this uses strongly the fact that U is 2-unipotent). Then $N(U) \leq N(S)$. Let $V = O_2°(C(i))$. Then V is a Sylow° 2-subgroup of $C(i)$ and thus contains U. Furthermore $N_V(U) \leq N(S)$ so $N_V°(U) \leq C_S°(i) = U$. It follows that $V = U$ and $N(V) \leq N(S)$. In particular $C(i) \leq N(S)$, as claimed. q.e.d.

Now we can prove Theorem 8 and in particular Theorem 5.

Proof of Theorem 8. Assuming the theorem fails, we find a group G of finite Morley rank satisfying the hypothesis $(*)$ above, and in particular a Sylow° 2-subgroup S of G is unipotent. Then $N(S)$ is strongly embedded in G. Setting $H = N°(S)$, we have $H = S \cdot C_H(S)$.

Now all involutions of $N(S)$ are conjugate under the action of $N(S)$ by Lemma 16. In particular they all lie in $Z(S)$. But $H = S \cdot C_H(S)$ acts trivially on $Z(S)$, and $N(S)/H$ is finite, so S contains only finitely many involutions. But S is 2-unipotent and hence contains an infinite elementary abelian subgroup. This is a contradiction. q.e.d. (Theorem 8)

4 Poizat's problem

As we have said, the original point of Theorem 5 was that it was needed to clarify a longstanding problem put forward by Poizat, concerning groups which are generically of finite exponent. Our best result in that direction is the following [B_0B_1C05].

Proposition 18. Let G be a connected group of finite Morley rank containing a definable generic subset whose elements are of order n for some fixed n. Then a Sylow 2-subgroup U of G is unipotent, $G = U * C_G(U)$ is a central product, and G/U is a group without involutions whose elements are generically of order n_0, where n_0 is the odd part of n.

For the proof, one begins by defining U not as a Sylow 2-subgroup of G but as a Sylow° 2-subgroup, and one proves the corresponding result; of course, one first gets G/U to be a group of degenerate type, and then one invokes Theorem 1, which is indeed the main result of [B_0B_1C05], to tie the final knots.

The connection between Poizat's problem and our Theorem 5 or 8 can be seen in the following result.

Proposition 19. Let G be a connected group of finite Morley rank containing a definable generic subset whose elements are of order n for some fixed n. Then G contains no nontrivial p-torus for any p.

Now this comes from a rather different, and more central, place in the theory of groups of finite Morley rank: it is really a statement about what happens when one *does* have nontrivial p-tori. Up to this point in the present article, p-tori have been conspicuous primarily by their absence. But of course the moral of our story is that we need good techniques for exploiting p-tori when we have them, and this is fortunately the case. What follows is a variant of the main result of [C05], which can easily be reduced to the form given there.

Theorem 20. Let T be a maximal p-torus in a connected group G of finite Morley rank, and let $H = C°(T)$, the connected centralizer. Then there is a definable subset $X \subseteq H$, generic in H, such that any two conjugates of X are either disjoint or equal, and such that the union of the conjugates of X is a generic subset of G.

This means that as far as rank computations are concerned, G can be replaced by a cartesian product $X \times Y$ with Y parametrizing the set of conjugates of X (in fact Y can be identified with the coset space $N(T) \backslash G$ in this situation).

The proof of this theorem involves all the rigidity properties of good tori and the connection between p-tori and good tori via decent tori. It was of course directly inspired by the situation in algebraic groups, and more particularly the theory of reductive algebraic groups, where $C°(T)$ reduces to a maximal torus of G.

There is a competing theory which is very useful in classification results, namely the theory of Carter subgroups. This theory provides another analog of conjugacy of maximal tori, valid in arbitrary solvable finite groups, and, suitably adjusted, also in arbitrary groups of finite Morley rank with no solvability hypothesis. The Carter theory and the theory of maximal decent tori have much in common. We will return to this point in the next section.

Now we can sketch the proof of Theorem 18. The main step from our present point of view is the proof of Proposition 19, that is the elimination of p-tori.

Proof of Proposition 19. Invoke Theorem 20. So the notation is as follows: T is a maximal p-torus, $H = C°(T)$, $X \subseteq H$ is the generic subset of H afforded by that result. Set $\hat{T} = d(T)$, the definable hull of T. One can then easily move the initial hypothesis around as follows:

Generically $x^n = 1$ in G.
Generically $x^n = 1$ in X.
Generically $x^n = 1$ in H.
Generically $x^n = 1$ in some coset of $\hat{T}$ in H.
Generically $x^n = 1$ in $\hat{T}$.
Identically, $x^n = 1$ in $\hat{T}$.
Identically, $x^n = 1$ in T.

And the last condition forces $T = 1$. An important point here is that the elements of any coset of $\hat{T}$ in H commute; this commutativity is what we were aiming at, as it allows the transition from a generic condition to an identical condition. q.e.d. (Proposition 19)

The hypothesis "$x^n = 1$ generically" passes to definable quotients but not to subgroups; but the conclusion on the absence of p-tori is fully inductive, and this gave us the necessary flexibility for our analysis.

5 Carter subgroups

In finite group theory, a Carter subgroup is a selfnormalizing nilpotent subgroup, and it is a remarkable fact that in solvable finite groups there is always a unique conjugacy class of Carter subgroups; neither existence nor uniqueness is evident. In groups of finite Morley rank it is more convenient to work with the connected analog of this notion: so now a Carter subgroup will be a connected definable nilpotent subgroup which is almost selfnormalizing, that is it coincides with the connected component of its normalizer. These behave as well in solvable groups of finite Morley rank as they do in the finite theory; but in fact *every* group of finite Morley rank has Carter subgroups in this sense [FJ05]. On the other hand, the issue of conjugacy is still open. There are other properties which are desirable in the context of finite Morley rank, notably the following.

Definition 21. A subgroup H of a connected group G of finite Morley rank is *generous* if the union of the conjugates of H is generic in G.

For example, centralizers of decent tori are generous; this is a variant of Theorem 20. If one considers centralizers of maximal decent tori one gets a sharper version as in that theorem, but even this looser version is useful.

In [J06] one finds the following.

Theorem 22. Let G be a connected group of finite Morley rank. Then there is at most one conjugacy class of generous Carter subgroups.

Here existence remains an open problem. Thus we have the existence of Carter subgroups and the uniqueness of generous Carter subgroups, and if the gap between these two could be closed we would have a theory which in important ways is sharper than the parallel theory for centralizers of maximal decent tori, which are not known to be nilpotent.

Of course, for T a maximal decent torus and $H = C°(T)$, the group H/T contains no nontrivial decent torus, that is it is of unipotent type, and thus H has a unipotent 2-subgroup U for which H/TU is of degenerate type as well as unipotent type. It is in fact unlikely that all such groups are nilpotent. Apart from the possible existence of simple groups of this type, it is quite likely that there are solvable nonnilpotent torsion-free groups of this type, a point tied up with the possible existence of "bad fields".

Frécon [F05] has shown that the two theories, based alternatively on Carter subgroups or centralizers of p-tori, essentially coincide in the *tame* case (with no "bad groups" or "bad fields" involved), so that one gets all the features of both in this case. As things now stand, in order to bring the theory based on tori to a comparable degree of completeness one would at a minimum need to eliminate simple groups of degenerate type. On the other hand, there is no such obvious obstruction to improving the Carter theory by proving the existence of generous Carter subgroups in full generality. If this could be done, it would then become appropriate to redefine "Carter subgroups" so as to include generosity.

References.

[A₀01] T. Altınel, Groupes de rang de Morley fini de type pair avec des sous-groupes faiblement inclus. Thèse d'habilitation, Université Claude Bernard, Lyon 1, 2001.

[A₀B₀C99] T. Altınel, A. Borovik, and G. Cherlin, On groups of finite Morley rank with weakly embedded subgroups. *J. Algebra*, 211:409–456, 1999.

[A₀B₀C∞] T. Altınel, A. V. Borovik, and G. Cherlin, Simple Groups of Finite Morley Rank, *book in preparation*.

[A₀C03] T. Altınel and G. Cherlin, On groups of finite Morley rank of even type. *J. Algebra*, 264:155-185, 2003.

[A₀C04] T. Altınel and G. Cherlin, Simple L^*-groups of even type with strongly embedded subgroups. *J. Algebra*, 272:95–127, 2004.

[A₀C05a] T. Altınel and G. Cherlin, Simple L^*-groups of even type with weakly embedded subgroups. *J. Algebra*, 286:76–106, 2005.

[A₀C05b] T. Altınel and G. Cherlin, Limoncello. *J. Algebra*, 291:371–413, 2005.

[A₁B₀01] C. Altseimer and A. Borovik, *Probabilistic recognition of orthogonal and symplectic groups*, in: Groups and Computation III, pp. 1–20, W. Kantor and A. Seress, eds., de Gruyter, Berlin, 2001; Corrections (and GAP code): math.GR/0110234.

[B₀B₁C05] A. Borovik, J. Burdges, and G. Cherlin, Involutions in groups of finite Morley rank of degenerate type. *Preprint*, September 2005.

[B₀P90] A. Borovik and B. Poizat, Tores et p-groupes, *J. Symbolic Logic*, 55:478–491, 1990.

[B₀N94] A. Borovik and A. Nesin, Groups of Finite Morley Rank. The Clarendon Press Oxford University Press, New York, 1994. Oxford Science Publications.

[B₁∞] J. Burdges, The Bender method in groups of finite Morley rank, *J. Algebra*, to appear.

[C05] G. Cherlin, Good tori in groups of finite Morley rank. *J. Group Theory*, 8:613–621, 2005.

[F05] O. Frécon, Carter subgroups in tame groups of finite Morley rank. *Preprint*, 2005.

[FJ05] O. Frécon and E. Jaligot, The existence of Carter subgroups in groups of finite Morley rank. *J. Group Theory*, 8:623–633, 2005.

[J06] E. Jaligot, Generix never gives up. *J. Symbolic Logic*, 71:599–610, 2006.

[PW00] B. Poizat and F. Wagner, Lift the Sylows! *J. Symbolic Logic*, 53:703–704, 2000.

[W01] F. Wagner, Fields of finite Morley rank. *J. Symbolic Logic*, 66:703–706, 2001.

Received: February 15, 2006;
In revised version: March 29, 2006;
Accepted by the editors: April 18, 2006.

Independence for Distributivity Numbers

Jörg Brendle

The Graduate School of Science and Technology
Kobe University
Rokko-dai 1-1, Nada-ku
Kobe 657-8501, Japan
E-mail: brendle@kurt.scitec.kobe-u.ac.jp

ABSTRACT. We show that it is consistent that $\mathfrak{h}(\mathbb{C}^\omega/\mathrm{fin}) < \min\{\mathfrak{h}, \mathrm{add}(\mathcal{M})\}$ where $\mathbb{C}$ is the Cohen algebra, $\mathfrak{h}$ ($\mathfrak{h}(\mathbb{C}^\omega/\mathrm{fin})$, respectively) denotes the distributivity number of $\mathcal{P}(\omega)/\mathrm{fin}$ ($\mathbb{C}^\omega/\mathrm{fin}$, resp.), and $\mathrm{add}(\mathcal{M})$ is the additivity of the meager ideal. We also prove that, consistently, $\mathfrak{h}_2 < \mathfrak{h}(\mathbb{C}^\omega/\mathrm{fin})$ where $\mathfrak{h}_2$ is the distributivity of number of the square of $\mathcal{P}(\omega)/\mathrm{fin}$. The first result answers a question of Balcar and Hrušák, and the second strengthens a result of Shelah and Spinas and answers a question of Dow.

Dedicated to Ulrich Felgner on the occasion of his 65th birthday

Introduction

We prove two independence results about the values of distributivity numbers of Boolean algebras which are regular open algebras of the Stone-Čech remainders of the natural numbers ω, the real numbers $\mathbb{R}$, or of their products.

Let $\mathbb{A}$ be an atomless Boolean algebra, and put $\mathbb{A}^+ := \mathbb{A} \setminus \{\mathbf{0}\}$. A set $D \subseteq \mathbb{A}^+$ is *dense* if for all $a \in \mathbb{A}^+$, there is $d \leq a$ belonging to D. A set D is *open* if $d \in D$ and $a \leq d$ imply $a \in D$. The *distributivity number* (or: *height*) $\mathfrak{h}(\mathbb{A})$ of $\mathbb{A}$ is the least size of a family $\mathcal{D}$ of dense open sets such that $\bigcap \mathcal{D}$ is not dense. Note that $\bigcap \mathcal{D}$ necessarily is open. The cardinal $\mathfrak{h}(\mathbb{A})$ is always regular. Equivalently, $\mathfrak{h}(\mathbb{A})$ is the smallest cardinality of a family of maximal antichains in $\mathbb{A}^+$ which has no common refinement. Here, an antichain B *refines* an antichain A if for all $b \in B$ there is $a \in A$ with $b \leq a$. If $\mathbb{A}$ is *homogeneous*, that is, $\mathbb{A}{\restriction}a := \{b \in \mathbb{A} : b \leq a\} \cong \mathbb{A}$ for all $a \in \mathbb{A}$, then $\mathfrak{h}(\mathbb{A})$ is the least size of a family $\mathcal{D}$ of dense open sets such that $\bigcap \mathcal{D} = \varnothing$. All Boolean algebras we shall study are homogeneous.

Partially supported by Grants-in-Aid for Scientific Research (C)(2) 15540120 and (C) 17540116, Japan Society for the Promotion of Science.

If we think of $\mathbb{A}$ as a forcing notion, $\mathfrak{h}(\mathbb{A})$ can also be characterized as the least κ such that for some $\mathbb{A}$-name $\dot{f}$ for a function from κ to the ground model V, some $a \in \mathbb{A}^+$ forces $\dot{f} \notin V$. In the homogeneous case, we may stipulate the trivial condition **1** forces $\dot{f} \notin V$. For Boolean algebras $\mathbb{A}$ and $\mathbb{B}$, write $\mathbb{A} <\circ \mathbb{B}$ if there is a *complete embedding* from $\mathbb{A}$ into $\mathbb{B}$. This immediately entails $\mathfrak{h}(\mathbb{A}) \geq \mathfrak{h}(\mathbb{B})$. Furthermore, $\mathbb{A} \times \mathbb{B}$ denotes the *free product* of $\mathbb{A}$ and $\mathbb{B}$, namely, $(\mathbb{A}^+ \times \mathbb{B}^+) \cup \{\mathbf{0}\}$. Since $\mathbb{A} <\circ \mathbb{A} \times \mathbb{B}$, $\mathfrak{h}(\mathbb{A}) \geq \mathfrak{h}(\mathbb{A} \times \mathbb{B})$ follows. For any Boolean algebra $\mathbb{A}$ let r.o.$(\mathbb{A})$ denote its (unique) *completion*. Clearly $\mathfrak{h}(\mathbb{A}) = \mathfrak{h}(\text{r.o.}(\mathbb{A}))$. The cardinal $\mathfrak{h}(\mathbb{A})$ has been thoroughly studied for a number of Boolean algebras, in particular for algebras which arise naturally in topology.

Consider the Boolean algebra $\mathcal{P}(\omega)/\text{fin} := \{[A] : A \subseteq \omega\}$ where $[A] := \{B \subseteq \omega : \text{the symmetric difference } A \triangle B \text{ is finite}\}$. The set $\mathcal{P}(\omega)/\text{fin}$ is ordered by: $[A] \leq [B]$ if A is *almost contained* in B ($A \subseteq^* B$, in symbols) if $A \setminus B$ is finite. More generally, given an arbitrary Boolean algebra $\mathbb{A}$, the *reduced power* is defined by $\mathbb{A}^\omega/\text{fin} := \{[f] : f \in \mathbb{A}^\omega\}$ where $[f] := \{g \in \mathbb{A}^\omega : \forall^\infty n (f(n) = g(n))\}$. The order on the Boolean algebra $\mathbb{A}^\omega/\text{fin}$ is given by: $[f] \leq [g]$ if $f(n) \leq g(n)$ for almost all n. Here, $\forall^\infty n$ stands for "for all but finitely many n" and is read as "for almost all n". Similarly, $\exists^\infty n$ means "for infinitely many n". Clearly, $\mathcal{P}(\omega)/\text{fin} \cong \{\mathbf{0}, \mathbf{1}\}^\omega/\text{fin}$. Thus $\mathcal{P}(\omega)/\text{fin} <\circ \mathbb{A}^\omega/\text{fin}$ and $\mathfrak{h}(\mathbb{A}^\omega/\text{fin}) \leq \mathfrak{h}$ for any Boolean algebra $\mathbb{A}$ where $\mathfrak{h} := \mathfrak{h}(\mathcal{P}(\omega)/\text{fin})$. Also $\mathfrak{h}_2 \leq \mathfrak{h}$ where $\mathfrak{h}_2 := \mathfrak{h}(\mathcal{P}(\omega)/\text{fin} \times \mathcal{P}(\omega)/\text{fin})$.

Recall that for a normal space X, the *Stone-Čech compactification* βX is the space of ultrafilters of closed subsets of X, and the *Stone-Čech remainder* $X^* := \beta X \setminus X$ is the space of free ultrafilters of closed sets where an ultrafilter $\mathcal{F}$ is *free* if $\bigcap \mathcal{F} = \varnothing$. For a topological space X, $O \subseteq X$ is *regular open* if $\text{Int}(\text{Cl}(O)) = O$. The collection of all regular open sets of X, r.o.(X), forms a complete Boolean algebra, called the *regular open algebra* of X. For example, r.o.$(\omega) = \text{r.o.}(\beta\omega) = \mathcal{P}(\omega)$ and r.o.(ω^*) is the completion of $\mathcal{P}(\omega)/\text{fin}$, r.o.$(\mathcal{P}(\omega)/\text{fin})$. Similarly, r.o.$(\omega^* \times \omega^*) = \text{r.o.}(\mathcal{P}(\omega)/\text{fin} \times \mathcal{P}(\omega)/\text{fin})$. Let $\mathbb{C}$ denote the *Cohen algebra*, that is, the algebra of clopen subsets of the Cantor space 2^ω. Recall that r.o.$(\mathbb{C}) = \mathcal{B}/\mathcal{M}$ where $\mathcal{B}$ denotes the Borel subsets of 2^ω and $\mathcal{M}$ is the ideal of meager sets. The algebra $\mathbb{C}$ is the forcing notion for adding a Cohen real. It is well-known and easy to see that r.o.$(\mathbb{R}^*) = \text{r.o.}(\mathbb{C}^\omega/\text{fin})$. Two topological spaces X and Y are said to be *coabsolute* if r.o.$(X) \cong \text{r.o.}(Y)$. By a classical result of Parovičenko's, $\mathcal{P}(\omega)/\text{fin}$, $\mathcal{P}(\omega)/\text{fin} \times \mathcal{P}(\omega)/\text{fin}$, $\mathbb{C}^\omega/\text{fin}$, etc., are all isomorphic under the continuum hypothesis CH so that ω^*, $\omega^* \times \omega^*$, and $\mathbb{R}^*$ etc. are all coabsolute.

Answering an old problem of Balcar, Pelant and Simon [B$_0$PS$_1$80], Shelah and Spinas [S$_0$S$_2$00] proved that in the *iterated Mathias model*, namely the model obtained by an ω_2-stage countable support iteration of Mathias forcing over a model for CH, $\mathfrak{h}_2 = \aleph_1$ while $\mathfrak{h} = \aleph_2$ so that ω^* and $\omega^* \times \omega^*$ are not coabsolute in this model. Similarly, Dow [D$_1$98] proved that ω^* and $\mathbb{R}^*$ are not coabsolute in the same model by showing $\mathfrak{h}(\mathbb{C}^\omega/\text{fin}) = \aleph_1$, thus answering an old problem of van Douwen [v90]. He conjectured that, in fact, $\mathfrak{h}(\mathbb{C}^\omega/\text{fin}) = \mathfrak{h}(\mathbb{C}^\omega/\text{fin} \times \mathbb{C}^\omega/\text{fin}) \leq \mathfrak{h}_2$ in ZFC. Balcar and Hrušák [B$_0$H$_1$05] gave an alternative, much shorter, proof of Dow's result by showing that $\mathfrak{h}(\mathbb{C}^\omega/\text{fin}) \leq \text{add}(\mathcal{M})$ in ZFC where $\text{add}(\mathcal{M})$, the *additivity of the meager ideal*, is the smallest size of a family $\mathcal{F} \subseteq \mathcal{M}$ with $\bigcup \mathcal{F} \notin \mathcal{M}$. It has been well-known (and is relatively simple to prove) that $\text{add}(\mathcal{M}) = \aleph_1$ in the Mathias model [B$_1$J95]. Since Shelah and Spinas [S$_0$S$_2$98] also proved the consistency of $\text{add}(\mathcal{M}) < \mathfrak{h}_2$, the consistency of $\mathfrak{h}(\mathbb{C}^\omega/\text{fin}) < \mathfrak{h}_2$ follows as well. In view of their result, Balcar and Hrušák [B$_0$H$_1$05] asked whether $\mathfrak{h}(\mathbb{C}^\omega/\text{fin}) < \min\{\mathfrak{h}, \text{add}(\mathcal{M})\}$ was consistent. We show this is indeed the case.

Theorem 1. $\mathfrak{h}(\mathbb{C}^\omega/\text{fin}) < \min\{\text{add}(\mathcal{M}), \mathfrak{h}\}$ *is consistent.*

We also solve Dow's conjecture [D$_1$98] negatively.

Theorem 2. $\mathfrak{h}_2 < \mathfrak{h}(\mathbb{C}^\omega/\text{fin})$ *is consistent.*

Notice this gives a new proof of the result of Shelah and Spinas [S$_0$S$_2$00] quoted above and also shows the consistency of $\mathfrak{h}(\mathbb{C}^\omega/\text{fin} \times \mathbb{C}^\omega/\text{fin}) < \mathfrak{h}(\mathbb{C}^\omega/\text{fin})$ because $\mathcal{P}(\omega)/\text{fin} \times \mathcal{P}(\omega)/\text{fin} <_\circ \mathbb{C}^\omega/\text{fin} \times \mathbb{C}^\omega/\text{fin}$ clearly entails that $\mathfrak{h}_2 \geq \mathfrak{h}(\mathbb{C}^\omega/\text{fin} \times \mathbb{C}^\omega/\text{fin})$ in ZFC.

Our proofs are different from previous proofs in this area in so far as we build finite support iterations of ccc forcing. Previous results have always used countable support iteration of proper forcing and, in fact, only the Mathias model, for the results quoted above. For obvious reasons, this model is not adequate for either of our results. From the technical point of view, the main ingredient is a careful investigation of names of subsets of ω in Laver forcing with a filter and its relatives, a theme originally devised in the closely related work in [B$_4$06b, §§3 & 4] and pursued in the more recent [B$_4\infty$]. Since some of the arguments are very similar, we are brief in places and refer to the somewhat more elaborate treatment in [B$_4$06b].

Our notation is standard. For forcing theory we refer to [B$_1$J95], for cardinal invariants of the continuum to [B$_3\infty$], and for Boolean algebras and distributivity numbers to [B$_0$S$_1$89]. For a survey on the results presented here and some related results as well as a more thorough explanation of the connection to topology, see [B$_4$06a]. Apart from the cardinals defined above, we shall use the *(un)bounding number* $\mathfrak{b}$, namely, the smallest cardinality of an unbounded family in the eventually dominating ordering $(\omega^\omega, \leq^*)$ on the Baire space ω^ω, and the *covering number of the meager ideal* $\text{cov}(\mathcal{M})$, that is, the least size of a family $\mathcal{F} \subseteq \mathcal{M}$ such that $\bigcup \mathcal{F} = 2^\omega$. It is well-known that add($\mathcal{M}$) = min{cov($\mathcal{M}$), $\mathfrak{b}$} [B$_1$J95] and that $\mathfrak{h} \leq \mathfrak{b}$ [B$_0$S$_1$89]. However, $\mathfrak{h}$ is independent of either add($\mathcal{M}$) or cov($\mathcal{M}$): $\mathfrak{h} <$ add($\mathcal{M}$) holds in the Hechler model while cov($\mathcal{M}$) $< \mathfrak{h}$ holds in the Mathias model as mentioned above [B$_1$J95].

1 Proof of Theorem 1

Laver forcing $\mathbb{L}_\mathcal{F}$ with a filter $\mathcal{F}$ is forcing with trees $T \subseteq \omega^{<\omega}$ such that for all $\sigma \in T$ which extend the stem of T, the set of successor nodes succ$_T(\sigma) = \{n : \sigma^\frown n \in T\}$ belongs to $\mathcal{F}$. The set $\mathbb{L}_\mathcal{F}$ is ordered by inclusion. It is a σ-centered forcing notion which adds a dominating real whose range diagonalizes the filter $\mathcal{F}$.

A free ultrafilter $\mathcal{U}$ on ω is a *Ramsey ultrafilter* if for all partitions $\langle X_n : n \in \omega \rangle$ of ω either there is n with $X_n \in \mathcal{U}$ or there is a *selector* $U \in \mathcal{U}$ (that is, $|X_n \cap U| \leq 1$ for all $n \in \omega$). It is well-known that for Ramsey ultrafilters $\mathcal{U}$, $\mathbb{L}_\mathcal{U}$ is forcing equivalent to the better known Mathias forcing $\mathbb{M}_\mathcal{U}$. However, for technical reasons, we shall stick with Laver forcing.

Let $\mathcal{I}_e$ be the ideal on $2^{<\omega}$ which is generated by functions in 2^ω, that is, $A \in \mathcal{I}_e$ if and only if there are $\ell \in \omega$ and $f_i \in 2^\omega$, $i < \ell$, such that $A \subseteq \{f_i \restriction n : i < \ell \text{ and } n \in \omega\}$. If we identify ω and $2^{<\omega}$ and let $\mathcal{F}_e$ denote the dual filter of $\mathcal{I}_e$, we may think of $\mathcal{F}_e$ as a filter on ω. *Eventually branching function forcing* $\mathbb{E}$ is $\mathbb{L}_{\mathcal{F}_e}$. It generically adds a function $e \in (2^{<\omega})^\omega$ such that for all $\phi : \omega \to 2^\omega$ from the ground model, $e(n) \not\subseteq \phi(n)$ for almost all n. The forcing $\mathbb{E}$ is a close relative of Hechler forcing $\mathbb{D}$.

Assume CH and $\Diamond_{S_1^2}$. Recall that $\Diamond_{S_1^2}$ means "there is a sequence $\{Z_\alpha :$ cf$(\alpha) = \omega_1$ and $\alpha < \omega_2\}$ such that for all $Z \subseteq \omega_2$, the set $\{\alpha < \omega_2 : $ cf$(\alpha) = \omega_1$ and $Z \cap \alpha = Z_\alpha\}$ is stationary". We make a finite support iteration $\langle \mathbb{P}_\alpha, \dot{\mathbb{Q}}_\alpha : \alpha < \omega_2 \rangle$ of ccc forcing such that

(A) if cf$(\alpha) = \omega_1$, then $\dot{\mathbb{Q}}_\alpha$ is Laver forcing $\mathbb{L}_{\dot{\mathcal{U}}_\alpha}$ with a Ramsey ultrafilter $\dot{\mathcal{U}}_\alpha$,

(B) if cf$(\alpha) \neq \omega_1$, then $\dot{\mathbb{Q}}_\alpha$ is eventually branching function forcing $\dot{\mathbb{E}}$.

The idea is that at limit stages α of cofinality ω_1, we use forcing (A) to kill potential witnesses for $\mathfrak{h} = \aleph_1$. This is a standard argument using $\Diamond_{S_1^2}$: at stage α, $\Diamond_{S_1^2}$ hands us down an initial segment for a potential witness for $\mathfrak{h} = \aleph_1$. We then build the Ramsey ultrafilter $\mathcal{U}_\alpha$ in ω_1 steps such that it diagonalizes the witness by taking care of the relevant requirements in, say, the even steps of the construction of $\mathcal{U}_\alpha$. This leaves us room for doing something else in the odd steps (see clause (i) below). We leave the details of the diamond argument to the reader (see [B$_4$06b, Lemma 3.7] for a very similar argument). Thus $\mathfrak{h} = \aleph_2$.

Since all the iterands add dominating reals, $\mathfrak{b} = \aleph_2$ follows. Because of the Cohen reals which arise in limit stages of cofinality ω, $\mathsf{cov}(\mathcal{M}) = \aleph_2$ follows. (Note that $\mathbb{E}$ also adds Cohen reals.) Thus $\mathsf{add}(\mathcal{M}) = \aleph_2$ and we are left with proving $\mathfrak{h}(\mathbb{C}^\omega/\text{fin}) = \aleph_1$.

Forcing (B) is used to build up families $\{\mathcal{F}_\beta : \beta < \omega_1\}$ such that

(a) each $\mathcal{F}_\beta$ is a maximal antichain in $\mathbb{C}^\omega/\text{fin}$,

(b) for $\beta < \beta'$, $\mathcal{F}_{\beta'}$ refines $\mathcal{F}_\beta$ (that is, for each $f \in \mathcal{F}_{\beta'}$ there is $g \in \mathcal{F}_\beta$ such that $f \leq^* g$ in $\mathbb{C}^\omega$).

To avoid having to deal with equivalence classes, we work with $\mathbb{C}^\omega$ instead of $\mathbb{C}^\omega/\text{fin}$. As a further reduction, instead of considering the Boolean algebra $\mathbb{C} = \mathcal{B}/\mathcal{M}$, we work with $\mathbb{C} = 2^{<\omega} \cup \{\mathbf{0}\}$ where $2^{<\omega}$ is the canonical dense subset, and $\mathbf{0}$ is the zero, of the Boolean algebra $\mathcal{B}/\mathcal{M}$. For $f \in \mathbb{C}^\omega$, let $\mathsf{supp}(f) = \{n : f(n) \neq \mathbf{0}\}$. The relation $f \leq^* g$ is defined as usual by "$f(n) \leq g(n)$ for almost all n" and then becomes equivalent to "$\mathsf{supp}(f) \subseteq^* \mathsf{supp}(g)$ and $f(n) \supseteq g(n)$ for almost all $n \in \mathsf{supp}(f)$."

Let $\mathcal{F}_\beta^{\leq \alpha} = \mathcal{F}_\beta \cap V_\alpha \in V_\alpha$ where V_α denotes the generic extension via $\mathbb{P}_\alpha$. Use book-keeping to produce $\alpha_{h,\beta}$ for each $h \in \mathbb{C}^\omega$ and each $\beta \in \omega_1$ such that $h \in \mathbb{C}^\omega \cap V_{\alpha_{h,\beta}}$ and the function $(h, \beta) \mapsto \alpha_{h,\beta}$ is one-to-one and onto ordinals of cofinality $< \omega_1$. For α with $\mathsf{cf}(\alpha) \neq \omega_1$ let h_α and β_α be the unique h and β with $\alpha = \alpha_{h,\beta}$. At each successor stage $\alpha + 1$ with $\mathsf{cf}(\alpha) \neq \omega_1$, countably many $\mathcal{F}_\beta$ will get a new element:

(c) if $\beta < \beta_\alpha$ and if h_α is incompatible with all $g \in \mathcal{F}_\beta^{\leq \alpha}$ then there is $f \leq^* h_\alpha$ belonging to $\mathcal{F}_\beta^{\leq \alpha+1}$,

(d) for $\beta < \beta' < \omega_1$, $\mathcal{F}_{\beta'}^{\leq \alpha+1}$ refines $\mathcal{F}_\beta^{\leq \alpha+1}$,

(e) if $\beta < \beta_\alpha$, $f \in \mathcal{F}_\beta^{\leq \alpha+1} \setminus \mathcal{F}_\beta^{\leq \alpha}$ and $\phi : \omega \to 2^\omega$ is a partial function belonging to V_α, then $f(n) \not\subseteq \phi(n)$ for almost all $n \in \mathsf{dom}(\phi) \cap \mathsf{supp}(f)$.

For $\beta \geq \beta_\alpha$, we will have $\mathcal{F}_\beta^{\leq \alpha+1} = \mathcal{F}_\beta^{\leq \alpha}$.

Fix α. To build $\mathcal{F}_\beta^{\leq \alpha+1} \setminus \mathcal{F}_\beta^{\leq \alpha}$, recursively construct $h^\beta \in \mathbb{C}^\omega \cap V_\alpha$ as follows: $h^0 = h_\alpha$, if β is limit, $h^\beta \leq^* h^{\beta'}$ for all $\beta' < \beta$, and if $\beta = \beta' + 1$ is successor and there is $g \in \mathcal{F}_{\beta'}^{\leq \alpha}$ compatible with $h^{\beta'}$, then h^β is a common extension. Let $\beta \leq \beta_\alpha$ be minimal such that h^β does not exist if there is such a β. Then $\beta = \beta' + 1$ is successor and all $g \in \mathcal{F}_{\beta'}^{\leq \alpha}$ are incompatible with $h^{\beta'}$. Let $f \leq^* h^{\beta'}$ be a $\mathbb{Q}_\alpha$-generic function. Recall that $\mathbb{Q}_\alpha = \mathbb{E}$ adds a new function $e \in \mathbb{C}^\omega$; by coding e below $h^{\beta'}$ we get f; in particular $\mathsf{supp}(f) = \mathsf{supp}(h^{\beta'})$. Put f into $\mathcal{F}_\gamma^{\leq \alpha+1}$ for all $\beta' \leq \gamma < \beta_\alpha$. (c) holds by construction and (e), by genericity. The recursive construction as well as the inductive hypothesis for α for (d) yield that (d) also holds for $\alpha + 1$. This completes the construction in stage α.

Note that (b) follows from (d), and (a) holds by (c) and the book-keeping. This ends the construction of the $\mathcal{F}_\beta$. In particular, $\mathcal{F}_\beta^{\leq \alpha+1} = \mathcal{F}_\beta^{\leq \alpha}$ if $\operatorname{cf}(\alpha) = \omega_1$ and $\mathcal{F}_\beta^{\leq \alpha} = \bigcup_{\gamma < \alpha} \mathcal{F}_\beta^{\leq \gamma}$ for limit ordinals α.

To see that $\{\mathcal{F}_\beta : \beta < \omega_1\}$ witnesses $\mathfrak{h}(\mathbb{C}^\omega / \operatorname{fin}) = \aleph_1$ we need to show that for all $\alpha < \omega_2$,

$$\forall f \in \mathbb{C}^\omega \cap V_\alpha \; \exists \beta < \omega_1 \; \forall g \in \mathcal{F}_\beta \; f \not\leq^* g. \tag{$+_\alpha$}$$

Note $f \not\leq^* g$ means there are infinitely many $n \in \operatorname{supp}(f)$ such that either $n \notin \operatorname{supp}(g)$ or $f(n) \not\supseteq g(n)$. We shall in fact prove that for all $\alpha < \omega_2$,

for all partial functions $\phi : \omega \to 2^\omega$ from $V_\alpha \; \exists \beta < \omega_1 \; \forall g \in \mathcal{F}_\beta$
$$\exists^\infty n \in \operatorname{dom}(\phi) \; (n \notin \operatorname{supp}(g) \text{ or } \phi(n) \not\supseteq g(n)). \tag{$\dagger_\alpha$}$$

Clearly $(\dagger_\alpha)$ implies $(+_\alpha)$: if $f \in \mathbb{C}^\omega \cap V_\alpha$ is given, for each $n \in \operatorname{supp}(f)$ find $\phi(n) \in 2^\omega$ such that $f(n) \subseteq \phi(n)$ and let $\operatorname{dom}(\phi) = \operatorname{supp}(f)$. The β which works for ϕ also works for f.

To get $(\dagger_\alpha)$ we shall prove by induction on $\alpha < \omega_2$ that

for all partial functions $\phi : \omega \to 2^\omega$ from $V_\alpha \; \exists \beta < \omega_1 \; \forall g \in \mathcal{F}_\beta^{\leq \alpha}$
$$\exists^\infty n \in \operatorname{dom}(\phi) \; (n \notin \operatorname{supp}(g) \text{ or } \phi(n) \not\supseteq g(n)). \tag{$\star_\alpha$}$$

This is enough because $(\dagger_\alpha)$ immediately follows from $(\star_\alpha)$ and property (e) above.

The preservation of $(\star_\alpha)$ in limit stages α of cofinality ω_1 is immediate by induction hypothesis and (e). For limit stages α of cofinality ω, the preservation of $(\star_\alpha)$ is a standard argument using the induction hypothesis and the fact that the $\mathcal{F}_\beta$ get no new elements in limit stages: $\mathcal{F}_\beta^{\leq \alpha} = \bigcup_{\gamma < \alpha} \mathcal{F}_\beta^{\leq \gamma}$. We leave the details to the reader (see [B$_4$06b, Lemma 3.2] for a very similar argument). The crux of the proof is to show $(\star_{\alpha+1})$ under the assumption $(\star_\alpha)$ holds. This is particularly troublesome in case $\operatorname{cf}(\alpha) = \omega_1$ in which case we need one more property:

for all partial functions $\phi : \omega \to 2^\omega$ from V_α there is an infinite $A \subseteq \operatorname{dom}(\phi)$ and there is a $\beta < \omega_1 \; \forall g \in \mathcal{F}_\beta^{\leq \alpha}$
$$\forall^\infty n \in A \cap \operatorname{supp}(g) \; (\phi(n) \not\supseteq g(n)). \tag{$\ddagger_\alpha$}$$

Property $(\ddagger_\alpha)$ strengthens $(\star_\alpha)$. However, if $\operatorname{cf}(\alpha) = \omega_1$, the two properties are equivalent.

Lemma 3. Assume $\operatorname{cf}(\alpha) = \omega_1$ and $(\star_\gamma)$ holds for $\gamma < \alpha$. Then $(\ddagger_\alpha)$ holds.

Proof. Let $\phi \in V_\alpha$. Since α has uncountable cofinality, there is $\gamma < \alpha$ such that $\phi \in V_\gamma$. Work in V_γ.

For $\beta < \omega_1$ and $g \in \mathcal{F}_\beta^{\leq \gamma}$ let $A_g = \{n \in \operatorname{dom}(\phi) \cap \operatorname{supp}(g) : \phi(n) \supseteq g(n)\}$. Since distinct $g, g' \in \mathcal{F}_\beta^{\leq \gamma}$ are incompatible, the corresponding A_g and $A_{g'}$ must be almost disjoint. Thus, for each β, we get an almost disjoint family $\mathcal{A}_\beta = \{A_g \in [\omega]^\omega : g \in \mathcal{F}_\beta^{\leq \gamma}\}$. We claim that for some β, $\mathcal{A}_\beta$ is not a finite maximal almost disjoint family below $\operatorname{dom}(\phi)$.

For assume for every β, $\mathcal{A}_\beta$ was finite maximal. Then there is n such that for cofinally many β, $|\mathcal{A}_\beta| = n$, say $\mathcal{A}_\beta = \{A_{g_{\beta,m}} : m < n\}$. Since the $\mathcal{F}_\beta^{\leq \gamma}$ refine each other (property (d)), for $\beta < \beta'$ and $g' \in \mathcal{F}_{\beta'}^{\leq \gamma}$ with $A_{g'} \in \mathcal{A}_{\beta'}$ there is $g \in \mathcal{F}_\beta^{\leq \gamma}$ with $A_g \in \mathcal{A}_\beta$ such that $A_{g'} \subseteq^* A_g$. Thus, without loss of generality, $A_{g_{\beta',m}} \subseteq^* A_{g_{\beta,m}}$ for all $m < n$. Since all $\mathcal{A}_\beta$ are maximal, we get in fact $A_{g_{\beta',m}} =^* A_{g_{\beta,m}}$ for all $m < n$. This means that any $\phi \restriction A_{g_{\beta,m}} \in V_\gamma$ witnesses the failure of $(\star_\gamma)$, a contradiction. Thus the claim is proved.

Fix $\beta \geq \beta_\gamma$ such that $\mathcal{A}_\beta$ is not a finite maximal almost disjoint family below $\operatorname{dom}(\phi)$. Since $\mathbb{Q}_\gamma$ adds a dominating real, $\mathcal{A}_\beta$ is not a maximal almost disjoint family in $V_{\gamma+1}$. Choose an infinite $A \subseteq \operatorname{dom}(\phi)$, $A \in V_{\gamma+1}$, almost disjoint from all members of $\mathcal{A}_\beta$. Since $\mathcal{F}_\beta^{\leq \gamma+1} = \mathcal{F}_\beta^{\leq \gamma}$, the conclusion of $(\ddagger_\alpha)$ holds for all $g \in \mathcal{F}_\beta^{\leq \gamma+1}$. Since $\phi \restriction A \in V_{\gamma+1}$, (e) entails it even holds for all $g \in \mathcal{F}_\beta^{\leq \alpha}$, and the proof is complete. q.e.d.

Using $(\ddagger_\alpha)$ and CH which holds in V_α, it is easy to build up the Ramsey ultrafilter $\mathcal{U}_\alpha$ such that

(i) for all partial functions $\phi : \omega \to 2^\omega$ and all partial one-to-one onto $f : \omega \to \operatorname{dom}(\phi)$ from V_α, if $\operatorname{dom}(f) \in \mathcal{U}_\alpha$, then there are $U \in \mathcal{U}_\alpha$ contained in $\operatorname{dom}(f)$ and $\beta < \omega_1$ such that $\phi(n) \not\supseteq g(n)$ for all $g \in \mathcal{F}_\beta^{\leq \alpha}$ and almost all $n \in f(U) \cap \operatorname{supp}(g)$.

To do this simply list all such ϕ, f in order type ω_1 and take care of them in, say, the odd steps of the recursive construction of $\mathcal{U}_\alpha$.

$$\star \star \star$$

Before getting into the details of the proof of $(\star_{\alpha+1})$, we develop some combinatorics for $\mathbb{L}_\mathcal{F}$; see [B₄06b, Section 4] (and also [B₂D₀85]) for closely related arguments.

Let $\mathcal{F}$ be a filter, and let φ be a statement of the $\mathbb{L}_\mathcal{F}$ forcing language. Recall that $A \subseteq \omega$ is *positive modulo* $\mathcal{F}$ if $\omega \setminus A \notin \mathcal{F}$. Say $\sigma \in \omega^{<\omega}$ *forces* φ

if $T \Vdash \varphi$ for some $T \in \mathbb{L}_{\mathcal{F}}$ with stem σ. Note that σ forces φ iff $\{n : \sigma^\frown n$ forces $\varphi\} \in \mathcal{F}$. Let $\text{rk}_\varphi(\sigma) = 0$ if σ forces φ. For $\alpha > 0$, say $\text{rk}_\varphi(\sigma) = \alpha$ if $\text{rk}_\varphi(\sigma) \neq \beta$ for any $\beta < \alpha$ and $\{n : \text{rk}_\varphi(\sigma^\frown n) < \alpha\}$ is positive modulo $\mathcal{F}$. Say σ *favors* φ if $\text{rk}_\varphi(\sigma) < \omega_1$. Note that σ favors φ iff $\{n : \sigma^\frown n$ favors $\varphi\}$ is positive modulo $\mathcal{F}$. Also, if $\mathcal{F}$ is an ultrafilter, then σ favors φ iff σ forces φ. For any filter $\mathcal{F}$, σ favors at least one of φ and $\neg\varphi$, and forces at most one of them.

Let $\dot{X}$ be an $\mathbb{L}_\mathcal{F}$-name for an infinite subset of ω. Say $A \subseteq \omega$ is *large (with respect to $\mathcal{F}$)* if *either* $\mathcal{F}$ is an ultrafilter and $A \in \mathcal{F}$ *or* every infinite subset of A is positive modulo $\mathcal{F}$. (This is equivalent to saying that A is a pseudointersection of $\mathcal{F}$.) Note A is large with respect to $\mathcal{F}_e$ if and only if it has finite intersection with every branch in $2^{<\omega}$. Call $\sigma \in \omega^{<\omega}$ *good* if there are large A and one-to-one $f : A \to \omega$ such that $\sigma^\frown n$ favors $f(n) \in \dot{X}$ for all $n \in A$. A sequence σ is *very good* if there is $X_0 \in [\omega]^\omega$ such that σ favors $a \in \dot{X}$ for all $a \in X_0$.

Lemma 4. Assume $\mathcal{F}$ is either $\mathcal{F}_e$ or a Ramsey ultrafilter. Then every very good σ is good.

Proof. Let X_0 witness that σ is very good. For each $a \in X_0$, $A_a = \{n : \sigma^\frown n$ favors $a \in \dot{X}\}$ is positive modulo $\mathcal{F}$.

If $\mathcal{F}$ is a Ramsey ultrafilter, we easily find $U \in \mathcal{F}$ and a bijection $f : U \to X_0$ such that $f^{-1}(a) \in A_a$ for all $a \in X_0$. Then U and f witness that σ is good.

If $\mathcal{F} = \mathcal{F}_e$, for each $a \in X_0$, we can find by compactness $f_a \in 2^\omega$ such that for all m there is $\tau \in A_a$ with $f_a \restriction m \subseteq \tau \not\subseteq f_a$. Next find f such that a subsequence of the f_a converges to f. Assume the subsequence is indexed by $X_1 \subseteq X_0$. It is then easy to find $\tau_a \in A_a$ for $a \in X_1$ such that $A = \{\tau_a : a \in X_1\}$ forms an antichain in $2^{<\omega}$. So the large A and $f : A \to X_1$ sending τ_a to a witness that σ is good. q.e.d.

Define a rank function ρ by recursion on the ordinals such that $\rho(\sigma) = 0$ if σ is good and, for $\alpha > 0$, $\rho(\sigma) = \alpha$ if $\rho(\sigma) \neq \beta$ for any $\beta < \alpha$ and $\{n : \rho(\sigma^\frown n) < \alpha\}$ is positive modulo $\mathcal{F}$.

Lemma 5. Assume $\mathcal{F}$ is either $\mathcal{F}_e$ or a Ramsey ultrafilter. For all $\sigma \in \omega^{<\omega}$, $\rho(\sigma) < \omega_1$.

Proof. Suppose $\rho(\sigma) = \infty$ (which means the rank is undefined). Let $X_0 = \{a : \sigma$ favors $a \in \dot{X}\}$. By the previous lemma and $\rho(\sigma) \neq 0$, X_0 must be finite. We recursively construct a tree $T \in \mathbb{L}_\mathcal{F}$ with stem σ such that for all $\tau \in T$ extending σ,

- $\{a : \tau$ favors $a \in \dot{X}\} \subseteq X_0$,

- $\rho(\tau) = \infty$.

Suppose $\tau \in T$ for some $\tau \supseteq \sigma$. We need to construct $\mathrm{succ}_T(\tau)$. By definition of ρ and the assumption on τ, $U_0 = \{n : \rho(\tau^\frown n) = \infty\} \in \mathcal{F}$. Also, for $a \notin X_0$, $A_a = \{n : \tau^\frown n \text{ favors } a \in \dot{X}\}$ is small modulo $\mathcal{F}$. Let $A = \bigcup\{A_a : a \notin X_0\}$.

Assume first there is a large $B \subseteq A$. If $\mathcal{F}$ is Ramsey, by pruning B if necessary, we may assume $|B \cap A_a| \leq 1$ for all $a \notin X_0$. If $\mathcal{F} = \mathcal{F}_e$, then, by definition of largeness, we must have that $B \cap A_a$ is finite for all $a \notin X_0$. Thus, again by pruning B if necessary, we may assume $|B \cap A_a| \leq 1$ for all $a \notin X_0$. It is then easy to construct $f : B \to \omega$ witnessing that τ is good, a contradiction.

Therefore there is no large $B \subseteq A$. This means A still belongs to the dual ideal of $\mathcal{F}$. Thus, if we let $\mathrm{succ}_T(\tau) = U_0 \setminus A$, the recursive requirements are satisfied. This completes the construction of the tree T.

Find $S \leq T$ and $a \notin X_0$ such that $S \Vdash a \in \dot{X}$. Then $\mathrm{stem}(S) \in T$ and $\mathrm{stem}(S)$ forces $a \in \dot{X}$, a contradiction. q.e.d.

Let $\dot{\phi} : \omega \to 2^\omega$ be an $\mathbb{L}_\mathcal{F}$-name for a partial function. Let $\dot{X} = \mathrm{dom}(\dot{\phi})$. For good σ, fix A_σ and $f_\sigma : A_\sigma \to \omega$ witnessing goodness. Let $X_\sigma = f_\sigma(A_\sigma)$. Define $\psi_\sigma : X_\sigma \to 2^\omega$ as follows. Let $a \in X_\sigma$ and $n = f_\sigma^{-1}(a)$. So $\sigma^\frown n$ favors $a \in \dot{X} = \mathrm{dom}(\dot{f})$. Find $\psi_\sigma(a)$ such that for all m, $\sigma^\frown n$ favors the conjunction of $a \in \dot{X}$ and $\dot\phi(a)\restriction m = \psi_\sigma(a)\restriction m$. This is possible because there are only finitely many possibilities for the value of $\dot\phi(a)\restriction m$, and at least one of them must be favored.

We are ready to complete the proof of the preservation of $(\star_\alpha)$ in the successor step.

Lemma 6. Assume $\mathrm{cf}(\alpha) = \omega_1$ and $(\ddagger_\alpha)$ holds. Then $(\star_{\alpha+1})$ holds as well.

Proof. Recall that in this case, $\mathbb{Q}_\alpha = \mathbb{L}_{\mathcal{U}_\alpha}$ where $\mathcal{F} = \mathcal{U}_\alpha$ is a Ramsey ultrafilter satisfying property (i) above.

This means that for all good σ we can find $U_\sigma \in \mathcal{U}_\alpha$ and β_σ satisfying the conclusion of (i) for the ψ_σ and f_σ defined above. Let β be the supremum of the β_σ. We claim that

$$\Vdash \forall g \in \mathcal{F}_\beta^{\leq\alpha} \; \exists^\infty n \in \dot X \; (n \notin \mathrm{supp}(g) \text{ or } \dot\phi(n) \not\supseteq g(n))$$

Since $\mathcal{F}_\beta^{\leq\alpha+1} = \mathcal{F}_\beta^{\leq\alpha}$, this is enough.

Assume the claim was false and there are $T \in \mathbb{L}_{\mathcal{U}_\alpha}$, $g \in \mathcal{F}_\beta^{\leq\alpha}$ and a_0 such that

$$T \Vdash \forall a \geq a_0 \; (\text{if } a \in \dot X \text{ then } a \in \mathrm{supp}(g) \text{ and } \dot\phi(a) \supseteq g(a))$$

By the previous lemma, we may assume without loss of generality that $\sigma =$ stem(T) is good. By (i), there is $U := U_\sigma \cap \mathrm{succ}_T(\sigma) \subseteq \mathrm{dom}(f_\sigma) \cap \mathrm{succ}_T(\sigma)$ belonging to $\mathcal{U}_\alpha$ such that $\psi_\sigma(a) \not\supseteq g(a)$ for all $a \in f_\sigma(U) \cap \mathrm{supp}(g)$. Fix any $n \in U$ such that $a = f_\sigma(n) \geq a_0$. Since $\sigma^\frown n$ favors $a \in \dot{X}$, we must have $a \in \mathrm{supp}(g)$. Find m such that $\psi_\sigma(a) \restriction m$ is incompatible with $g(a)$. Since $\sigma^\frown n$ favors $\dot{\phi}(a) \restriction m = \psi_\sigma(a) \restriction m$, we easily find $S \leq T$ with stem$(S) = \sigma^\frown n$ such that $S \Vdash \dot{\phi}(a) \restriction m = \psi_\sigma(a) \restriction m$. Thus $S \Vdash \dot{\phi}(a) \not\supseteq g(a)$, a contradiction.

q.e.d.

Lemma 7. Assume cf$(\alpha) \neq \omega_1$ and $(\star_\alpha)$ holds. Then $(\star_{\alpha+1})$ holds as well.

Proof. Recall that in this case, $\mathbb{Q}_\alpha = \mathbb{E} = \mathbb{L}_{\mathcal{F}_e}$ is eventually branching function forcing. The argument is very similar to the previous proof. However, since there are a couple of subtle differences, we present the details.

For all good σ we can find β_σ satisfying the conclusion of $(\star_\alpha)$ for the ψ_σ defined above. Let $\beta \geq \beta_\alpha$ be larger than all the β_σ. We claim that

$$\Vdash \forall g \in \mathcal{F}_\beta^{\leq \alpha} \ \exists^\infty n \in \dot{X} \ (n \notin \mathrm{supp}(g) \text{ or } \dot{\phi}(n) \not\supseteq g(n)).$$

Since $\mathcal{F}_\beta^{\leq \alpha+1} = \mathcal{F}_\beta^{\leq \alpha}$, this is enough.

Assume the claim was false and there are $T \in \mathbb{E}$, $g \in \mathcal{F}_\beta^{\leq \alpha}$ and a_0 such that

$$T \Vdash \forall a \geq a_0 \ (\text{if } a \in \dot{X} \text{ then } a \in \mathrm{supp}(g) \text{ and } \dot{\phi}(a) \supseteq g(a)).$$

By Lemma 5, we may assume without loss of generality that $\sigma =$ stem(T) is good. Since A_σ is large and $\mathrm{succ}_T(\sigma) \in \mathcal{F}_e$, we must have $A_\sigma \subseteq^* \mathrm{succ}_T(\sigma)$. By $(\star_\alpha)$, we can find $n \in A_\sigma \cap \mathrm{succ}_T(\sigma)$ such that, letting $a = f_\sigma(n) \in X_\sigma = \mathrm{dom}(\psi_\sigma)$, we have $a \geq a_0$ and $(a \notin \mathrm{supp}(g)$ or $\psi_\sigma(a) \not\supseteq g(a))$. Since $\sigma^\frown n$ favors $a \in \dot{X}$, we must have $a \in \mathrm{supp}(g)$. Thus $\psi_\sigma(a) \not\supseteq g(a)$. Find m such that $\psi_\sigma(a) \restriction m$ is incompatible with $g(a)$. Since $\sigma^\frown n$ favors $\dot{\phi}(a) \restriction m = \psi_\sigma(a) \restriction m$, we can recursively build $S \leq T$ with stem$(S) \supseteq \sigma^\frown n$ such that $S \Vdash \dot{\phi}(a) \restriction m = \psi_\sigma(a) \restriction m$. Thus $S \Vdash \dot{\phi}(a) \not\supseteq g(a)$, a contradiction.

q.e.d.

This completes the proof of Theorem 1.

⋆ ⋆ ⋆

We believe that a modification of the proof of Theorem 1 can be used to solve the following other old problem of van Douwen [v90] (see also [$D_1$98]).

For a topological space X without isolated points, the *Baire number* (or: *Novák number*) $\mathfrak{n}(X)$ of X is the least size of a family of nowhere dense sets covering X. So $\mathfrak{n}(\mathbb{R}) = \text{cov}(\mathcal{M})$. Let $\mathfrak{n} := \mathfrak{n}(\omega^*)$. It is easy to see that $\mathfrak{n}(\mathbb{R}^*) \leq \mathfrak{n}$. Van Douwen asked whether $\mathfrak{n}(\mathbb{R}^*) = \mathfrak{n}$. If we could strengthen the diamond argument so as to destroy all witnesses for $\mathfrak{n} = \aleph_2$ while still preserving $\mathfrak{h}(\mathbb{C}^\omega/\text{fin}) = \aleph_1$, we would get the consistency of $\mathfrak{n}(\mathbb{R}^*) < \mathfrak{n}$. The point is that by results of Balcar, Pelant, and Simon [B₀PS₁80] (see also [B₀S₁89, Theorem 3.10]), the model of Theorem 1 must satisfy $\mathfrak{n}(\mathbb{R}^*) = \aleph_2$. Unfortunately, we have been unable to do this so far.

⋆ ⋆ ⋆

We also notice that the argument of the proof can be generalized to show the consistency of $\mathfrak{h}(\mathbb{C}^\omega/\text{fin}) = \kappa$ and $\mathfrak{h} = \text{add}(\mathcal{M}) = \mathfrak{c} = \kappa^+$ for arbitrary regular κ. This is a standard argument and we confine ourselves to briefly mentioning the main adjustments. Use $\diamondsuit_{S_\kappa^{\kappa^+}}$ instead of $\diamondsuit_{S_1^2}$. Make a finite support iteration of length κ^+, doing (A) in limit stages of cofinality κ and replacing (B) by

(B*) if $\text{cf}(\alpha) \neq \kappa$, then $\dot{\mathbb{Q}}_\alpha$ is the two-step iteration of $\dot{\mathbb{E}}$ and a κ-stage finite support iteration of ccc p.o.'s of size $< \kappa$ which forces $\text{MA}_{<\kappa}(\sigma$-centered).

Note this entails that $\text{MA}_{<\kappa}(\sigma$-centered) will hold in each successor step as well as in each limit step of cofinality $\geq \kappa$. Using this, we can still build the Ramsey ultrafilters $\mathcal{U}_\alpha$ in κ steps such that $\mathcal{U}_\alpha$ is generated by a decreasing κ-chain. Thus $\mathfrak{h} = \kappa^+$. This time $< \kappa$ many $\mathcal{F}_\beta$ will get a new element in certain successor stages: use again $\text{MA}_{<\kappa}(\sigma$-centered) to carry out the construction. In Lemmata 3 and 6, replace $\text{cf}(\alpha) = \omega_1$ by $\text{cf}(\alpha) = \kappa$. In Lemma 7, we also need to consider the forcing which forces $\text{MA}_{<\kappa}(\sigma$-centered): by previous arguments, it is enough to look at the single step forcing; since this forcing has size $< \kappa$, and since the $\mathcal{F}_\beta^{\leq \alpha}$ form a decreasing chain under refinement, preservation of $(\star_\alpha)$ is immediate by standard arguments. Thus $\mathfrak{h}(\mathbb{C}^\omega/\text{fin}) \leq \kappa$. Finally notice that $\mathfrak{h}(\mathbb{C}^\omega/\text{fin}) \geq \kappa$ because $\text{MA}_{<\kappa}(\sigma$-centered) holds in the final model.

We do not know, however, how to prove the consistency of $\mathfrak{h}(\mathbb{C}^\omega/\text{fin})^+ < \mathfrak{h}$. Note that by the results of Balcar, Pelant, and Simon mentioned above, this consistency would solve van Douwen's problem as well.

2 Laver forcing for $\mathbb{C}^\omega/\text{fin}$

In this section, we develop the combinatorics of a Laver-like forcing which naturally increases the distributivity number $\mathfrak{h}(\mathbb{C}^\omega/\text{fin})$. This will be put to good use in the proof of Theorem 2 in the next section. For a Mathias-like forcing notion which naturally increases $\mathfrak{h}(\mathbb{C}^\omega/\text{fin})$, see [H$_0$05].

As before, for $f \in \mathbb{C}^\omega$, let $\text{supp}(f) = \{i : f(i) \neq \mathbf{0}\}$. A set $\mathcal{F} \subseteq \mathbb{C}^\omega$ is a $\mathbb{C}^\omega$-*filter* if

- all members of $\mathcal{F}$ have infinite support,
- for all $f \in \mathcal{F}$ and $g \in \mathbb{C}^\omega$, if $f \leq^* g$ then $g \in \mathcal{F}$,
- for all $f, g \in \mathcal{F}$, $h \in \mathcal{F}$ where $h(i) = f(i) \cap g(i)$ for all i.

$\mathcal{F}$ is *maximal* if for any $f \in \mathbb{C}^\omega \setminus \mathcal{F}$ there is $g \in \mathcal{F}$ such that $\{i \in \text{supp}(f) \cap \text{supp}(g) : f(i) \cap g(i) > \mathbf{0}\}$ is finite. If $\mathcal{F}$ is a maximal $\mathbb{C}^\omega$-filter, then its trace on $\mathcal{P}(\omega)$, $\text{trace}(\mathcal{F}) = \{\text{supp}(f) : f \in \mathcal{F}\}$, is an ultrafilter. Say $\mathcal{F}$ is *Ramsey* if its trace on $\mathcal{P}(\omega)$ is a Ramsey ultrafilter. A filter $\mathcal{F}$ is *concentrated on* $2^{<\omega}$ if for all $f \in \mathcal{F}$ there is $g \in \mathcal{F}$ with $g \leq^* f$ and $g(i) \in 2^{<\omega}$ for all $i \in \text{supp}(g)$. Call $\mathcal{F}$ *selective* if given any $f \in \mathcal{F}$ and any family $\{a_{n,i} : n \in \omega\}, i \in \text{supp}(f)$, of maximal antichains below $f(i)$, there is a selector $g \in \mathcal{F}$. This means that $\text{supp}(g) = \text{supp}(f)$ and that for all $i \in \text{supp}(f)$ there is n with $g(i) = a_{n,i}$. If $\mathcal{F}$ is selective then it is necessarily concentrated on $2^{<\omega}$. Also, if $\mathcal{F}$ is selective and its trace is an ultrafilter, then $\mathcal{F}$ is a maximal $\mathbb{C}^\omega$-filter.

For a $\mathbb{C}^\omega$-filter $\mathcal{F}$ concentrated on $2^{<\omega}$, define *tree forcing with* $\mathcal{F}$, $\mathbb{Q}_\mathcal{F}$. Let Σ be the collection of all finite sequences σ with $\sigma(k) \in \omega \times 2^{<\omega}$ for $k \in |\sigma|$. For such σ, let $\sigma_0 \in \omega^{|\sigma|}$ and $\sigma_1 \in (2^{<\omega})^{|\sigma|}$ denote the projections onto the two coordinates. Order Σ by $\tau \leq \sigma$ if $|\tau| = |\sigma|$, $\tau_0 = \sigma_0$ and $\tau_1(k) \leq \sigma_1(k)$ for all $k \in |\sigma|$. (Note the latter means $\tau_1(k) \supseteq \sigma_1(k)$ if we think of $\tau_1(k)$ and $\sigma_1(k)$ as finite sequences.) Elements of $\mathbb{Q}_\mathcal{F}$ are trees $T \subseteq \Sigma$ such that for $\tau \in T$ with $\text{stem}(T) \subseteq \tau$, there is $f \in \mathcal{F}$ such that $\{\langle i, a \rangle : \tau^\frown \langle i, a \rangle \in T\} = f \upharpoonright \text{supp}(f)$. Note that the restriction of the projection $\sigma \mapsto \sigma_0$ to any condition T is one-to-one. The ordering is given by $T \leq S$ if for all $\tau \in T$ there is a (necessarily unique) $\sigma \in S$ such that $\tau \leq \sigma$. Clearly $\mathbb{Q}_\mathcal{F}$ is a σ-centered forcing notion.

The forcing $\mathbb{Q}_\mathcal{F}$ generically adjoins a function $\Phi : \omega \to \omega \times 2^\omega$, namely, $\Phi(n) = \langle i_n, c_n \rangle = \langle i, c \rangle$ if for each m there is $T \in G$ with $|\text{stem}(T)| \geq n + 1$ and $\text{stem}(T)(n) = \langle i, c \upharpoonright m \rangle$ where G is the $\mathbb{Q}_\mathcal{F}$-generic filter over V. It is easy to see that $\Phi_0 \in \omega^\omega$, the projection of Φ onto the first coordinate given by $\Phi_0(n) = i_n$, is a dominating real over V, and that all $\Phi_1(n) = c_n$ are Cohen reals over V. Define $f_\Phi \in \mathbb{C}^\omega$ by $\text{supp}(f_\Phi) = \text{ran}(\Phi_0) = \{i_n : n \in \omega\}$ and $f_\Phi(i_n) = c_n \upharpoonright i_{n+1}$.

Lemma 8. *The function f_Φ diagonalizes the filter $\mathcal{F}$, i.e., $f_\Phi \leq^* f$ for all $f \in \mathcal{F}$.*

Proof. Let $f \in \mathcal{F}$. Since $\mathcal{F}$ is concentrated on $2^{<\omega}$ we may assume $f(i) \in 2^{<\omega}$ for all $i \in \text{supp}(f)$. Since $\{T \in \mathbb{Q}_\mathcal{F} :$ for all $\tau \in T$ with $\text{stem}(T) \subseteq \tau$, $\{i :$ there is a with $\tau^\frown \langle i, a \rangle \in T\} \subseteq \text{supp}(f)\}$ is dense in $\mathbb{Q}_\mathcal{F}$, $\text{supp}(f_\Phi) \subseteq^* \text{supp}(f)$ follows. Similarly, $\{T \in \mathbb{Q}_\mathcal{F} :$ for all $\tau \in T$ with $\text{stem}(T) \subseteq \tau$, if $\tau^\frown \langle i, a \rangle^\frown \langle j, b \rangle \in T$, then $a \leq f(i)$ and $j \geq |f(i)|\}$ is dense in $\mathbb{Q}_\mathcal{F}$ where $|f(i)|$ denotes the length of the finite sequence $f(i)$. Thus, by a genericity argument, $c_n \supset f(i_n)$ and even $c_n \restriction i_{n+1} \supset f(i_n)$, i.e., $f_\Phi(i_n) \leq f(i_n)$, for almost all n, as required. q.e.d.

We note that an analogous forcing can be defined for Boolean algebras $\mathbb{A}$ other than $\mathbb{C}$ as well. That is, one works with $\mathbb{A}^\omega$-filters $\mathcal{F}$, considers finite sequences σ with $\sigma(k) \in \omega \times \mathbb{A}$ for $k \in |\sigma|$, and redefines $\mathbb{Q}_\mathcal{F}$ accordingly. To mention a specific case, let $\mathbb{A} = \mathbb{B}$ be random forcing. Then $\mathbb{Q}_\mathcal{F}$ is still σ-linked (because random forcing is), and adds Φ as above such that Φ_0 is a dominating real, and $\{\Phi_1(n) : n \in \omega\}$, a sequence of random reals. However, since $\mathbb{B}$ has no countable dense subset, there is no notion analogous to being concentrated. We do not know whether $\mathbb{Q}_\mathcal{F}$ adds a diagonalization to $\mathcal{F}$ in this case, and we have no use for the forcing $\mathbb{Q}_\mathcal{F}$. The subsequent rank analysis, however, goes through in this more general case as well (see Lemmata 9 through 14 below).

As for $\mathbb{L}_\mathcal{F}$ in Section 1, we make a rank analysis for $\mathbb{Q}_\mathcal{F}$. Assume $\mathcal{F}$ is a maximal $\mathbb{C}^\omega$-filter. Let $\sigma \in \Sigma$, and let φ be a statement in the $\mathbb{Q}_\mathcal{F}$ forcing language. Say σ *forces* φ if there is $T \in \mathbb{Q}_\mathcal{F}$ with stem σ such that $T \Vdash \varphi$. Clearly, σ can force at most one of φ and $\neg\varphi$. Also, if σ forces φ and $\tau \leq \sigma$, then τ forces φ. Say

- $\text{rk}_\varphi(\sigma) = 0$ if there is $\tau \leq \sigma$ which forces φ,

- (for $\alpha > 0$) $\text{rk}_\varphi(\sigma) \leq \alpha$ if for all $g \in \mathcal{F}$ there is $\langle i, a \rangle \in g$ such that $\text{rk}_\varphi(\sigma^\frown \langle i, a \rangle) < \alpha$.

Say $\text{rk}_\varphi(\sigma) = \infty$ if $\text{rk}_\varphi(\sigma)$ is undefined.

Lemma 9. *If $\tau \leq \sigma$, then $\text{rk}_\varphi(\tau) \geq \text{rk}_\varphi(\sigma)$.*

Proof. If $\text{rk}_\varphi(\tau) = \infty$, there is nothing to show. So make induction on $\text{rk}_\varphi(\tau)$. If $\text{rk}_\varphi(\tau) = 0$, $\text{rk}_\varphi(\sigma) = 0$ is clear by definition. So suppose $\text{rk}_\varphi(\tau) = \alpha > 0$. Let $g \in \mathcal{F}$ be arbitrary. Find $\langle i, a \rangle \in g$ such that $\text{rk}_\varphi(\tau^\frown \langle i, a \rangle) < \alpha$. By induction hypothesis $\text{rk}_\varphi(\sigma^\frown \langle i, a \rangle) \leq \text{rk}_\varphi(\tau^\frown \langle i, a \rangle) < \alpha$. Therefore $\text{rk}_\varphi(\sigma) \leq \alpha$. q.e.d.

Lemma 10. *We have that $\text{rk}_\varphi(\sigma) \leq \alpha$ if and only if for all $g \in \mathcal{F}$ there is $f \in \mathcal{F}$ with $f \leq^* g$ such that $\text{rk}_\varphi(\sigma^\frown \langle i, f(i) \rangle) < \alpha$ for all $i \in \text{supp}(f)$.*

Proof. ($\Longleftarrow$) This is immediate by the previous lemma.

($\Longrightarrow$) Given $g \in \mathcal{F}$, define $\text{supp}(f) = \{i \in \text{supp}(g) : \text{rk}_\varphi(\sigma^\frown \langle i, g(i) \rangle) < \alpha\}$, $\text{supp}(h) = \{i \in \text{supp}(g) : \text{rk}_\varphi(\sigma^\frown \langle i, g(i) \rangle) \not< \alpha\}$, $f \restriction \text{supp}(f) = g \restriction \text{supp}(f)$ and $h \restriction \text{supp}(h) = g \restriction \text{supp}(h)$. By maximality of $\mathcal{F}$, either f or h belongs to $\mathcal{F}$. By assumption, h cannot belong to $\mathcal{F}$. Thus f belongs to $\mathcal{F}$, and we are done. q.e.d.

Lemma 11. For all σ, either $\text{rk}_\varphi(\sigma) < \infty$ or $\text{rk}_{\neg\varphi}(\sigma) < \infty$.

Proof. Assume both $\text{rk}_\varphi(\sigma) = \infty$ and $\text{rk}_{\neg\varphi}(\sigma) = \infty$. Then we can easily construct $T \in \mathbb{Q}_\mathcal{F}$ with $\text{stem}(T) = \sigma$ and such that for all $\tau \supseteq \sigma$ belonging to T, we have $\text{rk}_\varphi(\tau) = \infty$ and $\text{rk}_{\neg\varphi}(\tau) = \infty$. Find $S \leq T$ deciding φ, and let $\tau = \text{stem}(S)$. Without loss of generality $S \Vdash \varphi$. Thus τ forces φ. There is $\tau' \in T$ such that $\tau \leq \tau'$. Hence $\text{rk}_\varphi(\tau') = 0$, a contradiction. q.e.d.

Say that σ *favors* φ if $\text{rk}_\varphi(\sigma) < \infty$. By the previous lemma, σ favors at least one of φ and $\neg\varphi$. Also, by Lemma 9, if $\tau \leq \sigma$ and τ favors φ, then σ favors φ. For $\sigma \in T$, let $T_\sigma := \{\tau \in T : \tau \subseteq \sigma \text{ or } \sigma \subseteq \tau\}$.

Lemma 12. A sequence σ favors φ if and only if for all $T \in \mathbb{Q}_\mathcal{F}$ with $\text{stem}(T) = \sigma$, there is $S \leq T$ such that $S \Vdash \varphi$.

Proof. ($\Longleftarrow$) Assume $\text{rk}_\varphi(\sigma) = \infty$, and use the argument of the previous proof to get some T no extension of which forces φ.

($\Longrightarrow$) Let T be given and make induction on $\text{rk}_\varphi(\sigma)$. If $\text{rk}_\varphi(\sigma) = 0$, find $\tau \leq \sigma$ forcing φ. Let S witness this, i.e., $\text{stem}(S) = \tau$ and $S \Vdash \varphi$. Clearly, S and T are compatible, and a common extension is as required. So assume $\alpha := \text{rk}_\varphi(\sigma) > 0$. There is $\langle i, a \rangle$ such that $\sigma^\frown \langle i, a \rangle \in T$ and $\text{rk}_\varphi(\sigma^\frown \langle i, a \rangle) < \alpha$. By induction hypothesis, there is $S \leq T_{\sigma^\frown \langle i,a \rangle} \leq T$ forcing φ. q.e.d.

Corollary 13. A sequence σ forces φ if and only if $\text{rk}_{\neg\varphi}(\sigma) = \infty$.

Proof. ($\Longrightarrow$) Let T witness that σ forces φ. By the previous lemma σ cannot favor $\neg\varphi$.

($\Longleftarrow$) By the previous lemma, there is T with $\text{stem}(T) = \sigma$ such that no extension of T forces $\neg\varphi$. Hence $T \Vdash \varphi$. Thus σ forces φ. q.e.d.

Let $\dot{X}$ be a $\mathbb{Q}_\mathcal{F}$-name for a subset of ω. Define

- $\rho_{\dot{X}}(\sigma) = 0$ if for all $g \in \mathcal{F}$ and all n_0 there are $n \geq n_0$ and $\langle i, a \rangle \in g$ such that $\sigma^\frown \langle i, a \rangle$ favors $n \in \dot{X}$,

- (for $\alpha > 0$) $\rho_{\dot{X}}(\sigma) \leq \alpha$ if for all $g \in \mathcal{F}$ there is $\langle i, a \rangle \in g$ such that $\rho_{\dot{X}}(\sigma^\frown \langle i, a \rangle) < \alpha$.

As in Lemma 9, we show that if $\sigma \leq \tau$ then $\rho_{\dot{X}}(\sigma) \geq \rho_{\dot{X}}(\tau)$.

Lemma 14. For all σ, we have $\rho_{\dot{X}}(\sigma) < \infty$.

Proof. Assume $\rho_{\dot{X}}(\sigma) = \infty$. Clearly, there is n_0 such that for all $n \geq n_0$, σ does not favor $n \in \dot{X}$. Recursively build a tree $T \in \mathbb{Q}_\mathcal{F}$ such that $\text{stem}(T) = \sigma$ and for all $\tau \supseteq \sigma$ belonging to T,

- $\rho_{\dot{X}}(\tau) = \infty$,

- for all $n \geq n_0$, τ does not favor $n \in \dot{X}$ (i.e., $\text{rk}_{n \in \dot{X}}(\tau) = \infty$).

The sequence σ clearly satisfies both requirements. Assume τ has been constructed as required, and we need to produce the successor level. There is $g_0 \in \mathcal{F}$ such that for all $\langle i, a \rangle \in g_0$, $\rho_{\dot{X}}(\tau^\frown \langle i, a \rangle) = \infty$. Also, there are $g_1 \in \mathcal{F}$ and n_τ such that for all $n \geq n_\tau$ and all $\langle i, a \rangle \in g_1$, $\tau^\frown \langle i, a \rangle$ does not favor $n \in \dot{X}$. Fix n with $n_0 \leq n < n_\tau$. Since τ does not favor $n \in \dot{X}$, there is $h_n \in \mathcal{F}$ such that for all $\langle i, a \rangle \in h_n$, $\tau^\frown \langle i, a \rangle$ does not favor $n \in \dot{X}$. Let $g \in \mathcal{F}$ be such that $\text{supp}(g) \subseteq \text{supp}(g_0) \cap \text{supp}(g_1) \cap \bigcap_{n_0 \leq n < n_\tau} \text{supp}(h_n)$ and $g(i) = g_0(i) \cup g_1(i) \cup \bigcup_{n_0 \leq n < n_\tau} h_n(i) \in 2^{<\omega}$ for all $i \in \text{supp}(g)$. Then g still satisfies that for all $\langle i, a \rangle \in g$, $\rho_{\dot{X}}(\tau^\frown \langle i, a \rangle) = \infty$ and for all $n \geq n_0$, $\tau^\frown \langle i, a \rangle$ does not favor $n \in \dot{X}$. Hence we obtained the successor level, and the construction of T is complete.

Since $\Vdash$ "$\dot{X}$ is infinite", there are $S \leq T$ and $n \geq n_0$ such that $S \Vdash n \in \dot{X}$. So $\tau := \text{stem}(S)$ forces $n \in \dot{X}$. Hence there is $\tau' \geq \tau$ belonging to T such that τ' favors $n \in \dot{X}$, a contradiction. q.e.d.

Lemma 15. Let $\mathcal{F}$ be additionally Ramsey and selective. Assume $\rho_{\dot{X}}(\sigma) = 0$. Then there is $f \in \mathcal{F}$ such that

Case 1 either there is one-to-one $\phi : \text{supp}(f) \to \omega$ such that for all $i \in \text{supp}(f)$ and all $a \leq f(i)$, $\sigma^\frown \langle i, a \rangle$ favors $\phi(i) \in \dot{X}$,

Case 2 or for all $i \in \text{supp}(f)$, all $a \leq f(i)$ and all n_0, there is $n \geq n_0$ such that $\sigma^\frown \langle i, a \rangle$ favors $n \in \dot{X}$.

Proof. Fix i. Let $\{a_{m,i} : m \in \omega\}$ be a maximal antichain such that for all m

(I) either $\sigma^\frown \langle i, a_{m,i} \rangle$ favors no $n \in \dot{X}$,

(II) or there is n such that for all $a \leq a_{m,i}$, $\sigma^\frown \langle i, a \rangle$ favors $n \in \dot{X}$ and $\sigma^\frown \langle i, a_{m,i} \rangle$ does not favor $n' \in \dot{X}$ for any $n' > n$,

(III) or for all $a \leq a_{m,i}$ and all n_0 there is $n \geq n_0$ such that $\sigma^\frown \langle i, a \rangle$ favors $n \in \dot{X}$.

To get such an antichain A, first note it is easy to get an antichain A_0 all of whose elements satisfy either (I) or (III) and which is maximal with this property. Say a favors n if $\sigma^\frown \langle i, a \rangle$ favors $n \in \dot{X}$. Given any a incompatible with all members of A_0, there must be $b \leq a$ which favors only finitely many n. Also every $c \leq b$ must favor at least one n. Therefore we can indeed find $c \leq b$ and n such that all $d \leq c$ favor n and c does not favor any number larger than n. Thus A_0 can indeed be extended to the required maximal antichain A.

Now, let $f \in \mathcal{F}$ be a selector. That is $f(i) = a_{m,i}$ for some m. By maximality of $\mathcal{F}$, we may assume that

- either for all $i \in \text{supp}(f)$, $f(i)$ is as in (I),
- or for all $i \in \text{supp}(f)$, $f(i)$ is as in (II),
- or for all $i \in \text{supp}(f)$, $f(i)$ is as in (III).

The first alternative is ruled out by $\rho_{\dot{X}}(\sigma) = 0$.

Assume the second alternative. For $i \in \text{supp}(f)$, let $\phi(i)$ be such that for all $a \leq f(i)$, $\sigma^\frown \langle i, a \rangle$ favors $\phi(i) \in \dot{X}$, and $\sigma^\frown \langle i, f(i) \rangle$ does not favor $n' \in \dot{X}$ for any $n' > \phi(i)$. Since $\text{trace}(\mathcal{F})$ is a Ramsey ultrafilter, we may assume without loss of generality that ϕ is either constant or one-to-one on $\text{dom}(\phi) = \text{supp}(f)$. The first case is again ruled out by $\rho_{\dot{X}}(\sigma) = 0$, and the second is just the first case of the lemma.

Finally, the third alternative gives the second case of the lemma.   q.e.d.

3   Proof of Theorem 2

Structurally, the proof of Theorem 2 closely follows the proof of Theorem 1 in Section 1. The differences are mainly on the combinatorial side. Accordingly, we shall be brief when dealing with arguments which are analogous, and elaborate where major adjustments occur.

We assume again CH and $\diamondsuit_{S_1^2}$, and perform a finite support iteration $\langle \mathbb{P}_\alpha, \dot{\mathbb{Q}}_\alpha : \alpha < \omega_2 \rangle$ of ccc forcing such that

(A') if $\text{cf}(\alpha) = \omega_1$, then $\dot{\mathbb{Q}}_\alpha$ is tree forcing $\mathbb{Q}_{\dot{\mathcal{F}}_\alpha}$ with a maximal $\mathbb{C}^\omega$-filter $\dot{\mathcal{F}}_\alpha$ which is Ramsey and selective (and thus concentrated on $2^{<\omega}$),

(B') if $\text{cf}(\alpha) \neq \omega_1$, then $\dot{\mathbb{Q}}_\alpha$ is Hechler forcing $\dot{\mathbb{D}}$.

Again, at limit stages α of cofinality ω_1, we use forcing (A') to kill potential witnesses for $\mathfrak{h}(\mathbb{C}^\omega/\text{fin}) = \aleph_1$ by building $\mathcal{F}_\alpha$ accordingly. The relevant requirements are dealt with in the even steps of the construction (*cf.* clause (i') below). Thus $\mathfrak{h}(\mathbb{C}^\omega/\text{fin}) = \aleph_2$ and we are left with proving $\mathfrak{h}_2 = \aleph_1$.

Using forcing (B'), we build up families $\{\mathcal{A}_\beta : \beta < \omega_1\}$ such that

(a') each $\mathcal{A}_\beta$ is a mad family of rectangles in $\omega \times \omega$,

(b') for $\beta < \beta'$, $\mathcal{A}_{\beta'}$ refines $\mathcal{A}_\beta$ (that is, for each $(A', B') \in \mathcal{A}_{\beta'}$ there is $(A, B) \in \mathcal{A}_\beta$ such that $A' \subseteq^* A$ and $B' \subseteq^* B$).

Again, let $\mathcal{A}_\beta^{\leq \alpha} = \mathcal{A}_\beta \cap V_\alpha \in V_\alpha$. The book-keeping this time gives us $\alpha_{(X,Y),\beta}$ for $(X,Y) \in ([\omega]^\omega)^2$ and $\beta \in \omega_1$. We define $(X,Y)_\alpha$ and β_α to be the unique (X,Y) and β with $\alpha = \alpha_{(X,Y),\beta}$. At each successor stage $\alpha + 1$ with $\text{cf}(\alpha) \neq \omega_1$, countably many $\mathcal{A}_\beta$ will get a new element:

(c') if $\beta < \beta_\alpha$ and if $(X,Y)_\alpha$ is almost disjoint from all members of $\mathcal{A}_\beta^{\leq \alpha}$ then there is $(A, B) \subseteq^* (X, Y)_\alpha$ belonging to $\mathcal{A}_\beta^{\leq \alpha + 1}$,

(d') for $\beta < \beta' < \omega_1$, $\mathcal{A}_{\beta'}^{\leq \alpha+1}$ refines $\mathcal{A}_\beta^{\leq \alpha+1}$,

(e') if $\beta < \beta_\alpha$, $(A, B) \in \mathcal{A}_\beta^{\leq \alpha+1} \setminus \mathcal{A}_\beta^{\leq \alpha}$, $A = \{a_n : n \in \omega\}$ and $B = \{b_n : n \in \omega\}$ are their increasing enumerations, and $\phi : \omega \to \omega$ is a partial one-to-one function belonging to V_α, then $\phi(a_n) < b_n < a_{n+1}$ and $\phi^{-1}(b_n) < a_{n+1} < b_{n+1}$ for almost all n (if the values are defined).

For $\beta \geq \beta_\alpha$, we again stipulate $\mathcal{A}_\beta^{\leq \alpha+1} = \mathcal{A}_\beta^{\leq \alpha}$.

For fixed α, the construction of $\mathcal{A}_\beta^{\leq \alpha+1} \setminus \mathcal{A}_\beta^{\leq \alpha}$ is like the construction of $\mathcal{F}_\beta^{\leq \alpha+1} \setminus \mathcal{F}_\beta^{\leq \alpha}$ in Section 1. To guarantee (e') holds use that $\mathbb{Q}_\alpha = \mathbb{D}$ adds a dominating real k over V_α, and construct $(A = \{a_n : n \in \omega\}, B = \{b_n : n \in \omega\}) \in \mathcal{A}_\beta^{\leq \alpha+1} \setminus \mathcal{A}_\beta^{\leq \alpha}$ such that $a_n < k(a_n) \leq b_n < k(b_n) \leq a_{n+1}$ for all n. Since, for $\phi \in V_\alpha$, $\phi <^* k$ ($\phi^{-1} <^* k$, resp.) on the domain of ϕ (ϕ^{-1}, resp.), (e') follows.

The statement (d') yields (b'), and (c') and the book-keeping together entail (a'). Also, we have $\mathcal{A}_\beta^{\leq \alpha+1} = \mathcal{A}_\beta^{\leq \alpha}$ if $\text{cf}(\alpha) = \omega_1$ and $\mathcal{A}_\beta^{\leq \alpha} = \bigcup_{\gamma < \alpha} \mathcal{A}_\beta^{\leq \gamma}$ for limit ordinals α. To argue that $\{\mathcal{A}_\beta : \beta < \omega_1\}$ witnesses $\mathfrak{h}_2 = \aleph_1$ we need to show that for all $\alpha < \omega_2$,

$$\forall (X,Y) \in ([\omega]^\omega)^2 \cap V_\alpha \; \exists \beta < \omega_1 \; \forall (A,B) \in \mathcal{A}_\beta \; (X \not\subseteq^* A \text{ or } Y \not\subseteq^* B) \tag{$++_\alpha$}$$

By (e'), it suffices to show by induction on $\alpha < \omega_2$ that

$$\forall (X,Y) \in ([\omega]^\omega)^2 \cap V_\alpha \ \exists \beta < \omega_1 \ \forall (A,B) \in \mathcal{A}_\beta^{\leq \alpha} \ (X \not\subseteq^* A \text{ or } Y \not\subseteq^* B) \tag{$\star\star_\alpha$}$$

In limit stages α of cofinality ω_1, we shall in fact require

$$\forall C \in [\omega]^\omega \ \forall \text{ partial 1-1 functions } \phi, \psi : C \to \omega \ \exists D \in [C]^\omega$$
$$\exists \beta < \omega_1 \ \forall (A,B) \in \mathcal{A}_\beta^{\leq \alpha} \ D \cap \phi^{-1}(A) \cap \psi^{-1}(B) \text{ is finite} \tag{$\ddagger\ddagger_\alpha$}$$

Clearly, ($\ddagger\ddagger_\alpha$) strengthens ($\star\star_\alpha$). (Given X, Y, let $C = \omega$, and let $\phi : \omega \to X$ and $\psi : \omega \to Y$ be bijections. Let D and β be as in ($\ddagger\ddagger_\alpha$). Fix $(A, B) \in \mathcal{A}_\beta^{\leq \alpha}$. Then $D \cap \phi^{-1}(A) \cap \psi^{-1}(B)$ is finite. If $X \subseteq^* A$, $\omega = \phi^{-1}(X) \subseteq^* \phi^{-1}(A)$, so $D \subseteq^* \phi^{-1}(A)$ and $D \cap \psi^{-1}(B)$ is finite. So $\psi(D) \cap B$ is finite as well and $\psi(D) \subseteq \psi(\omega) = Y$. Thus $Y \not\subseteq^* B$.) In limit stages α of cofinality ω_1, we get ($\ddagger\ddagger_\alpha$) for free.

Lemma 16. Assume $\text{cf}(\alpha) = \omega_1$ and ($\star\star_\gamma$) holds for $\gamma < \alpha$. Then ($\ddagger\ddagger_\alpha$) holds.

Proof. The argument is similar to the proof of Lemma 3. Let $C, \phi, \psi \in V_\alpha$. There is $\gamma < \alpha$ such that $C, \phi, \psi \in V_\gamma$. Work in V_γ.

For $A, B \subseteq \omega$, consider $E_{A,B} = C \cap \phi^{-1}(A) \cap \psi^{-1}(B)$. Note that $\mathcal{E}_\beta = \{E_{A,B} \in [\omega]^\omega : (A,B) \in \mathcal{A}_\beta^{\leq \gamma}\}$ is an almost disjoint family for any β. We claim that for some β it is not a finite maximal almost disjoint family below C.

Indeed, if all $\mathcal{E}_\beta$ were finite maximal, property (d') would entail that their members are equal mod finite on a terminal segment of β's. Let $E_0, ..., E_{n-1}$ be these members. Find $(A_i^\beta, B_i^\beta) \in \mathcal{A}_\beta^{\leq \gamma}$ such that $E_i =^* C \cap \phi^{-1}(A_i^\beta) \cap \psi^{-1}(B_i^\beta)$. Thus $\phi(E_i) \subseteq^* A_i^\beta$ and $\psi(E_i) \subseteq^* B_i^\beta$. Therefore $(\phi(E_i), \psi(E_i)) \in V_\gamma$ witnesses the failure of ($\star\star_\gamma$), proving the claim.

Let $\beta \geq \beta_\gamma$ such that $\mathcal{E}_\beta$ is not a finite maximal almost disjoint family below C. Since $\mathbb{Q}_\gamma = \mathbb{D}$ adds a dominating real, $\mathcal{E}_\beta$ is not maximal in $V_{\gamma+1}$, and there is $D \subseteq C$ almost disjoint from all members of $\mathcal{E}_\beta$. Thus the conclusion of ($\ddagger\ddagger_\alpha$) holds for all $(A,B) \in \mathcal{A}_\beta^{\leq \gamma} = \mathcal{A}_\beta^{\leq \gamma+1}$.

Finally use (e') to show it also holds for all $(A, B) \in \mathcal{A}_\beta^{\leq \alpha} \setminus \mathcal{A}_\beta^{\leq \gamma}$: given such (A, B), let $A = \{a_n : n \in \omega\}$ and $B = \{b_n : n \in \omega\}$ be their increasing enumerations. By (e') there is n_0 such that $\psi \circ \phi^{-1}(a_n) < b_n$ and $\phi \circ \psi^{-1}(b_n) < a_{n+1}$ for all $n \geq n_0$. Thus $b_n \notin \psi \circ \phi^{-1}(A)$ for $n \geq n_0$. This means $\phi^{-1}(A) \cap \phi^{-1}(B)$ is finite, as required. q.e.d.

This shows in particular the preservation of ($\star\star_\alpha$) in limit stages of cofinality ω_1. Preservation in limit stages of cofinality ω is again a standard argument using $\mathcal{A}_{\bar{\beta}}^{\leq \alpha} = \bigcup_{\gamma < \alpha} \mathcal{A}_{\bar{\beta}}^{\leq \gamma}$.

Using ($\ddagger\ddagger_\alpha$) and CH which holds in V_α, it is easy to build up the maximal $\mathbb{C}^\omega$-filter $\mathcal{F}_\alpha$ such that

(i') for all partial one-to-one functions $\phi, \psi : \omega \to 2^\omega$ from V_α, if $\mathrm{dom}(\phi)$ and $\mathrm{dom}(\psi)$ belong to the trace of $\mathcal{F}_\alpha$, then there are $D \in \mathrm{trace}(\mathcal{F}_\alpha)$ and $\beta < \omega_1$ such that $D \cap \phi^{-1}(A) \cap \psi^{-1}(B)$ is finite for all $(A, B) \in \mathcal{A}_{\bar{\beta}}^{\leq \alpha}$.

For this simply list all such ϕ, ψ in order type ω_1 and go through them in the odd steps of the recursive construction of $\mathcal{F}_\alpha$.

$$\star \star \star$$

We are ready to deal with the preservation of $(\star\star_\alpha)$ in the successor step. First deal with the case $\mathrm{cf}(\alpha) = \omega_1$. Recall $\mathbb{Q}_\alpha = \mathbb{Q}_{\mathcal{F}_\alpha}$. Let $(\dot{X}, \dot{Y})$ be a $\mathbb{Q}_{\mathcal{F}_\alpha}$-name for a pair of infinite subsets of ω.

Fix σ such that $\rho_{\dot{X}}(\sigma) = 0$. Let $f_\sigma \in \mathcal{F}_\alpha$ be as in Lemma 15. If we are in Case 1, also fix one-to-one $\phi_\sigma : \mathrm{supp}(f_\sigma) \to \omega$ accordingly. In this case set $X_\sigma = \mathrm{ran}(\phi_\sigma)$. If we are in Case 2 of Lemma 15, for all $i \in \mathrm{supp}(f_\sigma)$ and all $a \leq f_\sigma(i)$, fix an infinite $X_\sigma^{i,a}$ such that $\sigma^\frown \langle i, a \rangle$ favors $n \in \dot{X}$ for all $n \in X_\sigma^{i,a}$.

For τ such that $\rho_{\dot{Y}}(\tau) = 0$ do the same thing: from Lemma 15, get $g_\tau \in \mathcal{F}_\alpha$ and either $\psi_\tau : \mathrm{supp}(g_\tau) \to \omega$ and $Y_\tau = \mathrm{ran}(\psi_\tau)$ such that the conclusion of Case 1 holds, or $Y_\tau^{j,b}$ for all $j \in \mathrm{supp}(g_\tau)$ and all $b \leq g_\tau(j)$ such that the conclusion of Case 2 holds.

Fix a pair (σ, τ) such that $\rho_{\dot{X}}(\sigma) = \rho_{\dot{Y}}(\tau) = 0$. If Case 1 holds for both σ and τ, then (i') gives us $D_{\sigma,\tau} \in \mathrm{trace}(\mathcal{F}_\alpha)$ and $\beta_{\sigma,\tau}$ such that $D_{\sigma,\tau} \cap \phi_\sigma^{-1}(A) \cap \psi_\tau^{-1}(B)$ is finite for all $(A, B) \in \mathcal{A}_{\bar{\beta}_{\sigma,\tau}}^{\leq \alpha}$.

If Case 2 holds for both σ and τ, applying property $(\star\star_\alpha)$ countably often gives us $\beta_{\sigma,\tau}$ such that for all $i \in \mathrm{supp}(f_\sigma), j \in \mathrm{supp}(g_\tau), a \leq f_\sigma(i)$ and $b \leq g_\tau(j)$, $X_\sigma^{i,a} \not\subseteq^* A$ or $Y_\tau^{j,b} \not\subseteq^* B$ for all $(A, B) \in \mathcal{A}_{\bar{\beta}_{\sigma,\tau}}^{\leq \alpha}$.

If Case 1 holds for σ and Case 2 for τ, first let $\psi_\tau^{j,b} : \omega \to Y_\tau^{j,b}$ be bijections for $j \in \mathrm{supp}(g_\tau)$ and $b \leq g_\tau(j)$, and then apply (i') countably many times to get $D_{\sigma,\tau} \in \mathrm{trace}(\mathcal{F}_\alpha)$ and $\beta_{\sigma,\tau}$ such that for all such j and b, $D_{\sigma,\tau} \cap \phi_\sigma^{-1}(A) \cap (\psi_\tau^{j,b})^{-1}(B)$ is finite for all $(A, B) \in \mathcal{A}_{\bar{\beta}_{\sigma,\tau}}^{\leq \alpha}$.

Similarly, if Case 2 holds for σ and Case 1 for τ, get bijections $\phi_\sigma^{i,a} : \omega \to X_\sigma^{i,a}$ for $i \in \mathrm{supp}(f_\sigma)$ and $a \leq f_\sigma(i)$. Then find again $D_{\sigma,\tau}$ and $\beta_{\sigma,\tau}$ accordingly.

Let β be the supremum of all $\beta_{\sigma,\tau}$ chosen, and let $D \in \mathrm{trace}(\mathcal{F}_\alpha)$ be a pseudointersection of the $D_{\sigma,\tau}$.

Lemma 17. $\Vdash_{\mathbb{Q}_{\mathcal{F}_\alpha}} \forall (A,B) \in \mathcal{A}_{\bar{\beta}}^{\leq\alpha}$ $(\dot{X} \not\subseteq^* A$ or $\dot{Y} \not\subseteq^* B)$.
Since $\mathcal{A}_{\bar{\beta}}^{\leq\alpha+1} = \mathcal{A}_{\bar{\beta}}^{\leq\alpha}$, this proves $(\star\star_{\alpha+1})$ in case $\mathrm{cf}(\alpha) = \omega_1$.

Proof. Assume this was false and there are $T \in \mathbb{Q}_{\mathcal{F}_\alpha}$, $(A,B) \in \mathcal{A}_{\bar{\beta}}^{\leq\alpha}$ and n_0 such that
$$T \Vdash \dot{X} \setminus n_0 \subseteq A \text{ and } \dot{Y} \setminus n_0 \subseteq B.$$
By Lemma 14, we may assume $\rho_{\dot{X}}(\sigma) = 0$ where $\sigma = \mathrm{stem}(T)$.

Assume first we are in Case 1 for σ. Let $f \in \mathcal{F}_\alpha$ be such that $\{\langle i,a\rangle : \sigma^\frown\langle i,a\rangle \in T\} = f\restriction\mathrm{supp}(f)$. Without loss of generality $\mathrm{supp}(f) \subseteq D \cap \mathrm{supp}(f_\sigma)$ and $f \leq f_\sigma$ (everywhere). If $\mathrm{supp}(f) \not\subseteq^* \phi_\sigma^{-1}(A)$, we find $i \in \mathrm{supp}(f)$ and $n = \phi_\sigma(i) \notin A$ with $n \geq n_0$. Since $f(i) \leq f_\sigma(i)$, we get that $\sigma^\frown\langle i,f(i)\rangle$ favors $n \in \dot{X}$ by the above construction (Case 1 of Lemma 15). By Lemma 12, we get $S \leq T_{\sigma^\frown\langle i,f(i)\rangle} \leq T$ forcing $n \in \dot{X} \setminus A$, a contradiction.

Thus $\mathrm{supp}(f) \subseteq^* \phi_\sigma^{-1}(A)$ and the latter set belongs to $\mathrm{trace}(\mathcal{F}_\alpha)$. By Lemma 14, $\rho_{\dot{Y}}(\sigma) < \infty$. Thus we find $S \leq T$ such that $\rho_{\dot{Y}}(\tau) = 0$ where $\tau = \mathrm{stem}(S)$. Assume first we are in Case 1 for τ. By construction, $D_{\sigma,\tau} \cap \phi_\sigma^{-1}(A) \cap \psi_\tau^{-1}(B)$ is finite and so $\psi_\tau^{-1}(B)$ cannot belong to the trace of $\mathcal{F}_\alpha$. We then reach a contradiction as in the second half of the previous paragraph. So assume we are in Case 2 for τ. Let $g \in \mathcal{F}_\alpha$ be such that $\{\langle j,b\rangle : \tau^\frown\langle j,b\rangle \in S\} = g\restriction\mathrm{supp}(g)$. Without loss $g \leq^* g_\tau$. Choose any j with $g(j) \leq g_\tau(j)$. By construction, $D_{\sigma,\tau} \cap \phi_\sigma^{-1}(A) \cap (\psi_\tau^{j,g(j)})^{-1}(B)$ is finite and so $(\psi_\tau^{j,g(j)})^{-1}(B)$ is not cofinite. Thus $Y_\tau^{j,g(j)} \not\subseteq^* B$ and we can find $n \geq n_0$ with $n \in Y_\tau^{j,g(j)} \setminus B$. Then $\tau^\frown\langle j,g(j)\rangle$ favors $n \in \dot{Y}$ by the above (see Case 2 of Lemma 15). Applying again Lemma 12, we get $S' \leq S_{\tau^\frown\langle j,g(j)\rangle} \leq S$ forcing $n \in \dot{Y} \setminus B$, a contradiction. This completes the proof for Case 1 for σ.

Thus assume Case 2 holds for σ. Again let $f \in \mathcal{F}_\alpha$ be such that $\{\langle i,a\rangle : \sigma^\frown\langle i,a\rangle \in T\} = f\restriction\mathrm{supp}(f)$. Again we may assume $\mathrm{supp}(f) \subseteq \mathrm{supp}(f_\sigma)$ and $f \leq f_\sigma$ (everywhere). The argument at the end of the last paragraph in fact shows that we must have $X_\sigma^{i,f(i)} \subseteq^* A$ for all $i \in \mathrm{supp}(f)$. Again find $S \leq T$ such that $\rho_{\dot{Y}}(\tau) = 0$ where $\tau = \mathrm{stem}(S)$. If we are in Case 2 for τ, the construction gives us again $Y_\tau^{j,g(j)} \not\subseteq^* B$ as at the end of the previous paragraph, a contradiction. If Case 1 holds for τ, $D_{\sigma,\tau} \cap (\phi_\sigma^{i,f(i)})^{-1}(A) \cap \psi_\tau^{-1}(B)$ is finite for any $i \in \mathrm{supp}(f)$. Since any such $(\phi_\sigma^{i,f(i)})^{-1}(A)$ is cofinite, $D_{\sigma,\tau} \cap \psi_\tau^{-1}(B)$ is in fact finite, and we reach a contradiction as in the second half of the second paragraph of this proof. This completes the argument.

q.e.d.

Lemma 18. Assume $\mathrm{cf}(\alpha) \neq \omega_1$ and that $(\star\star_\alpha)$ holds. Then $(\star\star_{\alpha+1})$ holds as well.

We leave the details of this proof to the reader, first because it is quite similar to but simpler than the previous argument (basically it corresponds to the situation where both σ and τ satisfy Case 2, a situation in which $(\ddag\ddag_\alpha)$ was not required and $(\star\star_\alpha)$ was enough (see above)), second because it proceeds along trodden paths, involving the classical and well-known rank analysis for Hechler forcing $\mathbb{D}$ due to Baumgartner and Dordal [B$_2$D$_0$85]. See [B$_4$06b, Section 4] for an argument which is very similar to the one needed here.

This completes the proof of Theorem 2.

$$\star\,\star\,\star$$

As for Theorem 1, the argument can be generalized to show the consistency of $\mathfrak{h}_2 = \kappa < \mathfrak{h}(\mathbb{C}^\omega/\mathrm{fin}) = \mathfrak{c} = \kappa^+$ for arbitrary regular κ. Again, this model will satisfy MA$_{<\kappa}(\sigma$-centered). We leave the details to the reader. Notice this also generalizes the original Shelah-Spinas result to the consistency of $\mathfrak{h}_2 = \kappa < \mathfrak{h} = \mathfrak{c} = \kappa^+$.

References.

[B$_0$H$_1$05]   B. Balcar and M. Hrušák, *Distributivity of the algebra of regular open subsets of $\beta\mathbb{R} \setminus \mathbb{R}$*, Top. Appl. 149 (2005), 1-7.

[B$_0$PS$_1$80]   B. Balcar, J. Pelant and P. Simon, *The space of ultrafilters on $\mathbb{N}$ covered by nowhere dense sets*, Fund. Math. 110 (1980), 11-24.

[B$_0$S$_1$89]   B. Balcar and P. Simon, *Disjoint refinement*, in: Handbook of Boolean algebra (J.D. Monk and R. Bonnet, eds.), North-Holland, Amsterdam (1989), 335-386.

[B$_1$J95]   T. Bartoszyński and H. Judah, *Set Theory, On the structure of the real line*, A K Peters, Wellesley, 1995.

[B$_2$D$_0$85]   J. Baumgartner and P. Dordal, *Adjoining dominating functions*, J. Symbolic Logic 50 (1985), 94-101.

[B$_3\infty$]   A. Blass, *Combinatorial cardinal characteristics of the continuum*, in: Handbook of Set Theory (A. Kanamori et al., eds.), to appear.

[B$_4$06a]   J. Brendle, *Distributivity numbers of $\mathcal{P}(\omega)/\mathrm{fin}$ and its friends*, Sūrikaiseki kenkyūsho kōkyūroku 1471 (2006), 9-18.

[B$_4$06b]   J. Brendle, *Van Douwen's diagram for dense sets of rationals*, Ann. Pure Appl. Logic, 143 (2006), 54-69.

[B$_4\infty$]   J. Brendle, *Distinguishing groupwise density numbers*, Monatshefte für Mathematik, to appear.

[D$_1$98]   A. Dow, *The regular open algebra of $\beta\mathbb{R}\setminus\mathbb{R}$ is not equal to the completion of $\mathcal{P}(\omega)/\mathrm{fin}$*, Fund. Math. 157 (1998), 33-41.

[H$_0$05]   F. Hernández-Hernández, *A tree π-base for $\mathbb{R}^*$ without cofinal branches*, Comment. Math. Univ. Carolinae 46 (2005), 721-734.

[S₀S₂98]   S. Shelah and O. Spinas, *The distributivity numbers of finite products of* $\mathcal{P}(\omega)/\text{fin}$, Fund. Math. 158 (1998), 81-93.

[S₀S₂00]   S. Shelah and O. Spinas, *The distributivity number of* $\mathcal{P}(\omega)/\text{fin}$ *and its square*, Trans. Amer. Math. Soc. 352 (2000), 2023-2047.

[v90]   E. K. van Douwen, *Transfer of information about* $\beta\mathbb{N} \setminus \mathbb{N}$ *via open remainder maps*, Illinois J. Math. 34 (1990), 769-792.

Received: February 21, 2006;
In revised version: May 2, 2006;
Accepted by the editors: May 12, 2006.

The Lower Part of Event Ontology

REGINE ECKARDT

Seminar für Englische Philologie
Georg-August-Universität Göttingen
Käte-Hamburger-Weg 3
37073 Göttingen, Germany
E-mail: regine.eckardt@phil.uni-goettingen.de

Introduction to the non-linguist reader

The art and science of formal logic presents one of the great success stories in mathematics. Largely unnoticed at many mathematical departments, formal logic has grown into a fundamental tool of research in artificial intelligence, cognitive science and linguistics. Many important insights in human reasoning and communication could not have been achieved, accumulated, compared, and improved without the help of formal logic (predicate logic, type theory, intuitionistic logic, and their model theory) as the common *lingua franca*.

Truth conditional semantics aims at modelling the meaning of words, phrases and sentences of natural languages on the basis of logic and model theory. Intriguingly, this research raises issues not only about words, sentence and grammar but also about our human perception of the world: Language talks about the world, and the grammatical categories, distinctions, patterns offer an indirect reflex of naive ontology—which frequently is, indeed, quite sophisticated. The present contribution argues that natural languages like English, German (and probably most others) contain constructions which suggest that naive ontology knows something like infinitesimally small objects. This is surprising, given that "naive" thinking is often equated with thinking in terms of discrete objects and finitely bounded steps. It seems that, sometimes, the seeds of advanced mathematical objects are already present in everyday speech.

It might be useful to briefly introduce the linguistic domain which is addressed in the paper. Most natural languages allow us to present an ongoing process or activity as a *telic* or an *atelic* one. A *telic* description (e.g., English: *Tom ate two sausages*) suggests that we think of the ongoing

---

This paper will also appear in the volume *Event Structure in Linguistic Form and Interpretation*, edited by Johannes Dölling, Tatjana Heyde-Zybatow, and Martin Schäfer (Walter de Gruyter, 2007).

activity as one which will be *finished* once the task is completed. An *atelic* description (e.g., *Tom ate chips*) suggests that the activity in question will at one point stop—for instance when Tom is sick of chips, or falls asleep—but it is inappropriate to suggest that the activity was "completed" or "finished" at that point. The paper adopts the view [K$_1$89] that *telicity* and *atelicity* are properties of the natural language description of an event, rather than of the event itself. One prominent motive for this assumption lies in the observation that an event can be described in different ways. One and the same activity of Tom can, for example, be described as *Tom drank a bottle of red wine* or as *Tom drank wine*. The first description implies that the activity was finished once the bottle was empty. The second description only suggests that Tom stopped drinking at some point or other (for example when the bottle was empty). There is no notion of completion inherent in the second description of the event. The compatibility of sentences with temporal adverbials like *in one hour* as contrasted with *for one hour* offers an independent test for the *telic/atelic* distinction. While *telic* descriptions must be modified with *in n time* adverbials and are ungrammatical with *for*-adverbials, *atelic* descriptions show the opposite pattern.

I want to thank Ulrich Felgner for introducing me to that field in mathematics where, to my eye, applications are more exciting and challenging than in any other branch. I hope that he doesn't count the ones like me as lost sons (or, daughters).

1 Conflicting applications of event ontology

The present paper contributes to the analysis of several linguistic phenomena where event ontology plays a role: the aspectual distinction between telic and atelic predicates on one side, and the analysis of negative polarity sensitive items on the other side. I will address an apparent conflict between two theories that both rest on a certain event ontology for natural language semantics. The conflict arises within the lower part of event ontology, and consists, briefly, in the following. On one hand, scholars who aim at modelling tense and aspect in the tradition of [K$_1$89] and [L$_1$83] commonly assume that certain properties of events are inherited by all their parts. These are called homogeneous properties and are assumed to be the logical counterpart of atelic predicates like *sleep* or *drink wine*. On the other hand, recent proposals to model negative polarity items (e.g., expressions like *lift a finger* or *sleep a wink*) have to assume that there is a level where the parts of events are so small that they can no longer reasonably inherit certain properties that hold true of their superevents.

Note that the perspective taken here is strictly theory-internal in that I will not defend the two approaches in question against competing other theories that aim at modelling the same natural language phenomena. There are more frameworks that treat the telic/atelic distinction, notably the recent event calculus by [v_0H_005], and the named conflict could in principle be resolved by resorting to one of these. There are other frameworks that treat negative polar items, and the conflict could be resolved by resorting to one of those for the second domain of phenomena. So the issue in question is, modestly, whether two specific kinds of event-based theories can be combined in a coherent manner. I think that this issue is still worthwhile for several reasons. The Link/Krifka approach to verbal aspect is well-developed and integrated in sophisticated analyses of other facts about language use, notably the important link between syntax (i.e., real sentences) and semantics (see [H_1K_098]). It would certainly be premature to dismiss a framework that offers these powerful links. The event-based analysis of negative polarity items in turn is hosted in the so-called scalar approaches that, unlike earlier treatments, acknowledge the strong rhetorical qualities of utterances like *Paul did not (even) lift a finger to help me* as the basis of the phenomenon. There is a wide agreement that such approaches are explanatorily more satisfactory, have been corroborated by independent neurolinguistic evidence, and also cover the data more adequately than the traditional competing treatments that can stay neutral with respect to claims about event ontology. [1] Finally, the assumption that two different linguistic phenomena offer independent insights into folk theories about time, space and action can lead to insights into the ways in which we conceptualize the world depending on current communicative interest. This makes the issue, in my eyes, worth following, in spite of the fact that the investigation of how humans *really* think about the world, and which underlying language of the mind *really* shapes our reasoning is fraught with methodological difficulties that have been discussed for centuries. All claims about human conceptualization of time, space and action that will be made in the following should be read with this hedge.

Having delimited the range of the enterprise, we can now turn to the core of the paper. In the following two sections, I shall recapitulate the respective positions and list some armchair assumptions about events that come along with either one. In §1.3, I will name two possible ways out of the dilemma which will be elaborated in this paper.

---

[1] Once again, there are excellent chances that these analyses can be integrated with alternative theories of aspect, like [v_0H_005], but it is fair to expect that the resulting theories will be ontologically more committed than in their initial version.

1.1 Aspect

The book [K$_1$89], following earlier work by Link [L$_1$83], can be seen as the groundbreaking elementary proposal to model the distinction between telic and atelic predicates on the basis of events. It is assumed that sentence radicals denote properties of events P. If the property P is quantised, the sentence makes a telic statement. If the property P is homogeneous, then the sentence makes an atelic statement. The simplest linguistic correlate to this distinction is the test of whether the duration of the eventuality described will be specified with an *in*-PP (telic) or a *for*-PP (atelic sentence). The simplest definitions of "being quantised" and "being homogeneous" are given in (1) and (2). Further refinements were discussed in subsequent literature but are of no immediate concern here. I use $\subset$ for the part-of relation in the domain of events.

(1) $\text{QUANT}(P) \Leftrightarrow \exists e(P(e) \ \& \ \forall e'(e' \subset e \rightarrow \neg P(e')))$.

(2) $\text{HOM}(P) \Leftrightarrow \forall e \forall e'(P(e) \ \& \ e' \subset e \rightarrow P(e'))$.

It is easy to see that these definitions rest crucially on further assumptions about event ontology. In particular, for QUANT to be meaningful we need to ensure that the event ontology as such does not have an atomic level. Events are conceived of much like the set of time intervals on the rationals, which does not have an atomic level of smallest intervals. Indeed, at least nonstative events e have a running time $\tau(e)$ which is an interval on the time line. The following assumptions, hence, seem to be uncontroversial for this kind of theory, even if single authors do not care to list them all explicitly. Note that the symbol $<$ stands for temporal precedence on the time line, as well as temporal precedence between two (temporally located) events. The relation $\subset$ stands for the part-of relation between two events as well as the subset relation between time intervals. Clearly, if $e \subset e'$ then $\tau(e) \subset \tau(e')$.

(3) There is no lower boundary to events:

$$\forall e \exists e'(e' < e)$$

(4) Boolean Structure: There is a summation operation $\oplus$ defined on events that adds up adjacent events (incl. overlapping events) to larger events:

$$\forall e \forall e'(\neg \exists e^*(\tau(e) < \tau(e^*) < \tau(e')) \rightarrow \exists f(e \oplus e' = f))$$

($<$ on time intervals is the partial ordering defined as $I < J$ iff $\forall i \forall j (i \in I \ \& \ j \in J \rightarrow i < j)$).

(5) Betweenness: Between any two events, there is another one.
$$\forall e \forall e'(e' \subset e \rightarrow \exists e^*(e' \subset e^* \subset e)).$$

(6) Differences: If e' is part of e, then there are non-overlapping e'', e''' that add up e' to e:

$\forall e \forall e'(e' \subset e \rightarrow$
$[\exists e''(e' \oplus e'' = e \ \& \ \neg \exists e^*(e^* \subset e' \ \& \ e^* \subset e''))$
$\lor \ \exists e'' e'''(e' \oplus e'' \oplus e''' = e \ \& \ \neg \exists e^*(e^* \subset e' \ \& \ e^* \subset e'') \ \&$
$\neg \exists e^*(e^* \subset e' \ \& \ e^* \subset e'''))].$

Atelic predicates are modelled as homogeneous predicates ([L$_1$83], [K$_1$89], [P00] and others). Link [L$_1$83] pointed out that, at least in the domain of real matter and things, these assumptions are in fact wrong, physically speaking. He notes that the matter *gold*, for instance, has an atomic level (namely, the level of single gold atoms) even though the natural language term '*gold*' behaves as if it denoted a homogeneous property. Link, as well as later authors, assumes that natural language ontology need not be isomorphic to a physically tenable model of the world, because the facts and phenomena in natural language have not been shaped by modern quantum physics but by folk views about nature and matter. Krifka [K$_1$89] makes similar remarks with respect to the lower end of event ontology. The general strategy hence seems to be, to use folk models of the world and time because we want to model the grammatical effects of folk notions about the world and time.

1.2 Negative Polarity Items

In a series of papers that go back to [F75], the behavior of negative polarity items is modeled based on the fact that they describe the weakest possible case of a set of salient alternatives [K$_1$95, L$_0$98, E03]. A sentence like (7) is predicted to be well formed because *lift a finger* denotes the smallest possible way of lending help.

(7) *Tom did not (even) lift a finger in order to help me.*

Interestingly, many languages distinguish between so-called *weak* and *strong* negative polarity items where strong NPIs are virtually restricted to negated contexts and rhetorical questions. The NPI in (7) is commonly assumed to be of the strong type. Informants agree that downward entailing contexts like *few*, *rarely*, etc. do not license it (see (8)) and that a question like (9) can only be used as a rhetorical question.

(8) *Few people (even) lifted a finger in order to help me.*

(9) *Did Tom even lift a finger in order to help you?*

There is a general consensus on how the data in (7)–(9) should be derived from the lexical meaning of the NPI, which is that expressions like *lift a finger, bat an eyelash, drink a drop* etc. denote irrelevantly small events or objects. This general idea has received different analyses by various authors. Krifka [K$_1$89] models it in probabilistic terms and stipulates the following strict inequality of probabilities:

(10) $p(\bigwedge_i$Tom did not do a_i in order to help me$) > p($Tom did not lift a finger$)$, where the conjunction $\bigwedge_i$ ranges over all possible alternative ways in which Tom could have helped me.

The distinct distribution of strong NPIs is derived from this inequality. The details of the theory will not be discussed further in this paper. While this probabilistic account for strong polarity sensitivity is logically consistent with other assumptions about event ontology, inequalities such as the one in (10) are hard to relate with intuitions about events and their subevents. One may suspect that a full model-theoretic account for (10) needs to address similar issues like the ones treated in this paper.

In a related but different vein, van Rooij [v$_1$03] explains the rhetorical quality of the question in (9) essentially by remarking that a positive answer to this question would be downright absurd, given that the question, according to his background theory, is such that a positive answer would pragmatically implicate that "lifting a finger" is also the maximal and hence the only thing that Tom undertook in order to help me. He states that this would not be a relevant act of helping, without further discussion. Eckardt ([E04, Chapter 4], [E05]) takes up this position and elaborates the distinction between weak and strong negative polarity items on the basic assumption that weak NPIs denote small objects and events but ones that are still reasonable things to do. Strong NPIs, in contrast, denote eventualities that are so small that they no longer fall in the right kind of category. For example, *lifting one's finger* may be a subevent of events of helping, but it is not an event of helping itself, and it can not occur in isolation (i.e., without an appropriate superevent of reasonable size). This relates to the observation that natural language terms like Dutch *ook maar* or German *auch nur*, if used in questions, lead to rhetorical questions that cannot possibly receive a positive answer.

The ontological implementation of the strong-weak distinction suggests the plausibility of axioms like the ones in (11) and (12). These hold in particular also for properties of events P that would be regarded as homogeneous in an aspectual theory.

(11) If you *really* look down into the lower end of ontology, some events e are just too small to count:
$\forall e(P(e) \to \exists \varepsilon(\varepsilon \subset e \ \& \ \neg P(\varepsilon)))$.

(12) Can we tell where?

   (a) $\exists e(P(e) \ \& \ \forall e'(e' \subset e \to \neg P(e')))$ ('yes'),
   (b) $\forall e(P(e) \to \exists e'(e' \subset e \ \& \ P(e)))$ ('no').

It is evident that assumptions like (11) are fatal for a predicate that an account of aspect would predict to be homogeneous. On the other hand, we might claim that negative polarity items which denote "minimal objects or events" do exactly what one should not do according to the general guidelines of aspect theory, which is to zoom into the lower part of event ontology which is simplified and idealized in this kind of modelling.

Note that the present conflict is not an easy one between folk theory and physical theories about the world. Both ways to view the lower end of event ontology are supported by "linguistic" facts. Hence, there seem to be two different folk theories about very small events. What kind of viewpoint shift is occuring here?

As the summary above already suggests, several analyses of "P-events too poor to mention" can be imagined. For present purposes, I will hypothetically adopt the

> **Strong position:** There are events e below P-events that are not themselves in P—even if P is intuitively a homogeneous predicate (for instance *'walk a single step'* is not something that is an event of *walking*.)

I will *not* defend the strong position as the best, or only possible one. The aim of this paper is to demonstrate how this position can be carried out. Before we turn to the details, I will lay out the roadmap of the paper in the following section.

1.3 Possible Solutions

What kind of "blindness" makes speakers prefer one kind of expectations on one occasion and another on another occasion? What kind of change in our world view takes place once we zoom in the lower end of ontology? Somewhat surprisingly, there are even two consistent answers to this question. The first one elaborates the idea that we make bold universal statements about events (like HOM) because we *ignore some events*. If we really take all events into account, we are forced to retract these strong universal statements. The step between view one and view two hence consists in increasing or reducing the underlying domain of events. From a superficial view, so to speak, we can not see all events and hence feel inclined to universal statements like HOM(P). The surprising part of this idea is that we seem to see a great many small events even before we took that closer look. How could we have overlooked so many of them? In §2, I offer an application of a model theoretic construction to the domain of events which shows that this is logically possible. In §3, I compute an actual example that might be useful as an illustration, or for concrete applications.

In §4, I turn to a second kind of explanation which rests on the assumption that we face an instance of the Sorites paradox. This view comes down to the claim that we make bold universal statements because we idealizingly *assume wrong properties* for some minor events. I will discuss one spellout of this view and turn to a final comparison in the last section.

2 Infinitesimal Events

Let $\mathcal{L}$ be a first order language that contains relations and functions appropriate to event ontology. Specifically, I will use a sortal distinction between events and time intervals (along with the classical sorts for individuals; I will ignore extensions to higher order logic in the subsequel). The unary function symbol τ will be interpreted as the function that maps each event onto its running time. The binary relation $\subset$ will be defined both on the set of events as well as the set of time intervals. The binary relation $\leq$ is defined primarily as the earlier-than relation on the set of time intervals. It can be shifted to the domain of events by assuming that $e_1 \leq e_2$ iff $\tau(e_1) \leq \tau(e_2)$. Finally, the binary function $\oplus$ is interpreted as event summation. For the present purposes I will assume that summation is restricted to temporally adjacent events. Nothing depends crucially on this assumption, but it is in the spirit of the general enterprise to see how two perfectly natural but contradictory views of event ontology relate to each other.

Let $\mathbf{E} = (E, \tau, \oplus, <, \subset)$ be an event structure for such a first order language, and one that specifically verifies the $\mathcal{L}$-axioms (3) to (6) above. We assume moreover that there is at least one homogeneous predicate P that lives on $\mathbf{E}$. The model theoretic construction will be spelled out with reference to P. It can easily be modified so as to extend to further homogeneous predicates.

Definition 1. Let $(e_i)_{i \in \mathbb{N}}$ be a sequence of events in $\mathbf{E}$. We call $(e_i)_{i \in \mathbb{N}}$ zero-convergent iff $\forall i \forall j (i < j \rightarrow e_j \subset e_i)$ and $\neg \exists \forall i (e \subset e_i)$. Let $\Phi_{\mathbf{E}}$ be the set of all zero-convergent sequences in $\mathbf{E}$. Let $\approx$ be the equivalence relation on $\Phi_{\mathbf{E}}$ defined by $(e_i)_{i \in \mathbb{N}} \approx (f_i)_{i \in \mathbb{N}}$ if and only if $\forall e_i \exists f_j (f_j \subset e_i)$ and $\forall f_k \exists e_r (e_r \subset f_k)$.

Next, we will augment $\mathcal{L}$ to a richer language $\mathcal{L}_\mathbf{E}$ by adding constant names for all events in $\mathbf{E}$. Formally, we could do so by taking the respective domain of $\mathbf{E}$, indexing all its elements e, for example, as e, in order to avoid confusion between objects and language, and add these indexed elements as new constant symbols to $\mathcal{L}$.

Now, consider the following sets of sentences in $\mathcal{L}_\mathbf{E}$:

(13) $\Phi[(e_i)_{i\in\mathbb{N}}] := \{x \subset e_i \mid i \in \mathbb{N}\} \cup \{\neg \forall e(x \subset e)\}$.

An object that would make all statements in $\Phi[(e_i)_{i\in\mathbb{N}}]$ true would be part of all events in $(e_i)_{i\in\mathbb{N}}$ without being the zero event. The next goal we need to achieve is

(i) to construct an event structure $\widehat{\mathbf{E}}$

(ii) that extends the original event structure $\mathbf{E}$ and

(iii) contains new elements ε such that

(iv) for each one of the sets Ψ of sentences as in (13), there is some ε for which all the formulae in Ψ hold true at once.

The construction we are aiming for should add such very small elements for all zero-convergent sequences in $\mathbf{E}$. However, we also must account for cases in which two such sequences converge to "the same point". In order to avoid contradictions, we need to ensure that only one infinitesimal element will be added in such cases. Therefore, we will first identify all co-convergent sequences.

First, we will choose some representative sequence $(e_i)_{i\in\mathbb{N}}$ for each of the equivalence classes modulo $\approx$ in $\Phi_\mathbf{E}$. Remember that this sequence now stands as the representative for all further sequences that consist of different events but eventually dove-tail with this representative in such a way as to converge to the same (so far: abstract) mini-event. For each one of these representatives, we now take the respective set of formulae $\Phi[(e_i)i \in \mathbb{N}]$ as in (13) and keep it in stock. We need to keep the free variables in each of the $\Phi[(e_i)_{i\in\mathbb{N}}]$ distinct. For the purpose of exposition here, I will use different letters x, y, z for different sets of formulae $\Phi[(e_i)_{i\in\mathbb{N}}]$, $\Phi[(f_i)_{i\in\mathbb{N}}]$. Generally, we must use variables in the definition of $\Phi[(e_i)_{i\in\mathbb{N}}]$ that are indexed with the respective sequence. I will not carry this out for obvious typographical reasons. Formally:

Definition 2. Choose a fixed set of representatives for the equivalence classes in $\Phi_\mathbf{E}/\approx$. Let $\Phi := \bigcup \{\Phi[(e_i)_{i\in N}] \mid (e_i)_{i\in\mathbb{N}}$ is a representative of some equivalence class in $\Phi_\mathbf{E}/\approx\}$.

We now need to conjoin these formulae with the elementary theory of **E**. Let therefore $\text{Th}(\mathbf{E}) := \{\psi \mid \psi$ is an atomic sentence in $\mathcal{L}_\mathbf{E}$ and $\mathbf{E} \models \psi\}$. Hence, $\text{Th}(\mathbf{E})$ offers a full description of all elements in **E**. Any model for $\text{Th}(\mathbf{E})$ will therefore contain a substructure that is isomorphic to the original structure **E**.

Finally, let us add the requirement that events between two *P*-events are again *P*-events:

$$\forall e \forall e' \forall e^*(e' \subset e^* \subset e \,\&\, P(e) \,\&\, P(e) \rightarrow P(e^*))$$

We can now turn to the construction of an event structure $\widehat{\mathbf{E}}$ which extends **E** in a conservative manner, which contains infinitesimal events, and where these infinitesimal events are not in the extension of *P* even though they might be parts of larger events that are in the extension of *P*.

First observe that the following set of formulae is finitely consistent:

$$\Phi \cup \text{Th}(\mathbf{E}) \cup \{\forall e \forall e' \forall e^*(e' \subset e^* \subset e \,\&\, P(e) \,\&\, P(e') \rightarrow P(e^*))\}$$

If we take any finite subset Δ of this set of formulae, we can prove its consistency by interpreting it in the old event structure **E** from which we started. More specifically, there is an interpretation I of the constant symbols and a variable assignment g for the free variables in Φ such that $\mathbf{E} \models^{Ig} \delta$ for all formulae in Δ. We can simply interpret all constant names $\dot{e}$ in Δ by the respective element e in **E** and interpret the variables y, x, z as events that are part of all those larger events that are mentioned in Δ, among the formulae collected in Φ. Because there are only a finite number of such statements in Δ and the decreasing sequences $(e_i)_{i \in \mathbb{N}}$ were assumed to be infinite, we can always find events that are smaller than a finite part of the infinitely decreasing sequence.

As all finitely consistent sets of formulae are also consistent, there exists a model $\widehat{\mathbf{E}}$ of $\Phi \cup \text{Th}(\mathbf{E})$. I will use ε, ε' etc. as meta-variables for elements that realize one of the types of infinitesimal objects, $\Phi[(e_i)_{i \in \mathbb{N}}]$.

As $\text{Th}(\mathbf{E})$ contains only atomic sentences in $\mathcal{L}_\mathbf{E}$ (i.e., importantly, not the clause about the homogeneity of *P*) we can moreover consistently assume that $\neg P(\varepsilon)$ for all infinitesimal objects. (This step of the construction will be formally legitimised below by a model construction that proves its consistency.)

We can now form the set of all infinitesimal small objects below *P* that are too small to be *P* themselves: $\lambda \varepsilon (\exists e (P(e) \,\&\, \varepsilon \subset e) \,\&\, \neg P(e))$. For convenient reference, let us call this area the infinitesimal part below *P*.

$$\text{INF}(P)(\varepsilon) \leftrightarrow \exists e(P(e) \,\&\, \varepsilon \subset e \,\&\, \neg P(\varepsilon))$$

The model $\widehat{\mathbf{E}}$ hence comes up to our expectations about "zooming into" the lower end of event ontology in the following way: We maintain everything that we believed about previously recognized events ($\mathbf{E}$ is a substructure of $\widehat{\mathbf{E}}$). All previous P-events as well as those that are between earlier P-events remain P-events. However, there is a lower level of previously unrecognized events that are not P.

3 An Example

In order to exemplify the above construction, I will repeat it on the basis of common mathematical structures. We will start with the real numbers $\mathbb{R}$ and the set of all *open* intervals over $\mathbb{R}$. Let us call this set E, in order to stress that we are not supposed to consider the internal structure of the objects in question from now on.

We can now take E to be the domain of events of our event structure and extend this set to a full event structure; specifically by adding the linearly ordered real numbers as our domain of time points. Let me define the basic relations and functions on events in $\mathbf{E}$ as follows:

- The timeline will consist of the real numbers $(\mathbb{R}, \leq)$.

- For all e in E, $\tau(e) := I$ iff $e = I$ (remembering e's internal structure for a moment).

- For all e, e' in E: $e \oplus e'$ is defined iff $e =]x; y[$ and $e' =]w; z[$ and either $]x; y[$ and $]w; z[$ have nonempty intersection or $y = w$.
  In case (i), $e \oplus e' :=$ the event represented by $]x; y[\cup]w; z[$.
  In case (ii), $e \oplus e' :=]x; z[$.
  (If desired, the operation $\oplus$ can be made commutative).

- An event e is a mereological part of another event e', written as "$e \subset e'$" iff, seen as intervals in $\mathbb{R}$, $e \subset e'$.

We can now choose the extension of a homogeneous predicate P in $\mathbf{E}$, starting from some maximal P-event e as $P(e^*)$ if and only if $e^* \subset e$.

Let us check that the structure $\mathbf{E}$ conforms to axioms (2) to (6) above:

(3) There is no lower boundary to events:

$$\forall e \exists e' (e' < e).$$

This holds true, because for each interval I in $\mathbb{R}$ there are more open intervals that are true parts of I.

(2) Homogeneous predicates P apply to events that consist of P-parts all the way down:

$$\text{HOM}(P) \Leftrightarrow \forall e \forall e'(P(e) \,\&\, e' \subset e \to P(e'))$$

This holds true due to definition of the extension of P.

(4) Boolean Structure: There is a summation operation $\oplus$ defined on events that adds temporally adjacent events (incl. overlapping events) to larger events:

$$\forall e \forall e'(\neg \exists e^*(\tau(e) < \tau(e^*) < \tau(e')) \to \exists f(e \oplus e' = f))$$

This holds true due to the definition of $\oplus$. Of interest to us are are events with a non-intersecting temporal extension ($]x;y[$ and $]y;z[$ with y in the temporal extension of neither). Here, the addition of events diverges from simple set union in $\mathbb{R}$.

(5) Betweenness: Between any two events, there is another one.
$\forall e \forall e'(e' \subset e \to \exists e^*(e' \subset e^* \subset e))$.

(6) Differences: If e' is part of e, then there are non-overlapping e'', e''' that add up e' to e:
$\forall e \forall e'(e' \subset e \to$
$[\exists e''(e' \oplus e'' = e \,\&\, \neg \exists e^*(e^* \subset e' \,\&\, e^* \subset e''))$
$\lor\; \exists e'' e'''(e' \oplus e'' \oplus e''' = e \,\&\, \neg \exists e^*(e^* \subset e' \,\&\, e^* \subset e'') \,\&\,$
$\neg \exists e^*(e^* \subset e' \,\&\, e^* e''')]$.

Both (5) and (6) hold true due to construction.

The construction of infinitesimal elements over this initial event structure **E** will result in the introduction of events that would correspond to single points in $\mathbb{R}$. Each zero-convergent sequence $(e_i)_{i \in \mathbb{N}}$ in **E** corresponds to a convergent sequence of intervals $(]x_i; y_i[)_{i \in \mathbb{N}}$ in $\mathbb{R}$. It is a theorem in $\mathbb{R}$ that the limit element of such sequences exist.

The event structure $\widehat{\mathbf{E}}$ arises from **E** by adding (closed) intervals that consist of one point only. The closure over these will result in $\widehat{\mathbf{E}}$ being all open and closed intervals over $\mathbb{R}$. We can consistently assume that all events ($[x;x]$) that correspond to single points in $\mathbb{R}$ are not in the extension of P.

The property P is therefore not homogeneous in the strong sense in $\widehat{\mathbf{E}}$ that *each and any* part of a P-event is again a P-event. However, homogeneity can be stated in the following weaker form:

$$\text{HOM}(P) \leftrightarrow [\forall e \forall e'(P(e) \,\&\, e' \subset e \to P(e') \lor \text{INF}(P)(e'))$$
$$\&\; \forall e(P(e) \to \exists e'(e' \subset e \,\&\, P(e')))].$$

Note that the structure as it is defined so far does not support axiom (3). Even though there is no lower limit to *P*-events, there are smallest events that have no proper parts, namely the events that correspond to single points.

In order to obtain a structure that supports (3), the construction would need to adopt the assumption that single points in fact hide another infinity of events. A concrete structure that illustrates this step can be built on the basis of tuples of real numbers. If we call E_0 the part in $\widehat{\mathbf{E}}$ below some infinitesimal event e, we can set:

$$E_0 := \{](x,a);(x,b)[\mid x,a,b \in \mathbb{R} \text{ and } a \leq b\}.$$

For all events e, e' in E_0:

$$e' < e :\Leftrightarrow e =](x,a);(x,b)[\text{ and } e' =](x,a');(x,b')[\text{ and }]a';b'[\subseteq]a;b[.$$

I will not further explore whether we can faithfully assume that the temporal extension of all these events comes down to the same point in time (plausibly). If we decide that we cannot, if we in other words maintain that events have unique temporal extension, then we are forced to add infinitesimal elements to the time line as well (see [R74, p. 244]).

As an aside, I would like to mention that the initial event structure in this example appears to shed light on a paradox about time that was posed by Sebastian Löbner [L₂97]. He pointed out that we have conflicting intuitions about time. On the one hand, we have a notion that there can be two immediately adjacent but nonintersecting time phases. On the other hand, we usually assume that between any two distinct time points there must be a third one, distinct from both (i.e., density). These intuitions are in fact not both supported in the same model, in the present construction. However, this model construction can explain how we shift between two possible conceptualizations of eventualities where one view supports assumption (i) and the other supports assumption (ii).

4 Solution Two: Sorites

The introduction of infinitesimally small events has turned out to be a consistent way to explain the conflicting intuitions listed in §1. However, you might object that the solution locates the "hazy phase" at the wrong point. You might maintain that the intuition that "*all walkings consist of smaller walkings*" does not come about by our failure to see small events. Let us take a closer look into this example. In fact, events for which we would cease to think about a *walking* are still quite macroscopic. Certainly, one step is not a walking. Certainly, two steps are not sufficient for a healthy *walking* either. Certainly, there seems to be some boundary somewhere between three and 100 steps (very loosely speaking) where the single steps end, and the real walking starts?

Looking at it from the upper end, we might likewise propose that the intuition that "*all walkings consist of smaller walkings*" comes about differently. Perhaps it means something like "if e is a walking, and if I take away one step of e to get to e', then e' will be a walking as well". We are not able to imagine the case where a walk e minus one step e' results in something too small for a walking. Cases like these have been discussed as the *heap paradox* in logic and philosophy. We can recast it as *weak homogeneity*. Consider the following condition.

(WH) $\quad \forall e \forall e' \forall e''(P(e) \& e = e' \oplus e'' \rightarrow (P(e') \vee P(e'')))$.

Condition (WH) is more cautious than full homogeneity. It reflects an intuition something like "if a P-event can be subdivided into two parts, then at least one (the larger one?) is again P". Let us assume that for each P, there is a uniform measure which distinguishes those parts of P-events that are too small to be P, e.g., '*step*' for walking, '*make a sound*' for '*say something*' etc. Assume moreover that all ordinary P-events in any ontology consist of a finite sequence of such STEP-P-events. Then, by induction, we will obtain instances of sorites sequences (see [G00]):

(14) $P(e)$.

(15) If $P(e)$ and $e = e_1 \oplus e_2$ and $\text{STEP}(P)(e_2)$ then $P(e_1)$ still.

(16) Any e in $P(e)$ is linked to some event ε such that there is a finite sequence $e = e_1, \ldots, e_k = \varepsilon$ where e_i and e_{i+1} are linked by the sorites relation in (ii), and such that $\text{STEP}(P)(\varepsilon)$.

This shows that (WH), even though it was a careful assumption about homogeneity, can not be maintained once we spell out all assumptions that are characteristic for a *heap paradox* case. Cases like these have received much discussion in the literature, and I refrain from recapitulating all the solutions that were proposed. Instead, I will base my discussion on work by Graff, specifically [G00]. She offers a solution to the heap paradox that rests in classical two-valued logic. This is advantageous for semantic modelling in the first case, because we need not burden semantic theory with controversial many-valued logics. More importantly, however, Graff's solution is particularly relevant to the present case insofar as it makes essential use of blind spots of the categorizing individual. Let me briefly outline her proposal.

Graff claims that, for any pair of objects (in our case: events e_1 and e_2) that are immediately linked by the sorites relation in question, the following cognitive effect occurs: Once we focus our attention on these two objects, their similarity is so salient that we cannot, subjectively, judge one to have property P but not the other. This is a subjective and essentially context-driven judgement, as Graff argues. If we decide for two events e_1 and e_2 where e_1 is a *walking* and e_2 is just one step shorter than e_1 that e_2 is likewise a *walking*, we tacitly expect that the two events e' and e'' which are a *walking*, a *non-walking* and separated by just one step are just somewhere lower on the scale of ever smaller events. This holds similarly for the dual case of two *non-walkings*. Globally speaking, therefore, there exists a borderline, i.e., two events e_1 and e_2 such that

$$\neg P(e_1) \;\&\; \neg P(e_2) \;\&\; P(e_1 \oplus e_2).$$

Looking at things locally, however—and this seems to be the kind of perspective that feeds our armchair intuitions about event ontology—we maintain principle (WH). Like for the previous solution, the condition on homogeneous predicates needs to be adapted:

$$\text{HOM}(P) \leftrightarrow [\forall e \forall e'(P(e) \;\&\; e' \subset e \rightarrow P(e') \vee \text{INF}(P)(e'))].$$

Note that in this case, we can not safely assume that all P-events have at least *some* parts that are again P. There is a strict boundary somewhere that separates P from $\text{INF}(P)$. We are just unable to locate it precisely:

$$\exists e(P(e) \;\&\; \forall e'(e' \subset e \rightarrow \neg P(e)).$$

Hence, the sorites solution and the infinitesimal construction, even though both capture our armchair intuitions about events, can be clearly distinguished by the logical truths that are supported by either kind of model.

5 Outlook and Summary

Both the construction of infinitesimal events and the sorites explanation appear to capture some of the essence of how we think about very small events. At present, I have no conclusive argument to favour one or the other treatment.

However, the existence of two logically distinct ways to fine-tune the notion of homogeneity could be put to work to distinguish cases that could not be differentiated by earlier theories. This is particularly interesting for cases where predicates appear to be homogeneous, but are not so perceived by speakers.

For example, Zucchi and White [Z01] investigate the so-called twigs and sequences puzzle. It has been observed that a sentence like (17) is ill-formed, although geometry tells us that the initial segment of a line is again a line and hence, each line consists of an infinity of shorter lines.

(17) *John drew a line for 2 minutes.

The difference between a case like (17) and a (well-formed) sentence like *John took a nap for 3 minutes* could be located in the different ways in which we think about smaller parts of a nap, and smaller parts of lines. For example, we could assume that objects like lines, sequences etc. are viewed as sorites-homogeneous but not infinitesimally homogeneous.

$$\text{HOM}(P) \leftrightarrow [\forall e \forall e'(P(e) \,\&\, e' \subset e \to P(e') \,\&\, \text{INF}(P)(e'))]$$
$$\exists e(P(e) \,\&\, \forall e'(e' \subset e \to \neg P(e)).$$

Atelic predicates in the sense of aspect semantics, by contrast, could be required to be homogeneous in the strict sense.

$$\text{HOM}(P) \leftrightarrow [\forall e \forall e'(P(e) \,\&\, e' \subset e \to P(e') \,\&\, \text{INF}(P)(e'))$$
$$\&\, \forall e(P(e) \to \exists e'(e' \subset e \,\&\, P(e')))].$$

This opens up a new possible line to distinguish between *John drew a line for 2 minutes* and *John ate beans for 10 minutes*, and hence could explain their different behaviour.

To summarize, in this paper I drew attention to conflicting assumptions about the lower end of event ontology that are suggested by different linguistic phenomena. Homogeneity (as required in the modelling of aspect) suggests that some properties P apply to large events and all their smaller parts, no matter how far down we look. Minimal-event-NPIs on the other hand suggest that events can indeed be too small to count as an element in the extension of P (for the same, or similar, properties P). I suggested that the dilemma can be resolved in two different ways.

The Infinitesimal Event construction rests on the assumption that the conceptual 'blind spot' of speakers that drives them to make inconsistent assumptions about event ontology on different occasions essentially consists in ignoring irrelevant material. As soon as we are forced to acknowledge the existence of extremely small events, we enrich our ontology, and readjust notions like HOM accordingly.

The starting point of the sorites solution is the hypothesis that we make idealised assumptions about the properties of very small events in everyday reasoning, just in order to keep matters simple. As soon as we are forced to think seriously about these minute eventualities, we acknowledge our idealisation as false, and readjust notions like HOM accordingly.

It appears very difficult to devise definite arguments in favour of one or the other of these two options. However, their joint existence opens up new perspectives in the investigation of aspect and related issues.

References.

[E03]　　R. Eckardt, Eine Runde im Jespersen-Zyklus: Negation, emphatische Negation, negative-polare Elemente im Altfranzösischen. KOPS (Konstanzer Online-Publikations-System) 991 (2003).
http://www.ub.uni-konstanz.de/kops/volltexte/2003/991.

[E04]　　R. Eckardt, Meaning Change under Reanalysis. Habilitationsschrift, Humboldt University Berlin 2004.
Published with revisions as "Meaning Change under Grammaticalization. An Inquiry into Semantic Reanalysis." Oxford: Oxford University Press.

[E05]　　R. Eckardt, Too poor to mention. Subminimal Eventualities and Negative Polarity Items. *In:* C. Maienborn, A. Wöllstein-Leisten (*eds.*), Events in Syntax, Semantics and Discourse. Tübingen: Niemeyer Verlag, 2005: 301–330.

[F75]　　G. Fauconnier, Pragmatic scales and logical structure. Linguistic Inquiry 6 (1975): 353-375.

[G00]　　D. Graff, Shifting Sands. An Interest-Relative Theory of Vagueness. Philosophical Topics 28.1 (2000):45–81.

[G03]　　D. Graff, Gap Principles, Penumbral Consequence, and Infinitely Higher-Order Vagueness. *In:* J. C. Beall (*ed.*), Liars and Heaps: New Essays on Paradox. Oxford: Oxford University Press, 2003.

[$H_1K_0$98]　　I. Heim, A. Kratzer, Semantics in Generative Grammar. Malden: Blackwell, 1998.

[$K_1$89]　　M. Krifka, Nominalreferenz und Zeitkonstitution: zur Semantik von Massentermen, Pluraltermen und Aspektklassen. München: Fink, 1989.

[$K_1$95]　　M. Krifka, The semantics and pragmatics of polarity items. Linguistic Analysis 25 (1995): 209–257.

[$L_0$98]　　U. Lahiri, Focus and Negative Polarity in Hindi. Natural Language Semantics 6 (1998): 57-123.

[$L_1$83]　　G. Link, The logical analysis of plurals and mass terms: A lattice theoretic approach. *In:* R. Bäuerle, U. Egli, C. Schwarze, A. von Stechow (*eds.*), Meaning, Use and Interpretation of Language. Berlin: de Gruyter, 1983.

[$L_2$97]　　S. Löbner, *private communication*, 1997.

[P00]　　C. Piñón, The syntax and semantics of *végig*. *In:* G. Alberti and I. Kenesi (*eds.*), Papers from the Pécs conference, Approaches to Hungarian. Szeged, Hungary, 2000: 201–236.

[R74]　　A. Robinson, Introduction to Model Theory and to the Metamathematics of Algebra. Amsterdam: North Holland, 1974.

[$v_0H_0$05]　　M. van Lambalgen, F. Hamm, The Proper Treatment of Events. Malden: Blackwell, 2005.

[v₁03]   R. van Rooy, Negative Polarity Items in Questions: Strength as Relevance. Journal of Semantics 20 (2003): 239–274.

[Z01]   S. Zucchi, Twigs, Sequences and the Temporal Constitution of Predicates. Linguistics and Philosophy 24.2 (2001): 223–270.

Received: February 13, 2006;
In revised version: September 25, 2006;
Accepted by the editors: December 10, 2006.

Existentially Closed Locally Finite CA-groups

ANDREAS ECKER

Heerstraße 69C
60488 Frankfurt a. M., Germany
E-mail: andreas_ecker@yahoo.com

ABSTRACT. We will show that existentially closed locally finite CA-groups are either isomorphic to $\mathrm{SL}_2(\tilde{\mathbb{F}}_2)$ (where $\tilde{\mathbb{F}}_2$ is the algebraic closure of the field with two elements) or to the semidirect product of an existentially closed abelian π-group by the direct sum of Prüfer groups of type q^∞ for $q \in \pi'$. But as this is only a necessary condition, we also give sufficient conditions for a group of this kind to be an e.c. locally finite CA-group.

1 Introduction

We call a group G a *CA-group* if the centralizer of every non-identity element in G is abelian. These groups are sometimes also called commutativity-transitive groups or CT-groups, because the relation that two elements of a CA-group G commute is transitive on $G - \{1\}$.

Examples for CA-groups are abelian groups, free groups, free solvable groups and the groups $\mathrm{SL}_2(K)$ for fields K of characteristic 2. The finite CA-groups were classified by L. Weisner [W$_0$25], M. Suzuki and G. E. Wall [S57, B$_1$SW58] (see also [W$_1$98]). They are either isomorphic to a group $\mathrm{SL}_2(2^n)$ for $n > 1$ (and thus simple) or to the semidirect product of an an abelian by a cyclic group. However a classification of all CA-groups seems nearly impossible. For example free groups and more generally free products of CA-groups are CA-groups. From the model-theoretic point of view it would be interesting to learn more about the existentially closed CA-groups. (As the class of CA-groups is inductive, every CA-group is embeddable in an existentially closed one.) But this problem also seems quite hard, since there is no global amalgamation theorem for CA-groups. A restriction to a subclass seems to be more promising. For the subclass of CSA-groups (see [GK$_0$M95, MR96]) examinations on the existentially closed groups have been done by E. Jaligot and A. O. Houcine in [JH$_2$04]. Some related results may be found in [E99]. We will restrict ourselves in this paper to the subclass of locally finite CA-groups. In this class we are able to use the classification of the finite CA-groups to get a classification of the existentially closed (e.c.) groups. For readers not familiar with the term *existentially closed* we will give a short introduction in Section 2. Most results in this paper are from the author's doctoral thesis [E99]. After finishing the thesis the author took notice of a paper from Y.-F. Wu [W$_1$98] where some related results are proven. We will refer to this article for several results. According to the classification of the finite CA-groups we will have to make a distinction between the solvable and the non-solvable groups. In the following, the class of solvable locally finite CA-groups will be denoted with $\mathcal{C}$.

A non-solvable existentially closed (e.c.) locally finite CA-group is isomorphic to $\mathrm{SL}_2(\bar{\mathbb{F}}_2)$ where $\bar{\mathbb{F}}_2$ is the algebraic closure of the field with 2 elements. We also prove in Theorem 21 that a solvable e.c. locally finite CA-group (i.e., an e.c. $\mathcal{C}$-group) is isomorphic to the semidirect product of an e.c. abelian π-group by $\bigoplus_{q \in \pi'} Z_{q^\infty}$. (Here π is a set of primes, and π' its complement in the set of all primes.) That raises the question whether the converse is true, i.e., whether all CA-groups of this type are e.c. in $\mathcal{C}$. This is not true, as we show in this paper. Moreover we give sufficient conditions for such a group to be e.c. in $\mathcal{C}$. Another question we want to consider is the number of non-isomorphic countable e.c. groups in $\mathcal{C}$.

2 Preliminaries

We start with a model-theoretic definition.

Definition 1. Let $\mathcal{L}$ be a first-order language. We say a structure $\mathfrak{M}$ is *existentially closed (e.c.) in* $\mathcal{K}$ if the following conditions hold:

(i) $\mathcal{K}$ is a class of $\mathcal{L}$-structures,

(ii) $\mathfrak{M} \in \mathcal{K}$,

(iii) for every quantifier-free formula $\Phi(\bar{x}, \bar{y})$ and for every tuple $\bar{a}$ in $\mathfrak{M}$: If there is a structure $\mathfrak{N} \in \mathcal{K}$, such that $\mathfrak{M} \subseteq \mathfrak{N}$ and $\mathfrak{N} \vDash \exists \bar{y} \, \Phi(\bar{a}, \bar{y})$, then already $\mathfrak{M} \vDash \exists \bar{y} \, \Phi(\bar{a}, \bar{y})$.

Existentially closed structures have a rich structure and are therefore often easier to classify. For example, the e.c. fields are exactly the algebraically closed fields. In inductive classes (i.e., classes that are closed under unions of chains and under isomorphisms) every structure can be embedded in an e.c. structure. For more model-theoretic details see [H$_1$93]. It is easy to see that the class of locally finite CA-groups is inductive. We want to gather some wellknown facts about CA-groups.

Fact 2. Let G be a finite CA-group. Then G is abelian-by-cyclic or isomorphic to a group $SL_2(2^n)$ for some n [W$_0$25, S57, B$_1$SW58].

Fact 3. The Fitting subgroup $\text{Fit}(G)$ of a CA-group G is abelian and the unique maximal normal abelian subgroup (see [W$_1$98, E99]).

Fact 4. Let G be a finite metabelian CA-group and A the Fitting subgroup of G. If $A \lneq G$, then G is a Frobenius group and A is the Frobenius kernel. The group A has a cyclic complement C in G and $|A| \equiv 1 \pmod{|C|}$ ([W$_0$25], see [W$_1$98] for a corrected proof).

Fact 5. Let G be a locally finite CA-group and N a normal subgroup of G. Then G/N is a CA-group (see [W$_1$98]).

From Fact 4 it follows that A is a Hall π-subgroup. Furthermore all complements of A are conjugate and maximal abelian subgroups and C acts fixed-point-free on A via conjugation. Notice that we use a stronger notion than sometimes in the literature.

Definition 6. We call a automorphism group $K \leq \text{Aut}(H)$ *fixed-point-free* if all $k \in K$, $k \neq \text{id}$ leave only the unit element of H fixed.

On the other hand, if H is an abelian group and $K \leq \text{Aut}(H)$ fixed-point-free and a CA-group, then $H \rtimes K$ is a CA-group (see Propositon 8).

Definition 7. We call a $\mathcal{C}$-group G *of type* π if π is a non-empty set of primes, the Fitting subgroup $F(G)$ of G is a π-group, and the complement of $F(G)$ in G is a π'-group.

3 Construction of CA-Groups

We will need some propositions on the construction of CA-groups. A special case of the following proposition (for N abelian) can be found in [W$_1$98, Lemma 7].

Proposition 8. Let G be a group and $N \triangleleft G$ such that

(i) N and G/N are CA-groups and

(ii) $\forall x \in N - \{1\} : C_G(x) \subseteq N$.

Then G is a CA-group.

Proof. Let $g \in G - \{1\}$. We have to show that $C_G(g)$ is abelian.
Case 1: $g \in N$.
Then $C_G(g) = C_N(g)$ because of (ii) and the claim follows because of (i).
Case 2: $g \in G - N$.
Let $h, k \in C_G(g)$. As G/N is a CA-group we have $hk = kh \cdot n$ for some

$n \in N$. We get $h, k, khn \in C_G(g)$ and thus $n \in C_G(g)$ and $g \in C_G(n)$. Because of (ii) this is only possible for $n = 1$. q.e.d.

We will use the proposition above for the construction of CA-groups. Let $G = A \rtimes Z$ be a CA-group, A an abelian p-group and Z locally cyclic. Then A can be considered as a module over the group ring $\mathbb{Z}_p Z$, where $\mathbb{Z}_p$ is the ring of p-adic integers. If $Z = \langle \{z\} \rangle$ is cyclic, then A is a $\mathbb{Z}_p[z]$-module in a natural way (let z act via conjugation on A). Therefore, if we want to construct a CA-group with Fitting subgroup A and a cyclic complement of order q, it suffices to define a fixed point free action of order q on A. Instead of a^z we will use the notation $z * a$ and consider A as an additive $\mathbb{Z}_p[z]$-module.

Proposition 9. Let $A \rtimes Z$ a non-abelian CA-group, where A is an abelian p-group, k not divisible by p and Z cyclic of order k. Furthermore let $a \in A$ and z a generating element of Z. Then there is a divisor g of the k-th cyclotomic polynomial F_k in $\mathbb{Z}_p[x]$ such that $g(z) * a = 0$.

Proof. It is sufficient to show that $F_k(z) * a = 0$ when $F_k(z)$ is the k-th cyclotomic polynomial. As k is the order of z, we have $(z^k - 1) * b = 0$ for all $b \in A$. Moreover there is a polynomial $H(x) \in \mathbb{Z}_p[x]$ such that $x^k - 1 = F_k(x) \cdot H(x)$. It suffices to show that the action of $H(z)$ is injective on A, because obviously $A \rtimes Z$ is locally finite. We show that for any irreducible divisor $f(x)$ of $H(x)$, $f(z)$ acts injective on A. Now let $f(x)$ be such an irreducible polynomial and $0 \neq b \in A$. Then f is the divisor of some l-th cyclotomic polynomial F_l, where l is a divisor of k. If $f(z) * b = 0$ then $(z^l - 1) * b = 0$, which means that b is fixed under the action of $z^l \neq 1$. This is not possible, as $z^l \neq 1$ acts fixed-point-free. q.e.d.

There is a converse to Proposition 9. As preparation we need the following proposition. It is a special case of a theorem from [H$_0$08, p. 63].

Proposition 10. Let $g, h \in \mathbb{Z}_p[x]$ be polynomials with no common divisors such that g and h have no common divisors modulo p. Then there exist $u, v \in \mathbb{Z}_p[x]$ such that $ug + vh = 1$.

Proof. Because $\mathbb{Q}_p[x]$ is a principal ideal ring, there are polynomials u', v' over the field of p-adic numbers $\mathbb{Q}_p$ such that $u'g + v'h = 1$. (Because a theorem of Gauss u' and v' also have no common divisors over $\mathbb{Q}_p[x]$.) The polynomials may be chosen in such a way that the degree of v' (respectively u') is smaller than that of g (respectively h). Now let m be the smallest natural number such that $u := p^m u'$ and $v := p^m v'$ are in $\mathbb{Z}_p[x]$. Then $ug + vh = p^m$ in $\mathbb{Z}_p$. We consider this equation modulo p. If $m > 0$, then

$ug + vh \equiv 0 \pmod{p}$. But this is not possible, because $u \equiv 0 \pmod{p}$ and $v \equiv 0 \pmod{p}$ would be a contradiction to the choice of m. As the degree of v (respectively u) is smaller than that of g (respectively h), g and h must have a common divisor modulo p. That is a contradiction to the assumption. Thus $m = 0$. q.e.d.

We will use the notation C_k for the cyclic group $\mathbb{Z}/(k \cdot \mathbb{Z})$.

Proposition 11. Let k be a natural number and p a prime not dividing k. Let $f(x) = x^n + \lambda_{n-1} \cdot x^{n-1} + \ldots + \lambda_1 \cdot x + \lambda_0$ be a divisor of the k-th cyclotomic polynomial over $\mathbb{Z}_p$ and A an abelian p-group. Then the group $A^{(n)} \rtimes C_k$ with the action (here let z be a generating element of C_k)

$$
\begin{aligned}
z * (a, 0, \ldots, 0) &:= (0, a, 0, \ldots, 0) \\
z^2 * (a, 0, \ldots, 0) = z * (0, a, 0, \ldots, 0) &:= (0, 0, a, 0, \ldots, 0) \\
&\vdots \\
z^n * (a, 0, \ldots, 0) = z * (0, \ldots, 0, a) &:= (-\lambda_0 \cdot a, \ldots, -\lambda_{n-1} \cdot a) \\
&= (z^n - f(z)) * (a, 0, \ldots, 0)
\end{aligned}
$$

is a CA-group.

Proof. The action of z may be described via the following function.

$$
\begin{aligned}
\varphi : A^n &\to A^n \\
(a_1, \ldots, a_n) &\mapsto (0, a_1, \ldots, a_{n-1}) + (-\lambda_0 a_n, \ldots, -\lambda_{n-1} a_n).
\end{aligned}
$$

This function is obviously an endomorphism of A^n. It is also injective

[Suppose $\varphi\big((a_1, \ldots, a_n)\big) = (0, \ldots, 0)$. Then $a_n = 0$, as λ_0 is invertible in $\mathbb{Z}$. Thus

$$(0, \ldots, 0) = \varphi\big((a_1, \ldots, a_n)\big) = (0, a_1, \ldots, a_{n-1})$$

and therefore $(a_1, \ldots, a_n) = (0, \ldots, 0)$.]

and surjective.

[Let $(b_1, \ldots, b_n) \in A^n$. We want to show that $(b_1, \ldots, b_n)$ is in the image of φ. For that purpose we have to show that the equational system

$$
\begin{aligned}
-\lambda_0 a_n &= b_1 \\
a_1 - \lambda_1 a_n &= b_2 \\
&\vdots \\
a_{n-1} - \lambda_{n-1} a_n &= b_n
\end{aligned}
$$

is solvable. As λ_0 is invertible, we get immediately

$$\begin{aligned} a_n &= -\lambda_0^{-1} b_1 \\ a_1 &= b_2 + \lambda_1 a_n = b_2 - \lambda_1 \lambda_0^{-1} b_1 \\ &\vdots \\ a_{n-1} &= b_n + \lambda_{n-1} a_n = b_n - \lambda_{n-1} \lambda_0^{-1} b_1.] \end{aligned}$$

We defined φ such that $\varphi^n(a, 0, \ldots, 0) = (\varphi^n - f(\varphi))(a, 0, \ldots, 0)$. As a direct conclusion we have $f(\varphi)(a, 0, \ldots, 0) = (0, \ldots, 0)$. Thus

$$\begin{aligned} \varphi^k(a, 0, \ldots, 0) &= ((\varphi^k - \mathrm{id}) + \mathrm{id})(a, 0, \ldots, 0) \\ &= (\varphi^k - \mathrm{id})(a, 0, \ldots, 0) + (a, 0, \ldots, 0). \end{aligned}$$

As $f(x)$ is a divisor of $x^k - 1$, we get $(\varphi^k - \mathrm{id})(a, 0, \ldots, 0) = (0, \ldots, 0)$. Analogously

$$\varphi^k\left(\varphi^l(a, 0, \ldots, 0)\right) = \varphi^l\left(\varphi^k(a, 0, \ldots, 0)\right) = \varphi^l(a, 0, \ldots, 0),$$

for $l \in \omega$. Therefore $\varphi^k(0, a, 0, \ldots, 0) = (0, a, 0, \ldots, 0)$ and so on. Thus k is the order of the action. To show that the action is fixed-point-free suppose $\varphi^l * (a_0, \ldots, a_{n-1}) = (a_0, \ldots, a_{n-1})$ for $(a_0, \ldots, a_{n-1}) \in A^{(n)}$ and some $l < k$. Then

$$(\varphi^l - \mathrm{id}) * (a_0, \ldots, a_{n-1}) = (0, \ldots, 0)$$

and

$$f(\varphi) * (a_0, \ldots, a_{n-1}) = (0, \ldots, 0).$$

The polynomials $f(x)$ and $x^l - 1$ have no common divisor in $\mathbb{Z}_p[x]$ and modulo p. Therefore there are (see Proposition 10) polynomials $u(x), v(x) \in \mathbb{Z}_p[x]$ such that

$$u(x) \cdot f(x) + v(x) \cdot (x^l - 1) = 1.$$

Thus

$$\begin{aligned} (a_0, \ldots, a_{n-1}) &= \left(u(\varphi) \cdot f(\varphi) + v(\varphi) \cdot (\varphi^l - 1)\right) * (a_0, \ldots, a_{n-1}) \\ &= (0, \ldots, 0). \end{aligned}$$

Thus there are no fixed points apart from 0. q.e.d.

4 A Decomposition

In the previous section we showed how cyclotomic polynomials may be used to construct abelian-by-finite CA-groups. In this section we will use cyclotomic polynomials in a similar way to decompose the Fitting subgroup of a locally finite solvable CA-group.

Proposition 12. Let $F_k(x) = g_1(x) \cdot \ldots \cdot g_n(x)$ be the decomposition of the k-th cyclotomic polynomial in its irreducible factors over $\mathbb{Z}_p$. Let $h_i(x) := \prod_{j \neq i} g_j(x)$ for $i = 1, \ldots, n$. (If $n = 1$ set $h_1(x) = 1$.) Then there are polynomials $u_i(x) \in \mathbb{Z}_p[x]$ ($i = 1, \ldots, n$) such that $u_1(x)h_1(x) + \ldots + u_n(x)h_n(x) = 1$.

Proof. We prove by induction over m that there exist polynomials $v_1^{(m)}, \ldots, v_m^{(m)} \in \mathbb{Z}_p[x]$ for $m \leq n$ such that

$$v_1^{(m)} h_1 + \ldots + v_m^{(m)} h_m = \prod_{j=m+1}^{n} g_j.$$

$m = 1$: Let $v_1^{(1)} := 1$.

$m \mapsto m+1$: $\prod_{j=1}^{m} g_j$ and g_{m+1} have no common divisor in $\mathbb{Z}_p[x]$ and modulo p. Therefore by Proposition 10 there exist $s_1, s_2 \in \mathbb{Z}_p[x]$ such that

$$s_1 \cdot g_{m+1} + s_2 \cdot \prod_{j=1}^{m} g_j = 1.$$

We multiply the equation with $\prod_{j=m+2}^{n} g_j$, replace $\prod_{j=m+1}^{n} g_j$ by $v_1^{(m)} h_1 + \ldots + v_m^{(m)} h_m$, and get

$$\prod_{j=m+2}^{n} g_j = s_1 \cdot \prod_{j=m+1}^{n} g_j + s_2 \cdot \prod_{\substack{j=1 \\ j \neq m+1}}^{n} g_j$$

$$= s_1 \cdot v_1^{(m)} h_1 + \ldots + s_1 \cdot v_m^{(m)} h_m + s_2 \cdot h_{m+1}.$$

Let $v_i^{(m+1)} := s_1 \cdot v_i^{(m)}$ for $i \leq m$ and $v_{m+1}^{(m+1)} := s_2$.

q.e.d.

Using the proposition we now can decompose a locally finite solvable CA-group $G = A \rtimes C_k$, where C_k is a cyclic group of order k.

Proposition 13. Let $A \rtimes C_k$ be a locally finite CA-group, A an abelian p-group and z a generating element of C_k. Let $F_k(x) = g_1(x) \cdot \ldots \cdot g_n(x)$ be the decomposition of the k-th cyclotomic polynomial into its irreducible factors over $\mathbb{Z}_p$. Define for $i = 1, \ldots, n$: $A_i := \{a \in A;\ g_i(z) * a = 0\}$. Then the following holds. The subgroups A_i ($i \leq n$) are invariant under the action of C_k, and $A = A_1 \oplus \ldots \oplus A_n$.

Proof. (1) The A_i are subgroups and invariant under the action.

[That is obvious.]

(2) The A_i are disjoint.

[Suppose $a \in A_i \cap \langle \bigcup_{j \neq i} A_j \rangle$. With Proposition 12 we get

$$\begin{aligned} a &= 1 * a &= \bigl(u_1(z)h_1(z) + \ldots + u_n(z)h_n(z)\bigr) * a \\ & &= u_1(z)h_1(z) * a + \ldots + u_n(z)h_n(z) * a. \end{aligned}$$

Because $a \in A_i$, we have $h_j(z) * a = 0$ for $j \neq i$. Hence $a = u_i(z)h_i(z) * a$. On the other hand $a \in \langle \bigcup_{j \neq i} A_j \rangle$, i.e., there are $b_j \in A_j$ for $j \leq n$ with $b_i = 0$ such that $a = b_1 + \ldots + b_n$. We get

$$a = u_i(z)h_i(z) * a = u_i(z)h_i(z) * b_1 + \ldots + u_i(z)h_i(z) * b_n = 0,$$

because for all $j \leq n$ we have $u_i(z)h_i(z) * b_i = 0$.]

(3) A is generated by the A_i.

[Let $a \in A$. With Proposition 12 we get

$$\begin{aligned} a &= 1 * a &= \bigl(u_1(z)h_1(z) + \ldots + u_n(z)h_n(z)\bigr) * a \\ & &= h_1(z)u_1(z) * a + \ldots + h_n(z)u_n(z) * a, \end{aligned}$$

with $h_i(z)u_i(z) * a \in A_i$ because of Proposition 9.]

q.e.d.

We want to decompose the blocks A_i into smaller ones. The A_i are modules over the ring $R_i := \mathbb{Z}_p[x]/I_i$ (let I_i be the ideal generated by g_i). The action of x on A_i corresponds to that of z. Notice that R_i is isomorphic to the ring $\mathbb{Z}_p$ with an adjointed primitive k-th root of unit. Therefore all the R_i are isomorphic. The ring R_i is a principal ideal domain, moreover the ideals in R_i are of the type (p^l) for some $l \in \mathbb{N}$. In what follows we will consider A_i as R_i-module. The A_i (for $i = 1, \ldots, n$) are divisible R_i-modules if and only if they are divisible as groups. This is because all elements not divisible by p are invertible in R_i.

The following proposition is proved according to that for divisible submodules (see for example [F70, Theorem 21.2]).

Proposition 14. Let the prerequisites be as in Proposition 13 and let $A = A_1 \oplus \ldots \oplus A_n$. Let p^m be the exponent of A_i and $C_{p^m}^{(s)} \cong X \leq A_i$, so that X is a submodule of A_i. Then X is a direct summand (as a R_i-module).

Proof. Let Y be a maximal R_i-submodule with $X \cap Y = \{0\}$. Suppose there is $z \in A_i - (X+Y)$. Because of the maximality of Y we get $X \cap \langle Y \cup \{z\} \rangle_{R_i} \neq \{0\}$. Thus $I := \{r \in R_i;\ rz \in X+Y\}$ is a non-trivial ideal in R_i. Hence I is generated by p^k for some $k \in \mathbb{N}$. Therefore there are $x \in X, y \in Y$ with $x = y + p^k z$. We show that x is divisible by p^k, which means that there is $x_1 \in X$ such that $p^k x_1 = x$. We have $p^{m-k} x = p^{m-k} y + p^m z = p^{m-k} y$. As $X \cap Y = \{0\}$ we get $p^{m-k} x = 0$. If p^t is the order of x, then x is divisible by p^{m-t}. As $m - k \geq t$, x is also divisible by $p^{m-(m-k)} = p^k$. Thus we can write $p^k x_1 = y + p^k z$, respectively $y = p^k(x_1 - z)$. Now consider $Y' := \langle Y \cup \{x_1 - z\} \rangle_{R_i}$. We claim that $X \cap Y' = \{0\}$. To prove this let $x_2 \in X, y_2 \in Y$ and $r \in R_i$ such that $x_2 = y_2 + r(x_1 - z)$. Then $rz \in X+Y$ and hence $r \in I$. Therefore there is $s \in R_i$ with $r(x_1 - z) = sp^k(x_1 - z) = sy \in Y$. Thus $x_2 = y_2 + sy \in Y$, which means that $x_2 = 0$. We found a contradiction, as $Y \subsetneq Y'$, but we did choose Y maximal. q.e.d.

Remark 15. One can choose Y in the proof such that Y contains a certain invariant subgroup that is disjoint to X.

Definition 16. Let M be a R-module. We call M *(directly) indecomposable* if there are no proper submodules $A, B \leq M$ such that $M = A \oplus B$.

Proposition 17. Let the conditions be as in Proposition 13, and R_i as defined above. Furthermore let A_i be finite. Then we can decompose A_i in a direct sum of indecomposable R_i-modules.

Proof. Let a be an element of maximal order in A_i and U the span of a as R_i-module (i.e., $U = \{r * a;\ r \in R_i\}$). We claim that U is indecomposable. Suppose $U = X_1 \oplus X_2$. Then $J_i := \{r \in R_i;\ r * a \in X_i\}$ for $i = 1, 2$ would be ideals in R_i that are not contained in each other. But this is not possible in R_i. The module U is of the same type as X in Proposition 14. Thus U is a direct summand. Let $A_i = U \oplus Y$, then we can find analogously a direct summand in Y. Inductively we get the desired decomposition. q.e.d.

In the proof of our main theorem (Theorem 22) we will have to extend a decomposition as in Proposition 17 to a decomposition of a group containing $A \rtimes C_k$. We will use the following proposition.

Proposition 18. Let $A \rtimes C_k \leq B \rtimes C_k$ be finite CA-groups such that $A \leq B \cong C_{p^m}^{(s)}$ are abelian p-groups. Let z be a generating element of C_k such that $g_i(z) * b = 0$ for all $b \in B$. Let $A = U_1 \oplus \ldots \oplus U_l$ be a decomposition of A in non-trivial indecomposable R_i-modules. Then there are indecomposable R_i-modules $V_1, \ldots, V_l$ and a R_i-module V such that $U_j \leq V_j$ (for $j = 1, \ldots, l$) and $B = V_1 \oplus \ldots \oplus V_l \oplus V$.

Proof. For $j \in \{1, \ldots, l\}$ let $v_j \in B$ be such that v_j is of order p^m, and $U_j \subseteq \langle \{v_j\}\rangle_{R_i}$. (Such v_j exist, because if x is of maximal order in U_j, then U_j is generated by x. Let v_j be a suitable root of x.) Let $V_j := \langle \{v_j\}\rangle_{R_i}$. We claim that $\langle V_j; j = 1, \ldots, l\rangle = V_1 \oplus \ldots \oplus V_l$. To prove this, suppose $z_1 + \ldots + z_l = 0$ with $z_j \in V_j$ for $j = 1, \ldots, l$. Choose k minimal such that $p^k z_j \in U_j$ for all $j = 1, \ldots, l$. Then $p^k z_j = 0$ (for $j = 1, \ldots, l$), as $A = U_1 \oplus \ldots \oplus U_l$. Suppose $k \geq 1$. Because z_j is in the span of v_j, there is a polynomial $h(x) \in \mathbb{Z}_p[x]$ of smaller degree than g_i with $z_j = h(z) * v_j$. As $p^k * z_j = 0$, the polynomial $h(x)$ is divisible by p^{m-k}. Thus $h(x) = p^{m-k} \cdot h'(x)$ for some polynomial $h'(x) \in \mathbb{Z}_p[x]$. On the other hand we have $p^{m-1} v_j \in U_j$, and therefore

$$p^{k-1} z_j = p^{k-1} p^{m-k} \cdot h'(z) * v_j = h'(z) * p^{m-1} v_j \in U_j.$$

That is a contradiction to the minimality of k. Thus $k = 0$ and $z_j = 0$ for $j = 1, \ldots, l$. It follows with Proposition 14 that $V_1 \oplus \ldots \oplus V_l$ is a direct summand. Hence there is some V with $B = V_1 \oplus \ldots \oplus V_l \oplus V$. q.e.d.

The previous results show how locally finite, abelian-by-finite CA-groups can be decomposed. On the other hand we can embed a finite solvable CA-group in a group similar to that we constructed in Proposition 11.

Proposition 19. Let $G = A \rtimes \langle\{z\}\rangle$ be a finite solvable CA-group, A abelian, and k be the order of z. Then G is embeddable in a group $B^{\phi(k)} \rtimes \langle\{z\}\rangle$, where the action of z is according to Proposition 11 and $f = F_k$, the k-th cyclotomic polynomial.

Proof. It is clear that it suffices to consider the case when A is a p-group. Let p^s be the exponent of A. According to Proposition 13 let $F_k(x) = g_1(x) \cdot \ldots \cdot g_n(x)$ and $A = A_1 \oplus \ldots \oplus A_n$. Let $m := \frac{\phi(k)}{n}$ (the degree of the polynomials $g_i(x)$). If we decompose A_i according to Proposition 17 in indecomposable R_i-modules, we get

$$A_i = C_{p^{i_1}}{}^m \oplus \cdots \oplus C_{p^{i_l}}{}^m.$$

We write $D_j := C_{p^{i_j}}{}^m$. We can embed D_j in $E_j := C_{p^s}{}^m$ as an abelian group. Choose $a \in D_j$ with order p^{i_j}. Let

$$b_0 \in E_j \quad \text{such that} \quad p^{s-i_j} * b_0 = a,$$
$$b_1 \in E_j \quad \text{such that} \quad p^{s-i_j} * b_1 = z * a,$$
$$\vdots$$
$$b_{m-1} \in E_j \quad \text{such that} \quad p^{s-i_j} * b_{m-1} = z^{m-1} * a.$$

It follows that E_j is (as a group) isomorphic to

$$\langle \{b_0\} \rangle \oplus \cdots \oplus \langle \{b_{m-1}\} \rangle.$$

Therefore one can define an action of z on E_j according to Proposition 11 with $f = g_i$. This action extends the action of z on D_j. If we do this for all j, we get an embedding

$$A_i \rtimes \langle \{z\} \rangle \to (C_{p^s}{}^m)^{r_i} \rtimes \langle \{z\} \rangle \text{ for some } r_i \in \mathbb{N}.$$

Let $r := \max_{1 \leq i \leq n}\{r_i\}$ and $B'_i := (C_{p^s}{}^m)^r$, then we can construct an embedding

$$A_i \rtimes \langle \{z\} \rangle \to B'_i \rtimes \langle \{z\} \rangle$$

in the same way. Now let $B := C_{p^s}{}^r$. When we decompose the group $B^{\phi(k)} \rtimes \langle \{z\} \rangle$ (with the action according to Proposition 11 and $f = F_k$) according to Proposition 13, we get

$$B^{\phi(k)} = B_1 \oplus \cdots \oplus B_n,$$

where the B_i are R_i-modules. It is easy to see that $B_i \cong B'_i$ as R_i-modules. Hence there is an embedding

$$A \rtimes \langle \{z\} \rangle \to B^{\phi(k)} \rtimes \langle \{z\} \rangle.$$

q.e.d.

5 An Amalgamation Lemma

As preparation for the characterisation of the existentially closed locally finite solvable CA-groups (Theorem 21) we need the following technical lemma.

Lemma 20 (Amalgamation Lemma). Let G be a locally finite solvable CA-group, A the Fitting subgroup of G, and Z a complement of A in G. Let π be a set of primes such that A is a π-group and Z is a π'-group. Let U and V be finite abelian π-groups, so that there are embeddings $\varphi_1 : V \to U$ and

$\psi_1 : V \to A$. Then there exists a locally finite CA-group $H = B \rtimes Y$ with the following properties. There are embeddings $\varphi_2 : U \to B$ and $\psi_2 : G \to H$ such that $\varphi_2 \circ \varphi_1 = \psi_2 \circ \psi_1$ and ψ_2 maps Z into $Y \cong \bigoplus_{q \in \pi'} \mathbb{Z}_{q^\infty}$.

Proof. We will define the action of Y on B in finite steps. The following (commuting) diagram shall give an idea. In this diagram, the arrows stand for embeddings.

$$\begin{array}{ccccccccc}
V & \to & A & \to & A \rtimes \langle\{z_1\}\rangle & \to & \cdots \to & A \rtimes \langle\{z_n\}\rangle & \to \cdots \dashrightarrow G \\
\downarrow & & \downarrow & & \downarrow & & & \downarrow & \downarrow \\
U & \to & B_0 & \to & B_1 \rtimes \langle\{y_1\}\rangle & \to & \cdots \to & B_n \rtimes \langle\{y_n\}\rangle & \to \cdots \dashrightarrow H
\end{array}$$

For easier notation we will write A and B_n additively as $\mathbb{Z}[z_n]$- respectively $\mathbb{Z}[y_n]$-modules for $n \in \omega$. The first step is an amalgamation of A und U with amalgamated subgroup V; we take the (outer) direct product of U and A and factorize modulo the subgroup $\{(\varphi_1(v), \psi_1(-v)); v \in V\}$. We will call this group B_0. From now on we will treat A as a subgroup of B_0. Let $\{q_n; n \in \omega\}$ be an enumeration of π' with the property that every $q \in \pi'$ occurs infinitely many times. We define $z_0 = y_0 = 1 \in \bigoplus_{q \in \pi'} \mathbb{Z}_{q^\infty}$ and $\eta_0 = \zeta_0 = 1 \in \mathbb{N}$. Let ζ_{n+1} be the biggest number dividing $\eta_{n+1} := \prod_{i=0}^{n} q_i$ such that there is an element of order ζ_{n+1} in Z. Let $z_{n+1} \in Z$ be an element of order ζ_{n+1} such that for $\xi := \frac{\zeta_{n+1}}{\zeta_n}$ the equation $z_{n+1}^\xi = z_n$ holds. Let $y_{n+1} \in \bigoplus_{q \in \pi'} \mathbb{Z}_{q^\infty}$ be an element of order η_{n+1} with $y_{n+1}^{q_n} = y_n$. Let ϕ be the Euler function ($\phi(n)$ is the number of primitive roots modulo n) and $K_n := B_0^{\phi(\eta_n)}$. Then $K_n \rtimes \langle\{y_n\}\rangle$ is a CA-group with the action described in Theorem 11. We construct B_n as a factor group of K_n, so that we get an embedding $A \rtimes \langle\{z_n\}\rangle \to B_n \rtimes \langle\{y_n\}\rangle$ in a natural way. We set $\gamma_n := \frac{\eta_n}{\zeta_n}$ and $k := \phi(\eta_n)$, and define

$$N_n := \{(z_n * a_1, z_n * a_2, \ldots, z_n * a_k) - y_n^{\gamma_n} * (a_1, a_2, \ldots, a_k);$$
$$a_1, a_2, \ldots, a_k \in A\}.$$

The group N_n is a subgroup of A^k. On A^k we have actions of y_n and z_n if we define

$$z_n * (a_1, \ldots, a_k) := (z_n * a_1, \ldots, z_n * a_k).$$

The actions of y_n and z_n commute obviously. Thus we get

$$N_n = (z_n - y_n^{\gamma_n}) * A^k.$$

It is easy to see that N_n is a normal subgroup of $K_n \rtimes \langle\{y_n\}\rangle$, i.e., N_n is invariant under the action of y_n. Let $B_n := K_n/N_n$. As $B_n \rtimes \langle\{y_n\}\rangle$ is a factor group of a locally finite CA-group, it is itself a CA-group (see Fact 5 in Section 2). We get the following (incomplete) diagram:

$$
\begin{array}{ccccccc}
\cdots & \to & A \rtimes \langle\{z_n\}\rangle & \to & A \rtimes \langle\{z_{n+1}\}\rangle & \to & \cdots \\
& & ? & & ? & & \\
\cdots & ? & B_n \rtimes \langle\{y_n\}\rangle & ? & B_{n+1} \rtimes \langle\{y_{n+1}\}\rangle & ? & \cdots \\
& & \uparrow & & \uparrow & & \\
\cdots & ? & K_n \rtimes \langle\{y_n\}\rangle & ? & K_{n+1} \rtimes \langle\{y_{n+1}\}\rangle & ? & \cdots
\end{array}
$$

It remains to show that we can replace the question marks in the diagram with embeddings. These embeddings shall make the diagram commutative. That means, we have to define mappings

I. $K_n \rtimes \langle\{y_n\}\rangle \to K_{n+1} \rtimes \langle\{y_{n+1}\}\rangle$,

II. $A \rtimes \langle\{z_n\}\rangle \to B_n \rtimes \langle\{y_n\}\rangle$,

III. $B_n \rtimes \langle\{y_n\}\rangle \to B_{n+1} \rtimes \langle\{y_{n+1}\}\rangle$,

and show that these mappings are (a) homomorphisms and (b) injective.

(I.) We start with the embedding $K_n \rtimes \langle\{y_n\}\rangle \to K_{n+1} \rtimes \langle\{y_{n+1}\}\rangle$. Here K_n is a $\mathbb{Z}[y_n]$-module and K_{n+1} is a $\mathbb{Z}[y_{n+1}]$-module and therefore (as $y_{n+1}^{q_n} = y_n$) also a $\mathbb{Z}[y_n]$-module. Thus it suffices to show that there is an injective $\mathbb{Z}[y_n]$-module homomorphism from K_n to K_{n+1}.

Let $k := \phi(\eta_n)$ and $l := \phi(\eta_{n+1})$. We define

$$
\begin{array}{rcl}
\theta_n : B_0^k & \to & B_0^l \\
(b_1, b_2, \ldots, b_k) & \mapsto & (b_1, 0, \ldots, 0) \\
& & + y_{n+1}^{q_n} * (b_2, 0, \ldots, 0) \\
& & + \ldots \\
& & + y_{n+1}^{(k-1)q_n} * (b_k, 0, \ldots, 0)
\end{array}
$$

(a) We have to show that this function is a homomorphism of $\mathbb{Z}[y_n]$-modules. We defined θ_n by

$$y_n^r * (b, 0, \ldots, 0) \mapsto y_{n+1}^{r \cdot q_n} * (b, 0, \ldots, 0) \text{ for } r < k.$$

We have to show that this condition also holds for $r \geq k$. Let $F_{\eta_n}(x)$ be the η_n-th cyclotomic polynomial. The action of y_n on B_0^k was defined in such a way that

$$F_{\eta_n}(y_n) * (b_1, \ldots, b_k) = (0, \ldots, 0) \text{ for all } (b_1, \ldots, b_k) \in K_n.$$

Therefore, if $x^r = P_1(x) \cdot F_{\eta_n}(x) + P_2(x)$ with $\deg(P_2) < k$, then

$$y_n{}^r * (b_1, \ldots, b_k) = P_2(y_n) * (b_1, \ldots, b_k).$$

On the other hand we have $x^{q_n \cdot r} = P_1(x^{q_n}) \cdot F_{\eta_n}(x^{q_n}) + P_2(x^{q_n})$. Now let $F_{\eta_{n+1}}$ be the η_{n+1}-th cyclotomic polynomial. Then $F_{\eta_{n+1}}(x)$ is a divisor of $F_{\eta_n}(x^{q_n})$ (more strictly: $F_{\eta_n}(x^{q_n}) = F_{\eta_{n+1}}(x) \cdot F_{\eta_n}(x)$ if $l = k \cdot (q_n - 1)$ and $F_{\eta_n}(x^{q_n}) = F_{\eta_{n+1}}(x)$ if $l = k \cdot q_n$). As

$$F_{\eta_{n+1}}(y_{n+1}) * (b_1, \ldots, b_l) = (0, \ldots, 0) \text{ for all } (b_1, \ldots, b_l) \in K_{n+1},$$

we get

$$y_{n+1}{}^{r \cdot q_n} * (b, 0, \ldots, 0) = P_2(y_{n+1}{}^{q_n}) * (b, 0, \ldots, 0) \text{ for all } b \in B.$$

Thus

$$\begin{aligned}
\theta_n(y_n{}^r * (b, 0, \ldots, 0)) &= \theta_n(P_2(y_n) * (b, 0, \ldots, 0)) \\
&= P_2(y_{n+1}{}^{q_n}) * (b, 0, \ldots, 0) \\
&= y_n{}^{r \cdot q_n} * (b, 0, \ldots, 0).
\end{aligned}$$

It follows easily that θ_n is a homomorphism.

(b) To show the injectivity of θ_n we have to consider two cases:

Case 1. q_n divides η_n.
Then $l := \phi(\eta_{n+1}) = \phi(\eta_n \cdot q_n) = q_n \cdot \phi(\eta_n) = q_n \cdot k$. Therefore

$$\theta_n(b_1, b_2, \ldots, b_k) = (\underbrace{b_1, 0, \ldots, 0}_{q_n}, b_2, \ldots, \underbrace{b_k, 0, \ldots, 0}_{q_n}),$$

and it is obvious that this is an embedding. We note for later purposes that K_{n+1} can equivalently be written as

$$\left\{ \begin{pmatrix} b_{11} & \cdots & b_{1k} \\ \vdots & & \vdots \\ b_{q_n 1} & \cdots & b_{q_n k} \end{pmatrix} ; b_{ij} \in B_0, 1 \le i \le q_n, 1 \le j \le k \right\},$$

where θ_n maps $(b_1, \ldots, b_k)$ to

$$\begin{pmatrix} b_1 & \cdots & b_k \\ 0 & \cdots & 0 \\ \vdots & & \vdots \\ 0 & \cdots & 0 \end{pmatrix}.$$

Case 2. q_n does not divide η_n.

Then $l = (q_n-1) \cdot k$. Let $c := y_{n+1}{}^{\eta_n}$ and $t := q_n - 1$. We construct a $\mathbb{Z}[y_{n+1}]$-module K'_{n+1} that is isomorphic to K_{n+1}. Here the elements of K'_{n+1} are written as $(t \times k)$-matrices over B_0. Let again $F_{\eta_{n+1}}(x) = x^k + \lambda_{k-1} + \ldots + \lambda_0$ be the η_n-th cyclotomic polynomial. We define fixed point free actions of $y_n = y_{n+1}{}^{q_n}$ and c on K'_{n+1}:

$$y_n * \begin{pmatrix} b_{11} & \ldots & b_{1k} \\ b_{21} & \ldots & b_{2k} \\ \vdots & & \vdots \\ b_{t1} & \ldots & b_{tk} \end{pmatrix} = \begin{pmatrix} 0 & b_{11} & \ldots & b_{1(k-1)} \\ 0 & b_{21} & \ldots & b_{2(k-1)} \\ \vdots & \vdots & & \vdots \\ 0 & b_{t1} & \ldots & b_{t(k-1)} \end{pmatrix} - \begin{pmatrix} \lambda_0 b_{1k} & \ldots & \lambda_{k-1} b_{1k} \\ \lambda_0 b_{2k} & \ldots & \lambda_{k-1} b_{2k} \\ \vdots & & \vdots \\ \lambda_0 b_{tk} & \ldots & \lambda_{k-1} b_{tk} \end{pmatrix}$$

and

$$c * \begin{pmatrix} b_{11} & \ldots & b_{1k} \\ b_{21} & \ldots & b_{2k} \\ \vdots & & \vdots \\ b_{t1} & \ldots & b_{tk} \end{pmatrix} = \begin{pmatrix} 0 & \ldots & 0 \\ b_{11} & \ldots & b_{1k} \\ \vdots & & \vdots \\ b_{(t-1)1} & \ldots & b_{(t-1)k} \end{pmatrix} - \begin{pmatrix} b_{t1} & \ldots & b_{tk} \\ b_{t1} & \ldots & b_{tk} \\ \vdots & & \vdots \\ b_{t1} & \ldots & b_{tk} \end{pmatrix}.$$

Notice that the definition is essentially the same as in Proposition 11. The element y_n acts on the columns and c acts on the rows (the q_n-th cyclotomic polynomial is $x^{q_n-1} + \ldots + x + 1$). As q_n does not divide η_n, there are natural numbers $0 < r < \eta_n$, $0 < s < q_n$ such that $r \cdot q_n + s \cdot \eta_n = 1$. Then $y_{n+1} = y_n{}^r \cdot c^s$ and we define

$$y_{n+1} * \begin{pmatrix} b_{11} & \ldots & b_{1k} \\ b_{21} & \ldots & b_{2k} \\ \vdots & & \vdots \\ b_{t1} & \ldots & b_{tk} \end{pmatrix} = (y_n{}^r \cdot c^s) * \begin{pmatrix} b_{11} & \ldots & b_{1k} \\ b_{21} & \ldots & b_{2k} \\ \vdots & & \vdots \\ b_{t1} & \ldots & b_{tk} \end{pmatrix}.$$

The action is welldefined and of order η_{n+1} because the actions of y_n and c commute. We now define the isomorphism σ between K_{n+1} and K'_{n+1} by

$$y_{n+1}{}^r * (b, 0, \ldots, 0) \mapsto y_{n+1}{}^r * \begin{pmatrix} b & 0 & \ldots & 0 \\ 0 & 0 & \ldots & 0 \\ \vdots & \vdots & \ddots & \vdots \\ 0 & 0 & \ldots & 0 \end{pmatrix} \quad \text{for } 0 \leq r \leq l-1.$$

Then $\sigma \circ \theta_n$ maps $(b_1, \ldots, b_k)$ to the matrix

$$\begin{pmatrix} b_1 & b_2 & \cdots & b_k \\ 0 & 0 & \cdots & 0 \\ \vdots & \vdots & \ddots & \vdots \\ 0 & 0 & \cdots & 0 \end{pmatrix}.$$

This mapping is obviously injective. Thus θ_n must be injective.

(II.) Now we show how $A \rtimes \langle \{z_n\} \rangle$ can be embedded in $B_n \rtimes \langle \{y_n\} \rangle$:

$$\begin{array}{rcl} A \rtimes \langle \{z_n\} \rangle & \to & B_n \rtimes \langle \{y_n\} \rangle \\ (a, z_n{}^r) & \mapsto & \left((a, 0, \ldots, 0) + N_n, y_n{}^{r \cdot \gamma_n} \right). \end{array}$$

Let $d_n := y_n{}^{\gamma_n}$.

(a) This function is a homomorphism. That may be concluded from

$$(z_n{}^r - d_n{}^r) * (a, 0, \ldots, 0) \in N_n \text{ for } r \in \mathbb{N},$$

which follows from

$$(z_n{}^r - d_n{}^r) = (z_n - d_n) \cdot (z_n{}^{r-1} + z_n{}^{r-2} d_n + \cdots + z_n d_n{}^{r-2} + d_n{}^{r-1}).$$

(b) The function is injective. Suppose $(a, 0, \ldots, 0) \in N_n$. Then

$$(a, 0, \ldots, 0) = (z_n - d_n) * (b_1, \ldots, b_k)$$

for some $b_1, \ldots, b_k \in A$. Let $F_{\zeta_n}(x) = x^\nu + \mu_{\nu-1} x^{\nu-1} + \cdots + \mu_0$ be the ζ_n-th cyclotomic polynomial. Then

$$\begin{aligned} F_{\zeta_n}(z_n) &= z_n{}^\nu + \mu_{\nu-1} z_n{}^{\nu-1} + \cdots + \mu_0 \\ &= (z_n - d_n) \cdot \left((z_n{}^{\nu-1} + \mu_{\nu-1} z_n{}^{\nu-2} + \cdots + \mu_1) \right. \\ &\quad + d_n \cdot (z_n{}^{\nu-2} + \mu_{\nu-1} z_n{}^{\nu-3} + \cdots + \mu_2) \\ &\quad + \cdots \\ &\quad \left. + d_n{}^{\nu-1} \right) \\ &\quad + F_{\zeta_n}(d_n). \end{aligned}$$

As

$$F_{\zeta_n}(z_n) * (a_1, \ldots, a_k) = F_{\zeta_n}(d_n) * (a_1, \ldots, a_k) = (0, \ldots, 0)$$

for all $a_1, \ldots, a_k \in A$, and $(a, 0, \ldots, 0) = (z_n - d_n) * (b_1, \ldots, b_k)$, we have

$$\begin{aligned} (0, \ldots, 0) &= \left((z_n{}^{\nu-1} + \mu_{\nu-1} z_n{}^{\nu-2} + \cdots + \mu_1) \right. \\ &\quad + d_n \cdot (z_n{}^{\nu-2} + \mu_{\nu-1} z_n{}^{\nu-3} + \cdots + \mu_2) \\ &\quad + \cdots \\ &\quad \left. + d_n{}^{\nu-1} \right) * (a, 0, \ldots, 0). \end{aligned}$$

The claim now follows with the same argument that we used to show the injectivity of θ_n. That means, because

$$\langle \{d_n{}^r *(b,0,\ldots,0);\ 0 \leq r \leq \nu - 1,\ b \in B_0\}\rangle$$
$$= \bigoplus_{r=0}^{\nu-1} \langle \{d_n{}^r * (b,0,\ldots,0);\ b \in B_0\}\rangle,$$

it follows that $a = 0$.

(III.) It is obvious that the embedding $B_n \rtimes \langle\{y_n\}\rangle \to B_{n+1} \rtimes \langle\{y_{n+1}\}\rangle$ must be defined in the following way to make the diagram commutative.

$$B_n \rtimes \langle\{y_n\}\rangle \quad \to \quad B_{n+1} \rtimes \langle\{y_{n+1}\}\rangle$$
$$((b_1,\ldots,b_k) + N_n, y_n{}^r) \quad \mapsto \quad \left(\theta_n(b_1,\ldots,b_k) + N_{n+1}, y_{n+1}{}^{r \cdot q_n}\right).$$

This is well-defined, because if $(b_1,\ldots,b_k) \in N_n$, then $\theta_n(b_1,\ldots,b_k) \in N_{n+1}$.

(a) It is clear that the mapping is a homomorphism, because θ_n is a homomorphism.

(b) It remains to show that the mapping is injective. Let $d_n := y_n{}^{\gamma_n}$ and $d_{n+1} := y_{n+1}{}^{\gamma_{n+1}}$. We have to show that if

$$\theta_n(b_1,\ldots,b_k) \in N_{n+1} = (z_{n+1} - d_{n+1}) * A^l,$$

then

$$(b_1,\ldots,b_k) \in N_n = (z_n - d_n) * A^k.$$

We have to consider several cases.

Case 1. $z_n = z_{n+1}$ and $d_n = y_n{}^{\gamma_n} = y_{n+1}{}^{\gamma_{n+1}} = d_{n+1}$.
Using the representation of the elements of K_{n+1} as matrices one can see that there is a y_n- and z_n-linear mapping $\Pi_n : K_{n+1} \to K_n$ such that $\Pi_n \circ \theta_n$ is the identity on K_n. Thus if

$$\theta_n(b_1,\ldots,b_k) = (z_{n+1} - d_{n+1}) * (a_1,\ldots,a_l)$$
$$= (z_n - d_n) * (a_1,\ldots,a_l),$$

then

$$(b_1,\ldots,b_k) = \Pi_n(\theta_n(b_1,\ldots,b_k))$$
$$= \Pi_n((z_n - d_n) * (a_1,\ldots,a_l))$$
$$= (z_n - d_n) * \Pi_n(a_1,\ldots,a_l) \in N_n.$$

Case 2. $z_n = z_{n+1}{}^{q_n}$ and $\gamma_n = \gamma_{n+1}$. (Then $d_n = d_{n+1}{}^{q_n}$.)
First notice that there is a $\mathbb{Z}[y_n]$-module L_n that is isomorphic to K_n and may be written as

$$L_n = \left\{ \begin{pmatrix} b_{11} & \cdots & b_{1v} \\ \vdots & & \vdots \\ b_{u1} & \cdots & b_{uv} \end{pmatrix} ;\ b_{ij} \in B_0 \right\},$$

where $v = \phi(\zeta_n)$, $u = \frac{\phi(\eta_n)}{\phi(\zeta_n)}$, and

$$d_n * \begin{pmatrix} b_{11} & \cdots & b_{1v} \\ \vdots & & \vdots \\ b_{u1} & \cdots & b_{uv} \end{pmatrix} =$$

$$\begin{pmatrix} 0 & b_{11} & \cdots & b_{1(v-1)} \\ \vdots & \vdots & & \vdots \\ 0 & b_{u1} & \cdots & b_{u(v-1)} \end{pmatrix} - \begin{pmatrix} \mu_0 b_{1v} & \cdots & \mu_{v-1} b_{1v} \\ \vdots & & \vdots \\ \mu_0 b_{uv} & \cdots & \mu_{v-1} b_{uv} \end{pmatrix}.$$

(Here let $F_{\zeta_n}(x) = x^v + \mu_{v-1} x^{v-1} + \cdots + \mu_0$ be again the ζ_n-th cyclotomic polynomial.) This can be done recursively in the way we introduced the matrices in (I.b). We do not say much about the action of y_n on L_n, but this is not important here. We will identify K_n with L_n and have therefore an embedding $\theta_n : L_n \to L_{n+1}$. We need to show the following claim.

Claim A.

$$N_n = \left\{ (z_n - d_n) * \begin{pmatrix} a_{11} & \cdots & a_{1v} \\ \vdots & & \vdots \\ a_{u1} & \cdots & a_{uv} \end{pmatrix} ;\ a_{ij} \in A \right\}$$

$$\stackrel{!}{=} \left\{ (z_n - d_n) * \begin{pmatrix} a_{11} & \cdots & a_{1(v-1)} & 0 \\ \vdots & & \vdots & \vdots \\ a_{u1} & \cdots & a_{u(v-1)} & 0 \end{pmatrix} ;\ a_{ij} \in A \right\}.$$

To prove Claim A it is obviously sufficient to show that for all $b_1, \ldots, b_u \in A$ the element

$$(z_n - d_n) * \begin{pmatrix} 0 & \cdots & 0 & b_1 \\ \vdots & & \vdots & \vdots \\ 0 & \cdots & 0 & b_u \end{pmatrix}$$

can be written as

$$(z_n - d_n) * \begin{pmatrix} a_{11} & \cdots & a_{1(v-1)} & 0 \\ \vdots & & \vdots & \vdots \\ a_{u1} & \cdots & a_{u(v-1)} & 0 \end{pmatrix},$$

with $a_{ij} \in A$ $(1 \leq i \leq u, 1 \leq j \leq v-1)$. Set

$$\begin{aligned} a_{w1} &:= \mu_0 z_n^{-1} * b_w \\ a_{w2} &:= (\mu_1 z_n^{-1} + \mu_0 z_n^{-2}) * b_w \\ &\vdots \\ a_{w(v-1)} &:= (\mu_{v-2} z_n^{-1} + \cdots + \mu_0 z_n^{-v+1}) * b_w \end{aligned}$$

for $1 \leq w \leq u$. Easy computation shows that

$$(z_n - d_n) * \begin{pmatrix} 0 & \cdots & 0 & b_1 \\ \vdots & & \vdots & \vdots \\ 0 & \cdots & 0 & b_u \end{pmatrix} = (z_n - d_n) * \begin{pmatrix} a_{11} & \cdots & a_{1(v-1)} & 0 \\ \vdots & & \vdots & \vdots \\ a_{u1} & \cdots & a_{u(v-1)} & 0 \end{pmatrix},$$

and Claim A is proven.

Subcase 2.1. q_n divides ζ_n. Then

$$\theta_n : \begin{pmatrix} a_{11} & \cdots & a_{1v} \\ \vdots & & \vdots \\ a_{u1} & \cdots & a_{uv} \end{pmatrix} \mapsto \begin{pmatrix} a_{11} & 0 & \cdots & 0 & a_{12} & \cdots \\ \vdots & \vdots & & \vdots & \vdots & \\ a_{u1} & 0 & \cdots & 0 & a_{u2} & \cdots \end{pmatrix}.$$

Now suppose

$$\theta_n \left(\begin{pmatrix} a_{11} & \cdots & a_{1v} \\ \vdots & & \vdots \\ a_{u1} & \cdots & a_{uv} \end{pmatrix} \right) = (z_{n+1} - d_{n+1}) * \begin{pmatrix} b_{11} & \cdots & b_{1(v-1)} & 0 \\ \vdots & & \vdots & \vdots \\ b_{u1} & \cdots & b_{u(v-1)} & 0 \end{pmatrix}.$$

The action we have to consider is only acting on the rows. Thus it suffices to treat the rows separately. For $1 \leq w \leq u$ we get

$$(\overbrace{a_{w1}, 0, \ldots, 0}^{q_n}, a_{w2}, \ldots) = (z_{n+1} * b_{w1}, z_{n+1} * b_{w2} - b_{w1}, \ldots, -b_{w(v-1)}).$$

Omitting the index w and using the replacements

$$\begin{aligned} \beta_{ij} &:= b_{i+(j-1)\cdot q_n} \\ q &:= q_n \\ z &:= z_{n+1} \end{aligned}$$

we can write this using matrices.

$$\begin{pmatrix} a_1 & \cdots & a_v \\ 0 & \cdots & 0 \\ \vdots & & \vdots \\ 0 & \cdots & 0 \end{pmatrix} = \begin{pmatrix} z*\beta_{11} & \cdots & z*\beta_{1v} - \beta_{q(v-1)} \\ z*\beta_{21} - \beta_{11} & \cdots & z*\beta_{2v} - \beta_{1v} \\ \vdots & & \vdots \\ z*\beta_{q1} - \beta_{(q-1)1} & \cdots & -\beta_{(q-1)v} \end{pmatrix}. \quad (1)$$

Set
$$x_1 := z^{-q+1} * \beta_{11} = z_{n+1}^{-q_n+1} * \beta_{11}$$
$$x_2 := z^{-q+1} * \beta_{12} = z_{n+1}^{-q_n+1} * \beta_{12}$$
$$\vdots$$
$$x_{v-1} := z^{-q+1} * \beta_{1(v-1)} = z_{n+1}^{-q_n+1} * \beta_{1(v-1)}$$

We claim that
$$(a_1, \ldots, a_v) = (z_n - d_n) * (x_1, \ldots, x_{v-1}, 0)$$
$$= (z_n * x_1, z_n * x_2 - x_1, \ldots, -x_{v-1})$$

and therefore
$$\begin{pmatrix} a_{11} & \cdots & a_{1v} \\ \vdots & & \vdots \\ a_{u1} & \cdots & a_{uv} \end{pmatrix} \in N_n.$$

Using the equation (1) above we get for $1 \leq r < v$
$$0 = z * \beta_{2r} - \beta 1r$$
$$0 = z * \beta_{3r} - \beta 2r$$
$$\vdots$$
$$0 = z * \beta_{qr} - \beta(q-1)r.$$

Thus
$$\beta qr = z^{-1} * \beta_{(q-1)r}$$
$$= z^{-2} * \beta_{(q-2)r}$$
$$\vdots$$
$$= z^{-q+1} * \beta_{1r} = z_{n+1}^{-q+1} * \beta_{1r}$$

and $\beta_{1v} = 0$. It follows that
$$a_1 = z * \beta_{11} = (z_{n+1}^{q_n} \cdot z_{n+1}^{-q_n+1}) * \beta_{11} = z_n * x_1$$
$$a_2 = z * \beta_{12} - \beta_{11} = z_n * x_2 - x_1$$
$$\vdots$$
$$a_{v-1} = z_n * x_{v-1} - x_{v-2}$$
$$a_v = z * \beta_{1v} - \beta_{q(v-1)} = 0 - z^{-q+1} * \beta_{1(v-1)} = -x_{-v-1}.$$

That is just what we had to show.

Subcase 2.2. q_n does not divide ζ_n. Analogously as in Subcase 2.1 we may treat the rows seperately. We again use matrices as in (I.) Case 2. and get

$$\begin{pmatrix} a_1 & \cdots & a_v \\ 0 & \cdots & 0 \\ \vdots & & \vdots \\ 0 & \cdots & 0 \end{pmatrix} = (z_{n+1} - d_{n+1}) * \begin{pmatrix} b_{11} & \cdots & b_{1v} \\ b_{21} & \cdots & b_{2v} \\ \vdots & & \vdots \\ b_{(q-1)1} & \cdots & b_{(q-1)v} \end{pmatrix}. \quad (2)$$

Set $q := q_n$ and $u := q - 1$. We make the following claim.
Let $b_{ij} \in A$ for $1 \leq i \leq u$, $1 \leq j \leq v$. Then there are $\alpha_{ij}, \beta_{ij} \in A$ for $1 \leq i \leq u$, $1 \leq j \leq v$ such that

$$(z_{n+1} - d_{n+1}) * \begin{pmatrix} b_{11} & \cdots & b_{1v} \\ \vdots & & \vdots \\ b_{u1} & \cdots & b_{uv} \end{pmatrix} = (z_n - d_n) * \begin{pmatrix} \alpha_{11} & \cdots & \alpha_{1v} \\ \vdots & & \vdots \\ \alpha_{u1} & \cdots & \alpha_{uv} \end{pmatrix} \quad (3)$$

$$+ (z_{n+1}^{\zeta_n} - d_{n+1}^{\zeta_n}) * \begin{pmatrix} \beta_{11} & \cdots & \beta_{1v} \\ \vdots & & \vdots \\ \beta_{u1} & \cdots & \beta_{uv} \end{pmatrix}. \quad (4)$$

To prove this, we take a look at the polynomials $x^q - 1$ and $x^{\zeta_n} - 1$. Their greatest common divisor in $\mathbb{Z}[x]$ and modulo p (i.e., in $\mathbb{F}_p[x]$) for all prime numbers p is $x - 1$, because q does not divide ζ_n. Using Proposition 10 there are polynomials $P_1(x), P_2(x) \in \mathbb{Z}[x]$ such that

$$x - 1 = (x^q - 1) \cdot P_1(x) + (x^{\zeta_n} - 1) \cdot P_2(x).$$

Replacing x by $z_{n+1} \cdot d_{n+1}^{-1}$ we get

$$z_{n+1} \cdot d_{n+1}^{-1} - 1 = ((z_{n+1} \cdot d_{n+1}^{-1})^q - 1) \cdot P_1(z_{n+1} \cdot d_{n+1}^{-1})$$
$$+ ((z_{n+1} \cdot d_{n+1}^{-1})^{\zeta_n} - 1) \cdot P_2(z_{n+1} \cdot d_{n+1}^{-1}),$$

respectively

$$z_{n+1} - d_{n+1} = (z_n - d_n) \cdot d_{n+1}^{1-q} \cdot P_1(z_{n+1} \cdot d_{n+1}^{-1})$$
$$+ (z_{n+1}^{\zeta_n} - d_{n+1}^{\zeta_n}) \cdot d_{n+1}^{1-\zeta_n} \cdot P_2(z_{n+1} \cdot d_{n+1}^{-1}).$$

The claim follows directly. With Claim A equation (2) turns to ($u := q - 1$)

$$\begin{pmatrix} a_1 & \cdots & a_v \\ 0 & \cdots & 0 \\ \vdots & & \vdots \\ 0 & \cdots & 0 \end{pmatrix} = (z_n - d_n) * \begin{pmatrix} \alpha_{11} & \cdots & \alpha_{1(v-1)} & 0 \\ \alpha_{21} & \cdots & \alpha_{2(v-1)} & 0 \\ \vdots & & \vdots & \vdots \\ \alpha_{u1} & \cdots & \alpha_{u(v-1)} & 0 \end{pmatrix}$$

$$+ (z_{n+1}^{\zeta_n} - d_{n+1}^{\zeta_n}) * \begin{pmatrix} \beta_{11} & \cdots & \beta_{1v} \\ \vdots & & \vdots \\ \beta_{(u-1)1} & \cdots & \beta_{(u-1)v} \\ 0 & \cdots & 0 \end{pmatrix}. \quad (5)$$

Let $x_i := z_n^{\zeta_n^{u-1}} * \alpha_{ui} + \cdots + z_n^{\zeta_n} * \alpha_{2i} + \alpha_{1i} = \sum_{j=1}^{u} z_n^{\zeta_n^{j-1}} * \alpha_{ji}$. Then

$$(a_1, \ldots, a_v) = (z_n - d_n) * (x_1, \ldots, x_{v-1}, 0)$$
$$= (z_n * x_1, z_n * x_2 - x_1, \ldots, -x_{v-1}).$$

To show this, let us consider the equations from (3). We get

$$\begin{aligned}
\beta_{(u-1)1} &= z_n * \alpha_{u1} \\
\beta_{(u-2)1} - z_{n+1}{}^{\zeta_n} * \beta_{(u-1)1} &= z_n * \alpha_{(u-1)1} \\
&\vdots \\
\beta_{11} - z_{n+1}{}^{\zeta_n} * \beta_{21} &= z_n * \alpha_{21}
\end{aligned}$$

and

$$\begin{aligned}
\beta_{(u-1)i} &= z_n * \alpha_{ui} - \alpha_{u(i-1)} \\
\beta_{(u-2)i} - z_{n+1}{}^{\zeta_n} * \beta_{(u-1)i} &= z_n * \alpha_{(u-1)i} - \alpha_{(u-1)(i-1)} \\
&\vdots \\
\beta_{1i} - z_{n+1}{}^{\zeta_n} * \beta_{2i} &= z_n * \alpha_{2i} - \alpha_{2(i-1)}
\end{aligned}$$

for $i < i < v$, and

$$\begin{aligned}
\beta_{(u-1)v} &= -\alpha_{u(v-1)} \\
\beta_{(u-2)v} - z_{n+1}{}^{\zeta_n} * \beta_{(u-1)v} &= -\alpha_{(u-1)(v-1)} \\
&\vdots \\
\beta_{1v} - z_{n+1}{}^{\zeta_n} * \beta_{2v} &= -\alpha_{2(v-1)}.
\end{aligned}$$

We get

$$\begin{aligned}
\beta_{11} &= z_n * \sum_{j=2}^{u} z_{n+1}{}^{\zeta_n{}^{j-2}} * \alpha_{j1} \\
\beta_{1i} &= z_n * \sum_{j=2}^{u} z_{n+1}{}^{\zeta_n{}^{j-2}} * \alpha_{ji} - \sum_{j=2}^{u} z_{n+1}{}^{\zeta_n{}^{j-2}} * \alpha_{j(i-1)} \quad (1 < i < v) \\
\beta_{1v} &= -\sum_{j=2}^{u} z_{n+1}{}^{\zeta_n{}^{j-2}} * \alpha_{j(v-1)}.
\end{aligned}$$

Moreover we get from (3)

$$\begin{aligned}
a_1 &= z_n * \alpha_{11} + z_{n+1}{}^{\zeta_n} * \beta_{11} \\
&= z_n * \sum_{j=1}^{u} z_{n+1}{}^{\zeta_n{}^{j-1}} * \alpha_{j1} \\
&= z_n * x_1,
\end{aligned}$$

and similarly

$$\begin{aligned}
a_i &= z_n * \alpha_{1i} - \alpha_{1(i-1)} + z_{n+1}{}^{\zeta_n} * \beta_{1i} \\
&= z_n * \sum_{j=1}^{u} z_{n+1}{}^{\zeta_n{}^{j-1}} * \alpha_{ji} - \sum_{j=1}^{u} z_{n+1}{}^{\zeta_n{}^{j-1}} * \alpha_{j(i-1)} \\
&= z_n * x_i - x_{i-1}
\end{aligned}$$

for $i < i < v$, and

$$\begin{aligned}
a_v &= -\alpha_{1(v-1)} + z_{n+1}\zeta_n * \beta_{1v} \\
&= -\sum_{j=1}^{u} z_{n+1}\zeta_n^{j-1} * \alpha_{j(v-1)} \\
&= -x_{v-1}.
\end{aligned}$$

We define B as the direct limit of $(B_n)_{n \in \mathbb{N}}$, Y as the direct limit of $(\langle\{y_n\}\rangle)_{n \in \mathbb{N}}$ and H as the direct limit of $(B_n \rtimes \langle\{y_n\}\rangle)_{n \in \mathbb{N}}$. Then $H = B \rtimes Y$ has the desired properties and the proof is complete.   q.e.d.

6   Characterization of the e. c. Groups in $\mathcal{C}$

We now state the main theorems of this paper.

Theorem 21. A countable existentially closed locally finite CA-group is isomorphic to

$$\mathrm{SL}_2(\tilde{\mathbb{F}}_2)$$

(where $\tilde{\mathbb{F}}_2$ is the algebraic closure of the field with 2 elements) or isomorphic to

$$\left(\bigoplus_{p \in \pi} \mathbb{Z}_{p^\infty}{}^{(\omega)}\right) \rtimes \left(\bigoplus_{q \in \pi'} \mathbb{Z}_{q^\infty}\right)$$

for some set $\pi \neq \emptyset$ of primes. An uncountable existentially closed locally finite CA-group is isomorphic to

$$A \rtimes \left(\bigoplus_{q \in \pi'} \mathbb{Z}_{q^\infty}\right),$$

where A is an existentially closed abelian π-group for some set $\pi \neq \emptyset$ of primes.

Proof. Let first G be a non-solvable existentially closed locally finite CA-group. It is wellknown (see Fact 2) that a non-solvable finite CA-group is isomorphic to $\mathrm{SL}_2(K)$, where K is a finite field of characteristic 2. It can be shown (see [E99]) that the class

$$\{G;\ G \text{ is isomorphic to } \mathrm{SL}_2(K) \text{ for some finite field } K \text{ of characteristic 2}\}$$

is inductive. It follows that a non-solvable existentially closed locally finite CA-group G is isomorphic to $\mathrm{SL}_2(K)$, where K is a locally finite field of characteristic 2. It is easy to see (see [K$_1$91]) that the field operations are definable in a group $\mathrm{SL}_2(K)$, when K is a perfect field of characteristic 2.

Therefore K has to be algebraically closed. As K also has to be locally finite, it follows that $K \cong \tilde{\mathbb{F}}_2$. We already made some remarks on the solvable case in Section 4. Let again $\Phi(\bar{x}, \bar{y})$ be a quantifier-free formula in the first-order language $\mathcal{L}$ of groups, and $\bar{a}$ a tuple of elements of A such that there is an abelian π-group $B \supseteq A$ in which $\exists \bar{y}\, \Phi(\bar{a}, \bar{y})$ holds. Let $\bar{b}$ be a satisfying tuple of elements in B and set $U := \langle \bar{a}, \bar{b} \rangle$. The Amalgamation Lemma 20 shows that there is a locally finite CA-group $H \geq G$ such that $U \leq \mathrm{Fit}(H)$. Let $1 \neq c \in \mathrm{Fit}(G)$. Then

$$H \models \exists \bar{y}\, (\Phi(\bar{x}, \bar{y}) \wedge \bigwedge_i [y_i, c] = 1).$$

As G is e.c., this sentence is also true in G, and hence $A \models \exists \bar{y}\, \Phi(\bar{a}, \bar{y})$. As seen in Lemma 20 the sentence $\exists x\, (x^n = 1 \wedge \bigwedge_{i<n} x^i \neq 1)$ can be fulfilled in a locally finite CA-group $H \geq G$ for every π'-number n, and is therefore also true in G.
q.e.d.

In Theorem 21 we gave necessary conditions for a group to be an e.c. locally finite CA-group. We will now examine which groups of this kind are existentially closed.

Theorem 22. For a locally finite CA-group $G = A \rtimes Z$ with

$$A \cong \bigoplus_{p \in \pi} Z_{p^\infty}{}^{(\kappa_p)} \quad \text{and} \quad Z \cong \bigoplus_{q \in \pi'} Z_{q^\infty}$$

are equivalent:

(i) G is existentially closed in the class of $\mathcal{C}$-groups (the class of locally finite solvable CA-groups).

(ii) Every finite solvable CA-group of type π is embeddable in G.

(iii) For all $p \in \pi$, $l \in \omega$ and every π'-number k the group $(C_p{}^{\phi(k)})^l \rtimes C_k$ can be embedded in G. Here the action of C_k on a block $C_p{}^{\phi(k)}$ is according to Theorem 11 with $f = F_k$, the k-th cyclotomic polynomial.

(iv) For all $p \in \pi$ and for every π'-number k the Sylow-p-group $A^{(p)}$ of A can be decomposed in the following way. Let $f(x) = g_1(x) \cdot \ldots \cdot g_n(x)$ the decomposition of the k-th cyclotomic polynomial in its irreducible factors over $\mathbb{Z}_p$. Furthermore let z be an element of order k in Z and $A_i := \{a \in A^{(p)};\ g_i(z) * a = 0\}$. Then for every $i = 1, \ldots, n$: $A_i \neq \{0\}$.

Proof. "(i) ⇒ (iii)". Let G be an e.c. $\mathcal{C}$-group, $p \in \pi$, $l \in \omega$, and k a π'-number. The proof of the Amalgamation Lemma 20 shows that there is a locally finite solvable CA-group H containing G and $(C_p^{\phi(k)})^l \rtimes C_k$. The operations in this group can be described through an existential sentence, which is fulfilled in H. Thus the sentence must also to be satisfied in G. That means $(C_p^{\phi(k)})^l \rtimes C_k$ is embeddable in G.

"(ii) ⇒ (iii)". This is trivial.

"(iii) ⇒ (ii)". Let $G' = A' \rtimes \langle\{z\}\rangle$ be a finite solvable CA-group of type π. According to Proposition 19 G' is embeddable in a group $B^{\phi(k)} \rtimes \langle\{z\}\rangle$. As B is a finite abelian group, we have $B = C_{p_1^{m_1}} \oplus \cdots \oplus C_{p_k^{m_k}}$. As $(C_{p_1} \oplus \cdots \oplus C_{p_k})^{\phi(k)} \rtimes \langle\{z\}\rangle$ is embeddable in G and A is divisible, G' must also be embeddable in G.

"(iii) ⇒ (iv)". $U \rtimes C_k = (C_p^{\phi(k)})^l \rtimes C_k$ is decomposable in $(U_1 \oplus \ldots \oplus U_n) \rtimes C_k$, where $U_i := \{a \in U;\ g_i(z) * a = 0\}$. Every U_i has $p^{\frac{\phi(k)}{n} \cdot l} - 1$ elements of order p (and thus is non-trivial) and is a subgroup of A_i.

"(iv) ⇒ (i)". Let $\exists \bar{x}\ \psi(\bar{g}, \bar{x})$ be an existential sentence with parameters $\bar{g}$ from G that is true in a group $H \geq G$. Let K be the finite subgroup of H that is generated by $\bar{g}$ and the fulfilling elements for $\bar{x}$. Let $U := G \cap K$. Let $Z' := C_H(Z)$. As Z is maximal as a locally cyclic π-group, it follows $Z' = Z$. We may assume that $K \cap Z \neq \{1\}$, because otherwise choose $1 \neq z \in Z$, and consider $\langle K \cup \{z\}\rangle$ instead of K. Let $A' := F(H)$ be the Fitting subgroup of H (thus $H = A' \rtimes Z$), $D := F(K)$ and $Y' := K \cap Z$ (thus $K = D \rtimes Y'$). Furthermore let $C := F(U)$ and $Y = U \cap Z$ (then $Y = Y'$, as $Y' = K \cap Z = K \cap (G \cap Z) = (K \cap G) \cap Z = U \cap Z = Y$). We get the following diagram.

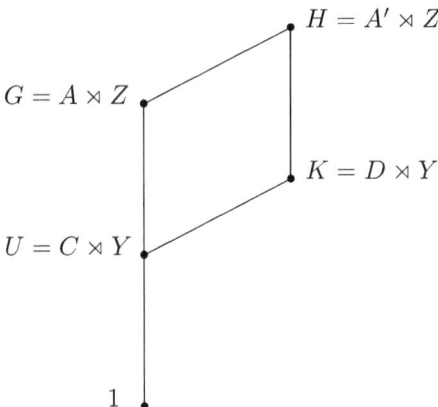

Let k be the order of Y. Let $p \in \pi$, and $C^{(p)}$ the p-subgroup of C, $D^{(p)}$ that of D. Then there are according to Proposition 13 decompositions

$C^{(p)} = C_1^{(p)} \oplus \ldots \oplus C_n^{(p)}$ and $D^{(p)} = D_1^{(p)} \oplus \ldots \oplus D_n^{(p)}$ with $C_i^{(p)} \leq D_i^{(p)}$ and $C_i^{(p)} < A_i$ for $i = 1, \ldots, n$. All A_i are divisible. (For $a \in A_i$ let $c \in A$ with $p \cdot c = a$. Then $c' := u_i(z)h_i(z) * c \in A_i$ and $p \cdot c' = a$.) Now let $i \in \{1, \ldots, n\}$. We have to find a subgroup in A_i that is isomorphic to $D_i^{(p)}$. The isomorphism shall fix $C_i^{(p)}$. We can assume that $D_i^{(p)} \cong C_{p^m}^{(s)}$ for some $m, s \in \omega$. That is possible, because H is embeddable in a divisible $\mathcal{C}$-group. With Proposition 17 we get a decomposition $C_i^{(p)} = U_1 \oplus \ldots \oplus U_l$ in indecomposable R_i-modules. According to Proposition 18 there are indecomposable R_i-modules $V_1, \ldots, V_l$ and a R_i-module V such that $U_j \leq V_j$ (for $j = 1, \ldots, l$) and $D_i^{(p)} = V_1 \oplus \ldots \oplus V_l \oplus V$. Now let v_j be a generating element of V_j ($j = 1, \ldots, l$) as a R_i-module. Let $u_j = p^{n_j} v_j$, with n_j minimal such that $u_j \in U_j$ and $w_j \in A_i$ with $u_j = p^{n_j} w_j$. The mapping

$$V_1 \oplus \ldots \oplus V_l \to \langle \{w_1\} \rangle \oplus \ldots \oplus \langle \{w_l\} \rangle$$

defined by $v_j \mapsto w_j$ for $j = 1, \ldots, l$ is a R_i-isomorphism. Because of (iv) there are infinitely many elements of order p in A_i. In particular there is a finite R_i-submodule of exponent p in A_i, which is disjoint to $\langle \{w_1\} \rangle \oplus \ldots \oplus \langle \{w_l\} \rangle$ and has cardinality $\geq |V|$. This submodule is embeddable in a finite submodule X of exponent $p^{|V|}$ in A_i, which is disjoint to $\langle \{w_1\} \rangle \oplus \ldots \oplus \langle \{w_l\} \rangle$. Obviously V is isomorphic to a submodule W of this module. Let W be a submodule of A_i that is isomorphic to V and disjoint to $\langle \{w_1\} \rangle \oplus \ldots \oplus \langle \{w_l\} \rangle$, then one gets an embedding

$$V_1 \oplus \ldots \oplus V_l \oplus V \to \langle \{w_1\} \rangle \oplus \ldots \oplus \langle \{w_l\} \rangle \oplus W \hookrightarrow A_i$$

The mapping fixes $C_i^{(p)}$. One can do the same for all $i = 1, \ldots, n$ and analogously for all prime numbers $p \in \pi$. Altogether one gets an embedding of K in G. That means that the sentence $\exists \bar{x}\, \psi(\bar{g}, \bar{x})$ is fulfilled in G. Hence G is existentially closed. q.e.d.

We give a second characterization of e.c. $\mathcal{C}$-groups.

Corollary 23. Let G be a $\mathcal{C}$-group. Then the following is equivalent:

(i) G is an e.c. $\mathcal{C}$-group.

(ii) G is divisible, and there is $\pi \subseteq \mathbb{P}$ such that every finite solvable CA-group of type π is embeddable in G.

Proof. "(i)$\Rightarrow$(ii)". The divisibility follows from Theorem 21. It follows from Theorem 22 that every finite solvable CA-group of type π is embeddable.

"(ii)⇒(i)". Let $A := [G, G]$. Then there is a locally cyclic subgroup $Z \leq G$ with $G = A \rtimes Z$. It is clear that G has to be of type π. For every $p \in \pi$ there must be infinitely many elements of order p in A. That is because for every π'-number k the group $(C_p)^{\phi(k)} \rtimes C_k$ is embeddable in G. As G and therefore also A are divisible, it follows $A \cong \bigoplus_{p \in \pi} Z_{p^\infty}{}^{(\kappa_p)}$. It is also easy to see that $Z \cong \bigoplus_{q \in \pi'} Z_{q^\infty}$. The rest follows with Theorem 22. q.e.d.

We now have determined the structure of existentially closed $\mathcal{C}$-groups quite clearly. The question now is how many non-isomorphic e.c. $\mathcal{C}$-groups of type π exist for fixed π. We will discuss this question in the following sections.

7   A Tree

Our next aim will be to determine the number of non-isomorphic countable e.c. $\mathcal{C}$-groups of type π. For that purpose we will construct a tree. The idea is to continue the decompositon introduced in Proposition 13. Let $G = A \rtimes Z$ be an e.c. $\mathcal{C}$-group of type π, and let k_1, k_2 be π'-numbers such that k_1 divides k_2. Let $z_i \in Z$ with order k_i for $i = 1, 2$. Then the decomposition of A with z_2 is a refinement of the decomposition with z_1. If we refine the decomposition step by step, we obtain a tree. We can also describe this tree with cyclotomic polynomials. We prefer this representation because it is independent of the chosen e.c. $\mathcal{C}$-group.

Definition 24. Let π be a non-empty set of primes, and $p \in \pi$. Let $\{q_n; n \in \omega\}$ be an enumeration of π' with the property that every prime occurs infinitely many times. Let $k_0 := 1$ and $k_n := \prod_{i<n} q_i$ for $n \geq 1$. Furthermore let L_n be the set of the normed irreducible divisors of the k_n-th cyclotomic polynomial over $\mathbb{Z}_p$ (respectively $\mathbb{Q}_p$). We define the following relation on $T := \bigcup_{n \in \omega} L_n$.

$$g \leq h \quad \Leftrightarrow \quad g \in L_n, h \in L_m, \text{ with } n \leq m,$$
$$\text{and } h(x) \text{ divides } g(x^{\frac{k_m}{k_n}}).$$

Let $\mathcal{T} = \mathcal{T}_\pi^{(p)} := \langle T, \leq \rangle$.

Proposition 25. The structure $\mathcal{T}$ is a tree, and L_n is the n-th level of $\mathcal{T}$.

Proof. We first show that $\mathcal{T}$ is partially ordered.

$\leq$ is reflexive. That is obvious.

$\leq$ is anti-symmetric. Let $g \leq h$ and $h \leq g$, $g \in L_n$, $h \in L_m$. Then $n = m$. That means that $g(x)$ divides $h(x)$, and $h(x)$ divides $g(x)$. As g and h are irreducible and normed, it follows that $g = h$.

$\leq$ is transitive. Let $g_1 \leq g_2$, $g_2 \leq g_3$, and $g_i \in L_{n_i}$ for $i = 1, 2, 3$. Let $r_1 := \frac{k_{n_2}}{k_{n_1}}$, $r_2 := \frac{k_{n_3}}{k_{n_2}}$, and $s := \frac{k_{n_3}}{k_{n_1}}$. Then there are polynomials $h_1, h_2 \in \mathbb{Z}_p[x]$ such that

$$g_1(x^{r_1}) = g_2(x) \cdot h_1(x) \quad \text{and} \quad g_2(x^{r_2}) = g_3(x) \cdot h_2(x).$$

It follows that

$$g_1(x^s) = g_1((x^{r_2})^{r_1}) = g_2(x^{r_2}) \cdot h_1(x^{r_2}) = g_3(x) \cdot h_2(x) \cdot h_1(x^{r_2}).$$

Hence $g_3(x)$ is a divisor of $g_1(x^s)$ and thus $g_1 \leq g_3$.

Now we prove that a element $h \in L_m$ has exactly one predecessor in every level L_n with $n < m$.

There is $g \in L_n$ such that $g \leq h$.

Let K be the splitting field of h over $\mathbb{Q}_p$ and $a \in K$ a zero of h. Then the multiplicative order of a is k_m. Let $r := \frac{k_m}{k_n}$. The element $b := a^r$ therefore has multiplicative order k_n, and hence is a zero of the k_n-th cyclotomic polynomial. Thus there is $g \in L_n$ with $g(b) = 0$. That means that a is a zero of $g(x^r)$. As a is also a zero of h, it follows that $h(x)$ is a divisor of $g(x^r)$, because h is irreducible.

Let $g_1, g_2 \in L_n$ with $g_1, g_2 \leq h$. Then $g_1 = g_2$.

Let again K be the splitting field of h over $\mathbb{Q}_p$ and $a \in K$ a zero of h. Let again $r := \frac{k_m}{k_n}$. Then $h(x)$ is a divisor of $g_1(x^r)$ and $g_2(x^r)$. Thus a is a zero of $g_1(x^r)$ and $g_2(x^r)$. The element $b := a^r$ is hence a zero of $g_1(x)$ and $g_2(x)$. If $g_1 \neq g_2$, then the polynoms can have no common zero. Therefore $g_1 = g_2$.

Thus the set of elements below an element h is well-ordered. The level L_0 consists only of the polynomial $x - 1$. This is the minimal element of the tree.

q.e.d.

We can consider three interpretations of the tree $\mathcal{T}$. As an order on the irreducible divisors of the k_n-th cyclotomic polynomials, as a decomposition of the Sylow-p-group of an e.c. $\mathcal{C}$-group of type π, or as an order on the k_n-th roots of unity in $\tilde{\mathbb{Q}}_p$.

First let us consider the relationship to the decomposition of the p-subgroup of an e.c. $\mathcal{C}$-group of type π. Let G be an e.c. $\mathcal{C}$-group of type π and $A^{(p)}$ the p-Sylow-group of G. Let $z_n \in G$ with order k_n and $z_{n+1} \in G$

with $z_{n+1}{}^{q_n} = z_n$. Furthermore let $g \in L_n$, $h \in L_{n+1}$. Let $V := \{a \in A^{(p)}; h(z_{n+1}) * a = 0\}$ and $U := \{a \in A^{(p)}; g(z_n) * a = 0\}$. Then

$$V \subseteq U \quad \Leftrightarrow \quad g \leq h.$$

That means that we can choose $\{z_n;\ n \in \omega\} \subseteq G$ such that z_n has order k_n and $z_{n+1}{}^{q_n} = z_n$ for all $n \in \omega$, and identify $g \in L_n$ with $U := \{a \in A^{(p)};\ g(z_n) * a = 0\}$. The ordering of $\mathcal{T}$ is then realized by the U in $A^{(p)}$. Notice that because of Theorem 22 none of the U is trivial.

We turn to the interpretation as an order on the roots of unity. Let $n \leq m$, let g be an irreducible divisor of the k_n-th and let h be an irreducible divisor of the k_m-th cyclotomic polynomial. Let $r := \frac{k_m}{k_n}$ and $\tilde{\mathbb{Q}}_p$ the algebraic closure of $\mathbb{Q}_p$. Then

$$g \leq h \quad \Leftrightarrow \quad \forall z \in \tilde{\mathbb{Q}}_p : h(z) = 0 \to g(z^r) = 0$$
$$\Leftrightarrow \quad \exists z \in \tilde{\mathbb{Q}}_p : h(z) = 0 \wedge g(z^r) = 0.$$

Let $E = E(\pi') := \{z \in \tilde{\mathbb{Q}}_p;\ \text{There is a } \pi'\text{-number } k \text{ with } z^k = 1\}$ the set of π'-roots of unity. As a multiplicative group E is isomorphic zu $\bigoplus_{q \in \pi'} Z_{q^\infty}$. On E we can define an equivalence relation R. For $z_1, z_2 \in E$ let

$z_1 \sim z_2 \quad \Leftrightarrow \quad$ There is a π'-number k and an irreducible divisor g
of the k-th cyclotomic polynomial with $g(z_1) = g(z_2) = 0$.

The equivalence classes of this relation can be mapped one-to-one to the set of irreducible divisors of the k-th cyclotomic polynomial for π'-numbers k. An order on E/R can be defined in the following way.

$$[z_1] \leq [z_2] \quad \Leftrightarrow \quad z_1 \text{ has order } r, z_2 \text{ has order } s,$$
$$r \text{ divides } s, \text{ and } z_2^{\frac{s}{r}} \in [z_1].$$

The order of the tree is a restriction of this order to the subset

$$\{[z];\ \text{The order of } z \text{ is } k_n \text{ for some } n \in \mathbb{N}\}.$$

Now let K be the closure of $\mathbb{Q}_p$ under the π'-roots of unity, i.e., $K = \mathbb{Q}_p(E)$. We consider E as a subgroup of the multiplicative group of the field K and define $\mathcal{A} := \text{Aut}(E)$. We next consider the set of field automorpisms of K that fix $\mathbb{Q}_p$, i.e., the Galois group $G(K : \mathbb{Q}_p)$ and define $\mathcal{U} := \{\phi \upharpoonright_E;\ \phi \in G(K : \mathbb{Q}_p)\}$ as a subgroup of $\mathcal{A}$. The equivalence classes of the relation R are exactly the orbits of E unter $\mathcal{U}$. Now the structure of $\mathcal{A}$ is easy to determine. As $E \cong \bigoplus_{q \in \pi'} Z_{q^\infty}$ and $\text{Aut}(Z_{q^\infty}) = \mathbb{Z}_q^*$ (the group of unities of q-adic integers), it follows that

$$\mathcal{A} \cong \prod_{q \in \pi'} \mathbb{Z}_q^*.$$

(Here $\prod$ is the cartesian product.) Hence $\mathcal{A}$ is abelian. One gets a homomorphism
$$f: \begin{array}{rcl} \mathcal{A}/\mathcal{U} & \to & \mathrm{Aut}(\mathcal{T}) \\ \phi \cdot \mathcal{U} & \mapsto & \phi': [z] \mapsto [\phi(z)]. \end{array}$$

As $\mathcal{A}$ is transitive on the sets of primitive k-th roots of unity, $\mathrm{Aut}(\mathcal{T})$ is also transitive on the single levels. In other words, we proved the following proposition.

Proposition 26. $\mathcal{T}$ is homogeneous, i.e., given two elements g_1, g_2 of a level, then g_1 can be mapped on g_2 by an isomorphism of the tree.   q.e.d.

We will need a number-theoretic proposition.

Proposition 27. Let π be a set of primes such that $\pi' := \mathbb{P} - \pi$ is finite, and $p \in \pi$. Let

$$k := \begin{cases} \prod\limits_{q \in \pi'} q & \text{if } 2 \notin \pi, \\ 2 \cdot \prod\limits_{q \in \pi'} q & \text{if } 2 \in \pi. \end{cases}$$

Let k' be a multiple of k and s the smallest natural number such that k' divides $p^s - 1$, and let r be the largest π'-number that divides $p^s - 1$. Then $q \cdot r$ is the largest π'-number that divides $p^{q \cdot s} - 1$. Moreover $q \cdot s$ is the smallest number t such that $q \cdot r$ is a divisor of $p^t - 1$.

Proof. Let $x \in \mathbb{N}$ with $p^s = 1 + xr$. It follows that

$$p^{qs} = (1 + xr)^q = \sum_{i=0}^{q} \binom{q}{i} x^i r^i,$$

respectively

$$p^{qs} - 1 = \sum_{i=1}^{q} \binom{q}{i} x^i r^i.$$

It follows directly that qr divides $qxr, r^2, \ldots, r^i, \ldots$, and thus also $\sum_{i=1}^{q} \binom{q}{i} x^i r^i$. That means that qr divides $p^{qs} - 1$. Now suppose that there is $q' \in \pi'$ such that $q'qr$ is a divisor of $p^{qs} - 1$. Suppose $q > 2$, then $q'qr$ is a divisor of $\frac{q(q-1)}{2} r^2$ and of r^i for $i > 2$. If $q = 2$, then according to our assumption 4 is also a divisor of r. It follows that $q'qr$ is a divisor of r^i for $i \geq 2$. Hence in both cases $q'qr$ must be a divisor of qxr, which means that q' is a divisor of x. That is a contradiction, because x is obviously a π-number. Now let t be minimal such that qr is a divisor of $p^t - 1$. According to the assumption s is the smallest number such that r is a divisor of $p^s - 1$, and hence $s < t$. Moreover s is a divisor of t and t a divisor of qs.

That is only possible for $t = qs$.   q.e.d.

Lemma 28. The tree $\mathcal{T}$ has finitely many branches if π' is finite, and $2^{\aleph_0}$ branches if π' is infinite.

Proof. To get a survey of the branches of $\mathcal{T}$, it is most convenient to examine the width of the tree. We make the following definitions. Let

$l_n := |L_n|$, the number of irreducible divisors of F_{k_n} over $\mathbb{Z}_p$, and

$\alpha_n := \min\{l \in \mathbb{N};\ k_n \text{ divides } p^l - 1\}$.

The number l_n can be calculated with the degree of an irreducible divisor. Every irreducible divisor of the k_n-th cyclotomic polynomial over $\mathbb{Q}_p$ has degree α_n. That is because the field $\text{GF}(p^{\alpha_n})$ with p^{α_n} elements is the smallest field of characteristic p that contains an element of multiplicative order k_n (and therefore the splitting field of the k_n-th cyclotomic polynomial over $\mathbb{F}_p$). Now let ϕ be the Euler function, then the number of irreducible divisors is $l_n := \frac{\phi(k_n)}{\alpha_n}$. We want to examine in which cases $l_{n+1} > l_n$. It is clear that $\alpha_n \leq \alpha_{n+1}$. Moreover α_n is a divisor of α_{n+1}. That is because the splitting field $\text{GF}(p^{\alpha_{n+1}})$ of the k_{n+1}-th cyclotomic polynomial is an extension of $\text{GF}(p^{\alpha_n})$.

We have to consider several cases.

Case 1. q_n divides k_n.
Then $\phi(k_{n+1}) = q_n \cdot \phi(k_n)$. Moreover α_{n+1} is a divisor of $q_n \cdot \alpha_n$. (That is analogous to Proposition 27.)

Subcase 1.a. $\alpha_{n+1} = \alpha_n$
Then k_{n+1} is already a divisor of $p^{\alpha_n} - 1$. It follows that $l_{n+1} > l_n$.

Subcase 1.b. $\alpha_{n+1} = q_n \cdot \alpha_n$
Then $l_{n+1} = \frac{\phi(k_{n+1})}{\alpha_{n+1}} = \frac{q_n \cdot \phi(k_n)}{q_n \cdot \alpha_n} = \frac{\phi(k_n)}{\alpha_n} = l_n$.

Case 2. q_n does not divide k_n.
Then $\phi(k_{n+1}) = (q_n - 1) \cdot \phi(k_n)$.

Subcase 2.a. α_n even and $q_n > 2$.
$k_{n+1} = q_n \cdot k_n$ divides $p^{\frac{(q_n-1)\cdot\alpha_n}{2}} - 1$, because $(p^{\frac{\alpha_n}{2}})^{q_n-1} \equiv 1 \pmod{q_n}$. Thus α_{n+1} is a divisor of $\frac{(q_n-1)\cdot\alpha_n}{2}$. We get $l_{n+1} = \frac{\phi(k_{n+1})}{\alpha_{n+1}} \geq \frac{(q_n-1)\cdot\phi(k_n)}{\frac{(q_n-1)\cdot\alpha_n}{2}} = \frac{2\cdot\phi(k_n)}{\alpha_n} > l_n$.

Subcase 2.b. α_n odd or $q_n = 2$.
From $l_{n+1} = l_n$ it follows that $\alpha_{n+1} = (q_n - 1) \cdot \alpha_n$. Thus α_{n+1} is even for $q_n > 2$.
Else $l_{n+1} > l_n$.

Now let π' be finite. Then for sufficiently large n, only Case 1 can occur. We show that Subcase 1.a. can happen only finitely often. Define $k' := k$, s and r as in Proposition 27. Let m be the smallest number such that r is a divisor of k_m. It follows inductively from Proposition 27 that $\alpha_n = \frac{k_n}{r} \cdot s$ for $n \geq m$. Hence $l_n = l_m$ for $n \geq m$.

For infinite π', Case 2 occurs infinitely often. It suffices to consider this case for $q_n > 2$. If $l_{n+1} = l_n$ and $q_n > 2$ in Subcase 2.b., then α_m is even for $m > n$. Hence we get for sufficiently large n always $l_{n+1} > l_n$, if Case 2 occurs.
q.e.d.

8 The Number of Non-Isomorphic e. c. $\mathcal{C}$-Groups

Theorem 29. If π' is infinite, then there are $2^{\aleph_0}$ non-isomorphic countable e.c. $\mathcal{C}$-groups of type π.

Proof. Let $p \in \pi$ and $\mathcal{T}$ be the tree defined in Definition 24. Let $\mathcal{S} = \langle S, \leq \rangle$ be a subtree of $\mathcal{T}$ without blind alleys and with the property that

$$\text{for all } a, b, c \in T, \text{ if } a \leq b \leq c \text{ and } a, c \in S, \text{ then } b \in S.$$

(In other words, every branch of $\mathcal{S}$ shall be a branch of $\mathcal{T}$.) Then we can construct a model-theoretic type $p^{\mathcal{S}}(x)$ in the following way. Let $n \in \omega$ and $g_{i_1}, \ldots, g_{i_m}$ the irreducible divisors of the k_n-th cyclotomic polynomial that lie in $\mathcal{S}$. Let $g(x), h_l(x) \in \mathbb{Z}[x]$ with $g(x) \equiv \prod_{j=1}^{m} g_{i_j}(x) \pmod{p}$ and $h_l(x) \equiv \prod_{j \neq l} g_{i_j}(x) \pmod{p}$ for $l = 1, \ldots, m$. Let $\phi_n^{\mathcal{S}}(x)$ be the formula

$$\phi_n^{\mathcal{S}}(x) \; :\rightleftharpoons \; x^p = 1 \; \wedge \; \exists y : y^{k_n} = 1 \; \wedge \; g(y) * x = 1 \; \wedge \; \bigwedge_{j=1}^{m} h_j(y) * x \neq 1.$$

For $g(y) = y^m + a_{m-1} y^{m-1} + \ldots + a_0$ the term $g(y) * x$ shall be an abbreviation for

$$(y^m)^{-1} \cdot x \cdot y^m \cdot \left((y^{m-1})^{-1} \cdot x \cdot y^{m-1} \right)^{a_{m-1}} \cdot \ldots \cdot x^{a_0}.$$

Set
$$p^{\mathcal{S}}(x) := \{\phi_n^{\mathcal{S}}(x); \; n \in \omega\}.$$

We have to show that $p^{\mathcal{S}}(x)$ is consistent. Let $n < m$. From $G \models \phi_m^{\mathcal{S}}(a)$ for some $a \in G$ it follows that $G \models \phi_n^{\mathcal{S}}(a)$. Let G be a $\mathcal{C}$-group and

$a \in G$ such that $G \vDash \phi_m^{\mathcal{S}}(a)$. Let z be a realization of y in $\phi_m^{\mathcal{S}}(a)$. Then $G \vDash \phi_n^{\mathcal{S}}(a)$, as z^{k_m/k_n} is a fulfilling element for y, because $\mathcal{S}$ is a subtree. The formula $\phi_n^{\mathcal{S}}(x)$ is realized in every e.c. $\mathcal{C}$-group. Let $z \in Z$ of order k_n and $A^{(p)} = A_1 \oplus \ldots \oplus A_l$ the decomposition of the p-Sylow-group according to Proposition 13. Choose $0 \neq a_i \in A_i$ with order p and set $a := a_{i_1} + \ldots + a_{i_m}$. Then $G \vDash \phi_n^{\mathcal{S}}(a)$, as z (inserted for y) satisfies the sentence.

Now let $\mathcal{S}_1 = \langle S_1, \leq \rangle$ and $\mathcal{S}_2 = \langle S_2, \leq \rangle$ be subtrees such that there is $n \in \omega$ with $|S_1 \cap L_n| \neq |S_2 \cap L_n|$. Then $p^{\mathcal{S}_1}(x)$ and $p^{\mathcal{S}_2}(x)$ cannot be realized by the same element. Suppose $G = A \rtimes Z$ is an e.c. $\mathcal{C}$-group of type π, and $a \in G$ realizes $p^{\mathcal{S}_1}(x)$ and $p^{\mathcal{S}_2}(x)$. Then there is $z_i \in Z$ that is a fulfilling element for $\phi_n^{\mathcal{S}_i}[a]$ ($i = 1, 2$). As z_1 and z_2 have the same order, there is a group automorphism of Z which maps z_1 on z_2. The induced tree automorphism of $\mathcal{T}$ then maps $S_1 \cap L_n$ on $S_2 \cap L_n$, a contradiction.

Let $\sigma := \{n \in \omega; a_n < a_{n+1}\}$, where a_n is defined as in Lemma 28. For $\tau \subseteq \sigma$ we define a subtree $\mathcal{S}_\tau$ recursively as follows. Suppose the m-th level $L_{\tau,m}$ of the tree is constructed for $m \leq n$. If $n \in \tau$, then the $n+1$-th level $L_{\tau,n+1}$ consists of all elements which are in $\mathcal{T}$ successors of an element of $L_{\tau,n}$. If $n \notin \tau$ choose for every element in $L_{\tau,n}$ exactly one successor. For any two of these subtrees there is obviously always a $n \in \omega$ such that the n-th level has distinct cardinality. As π' is infinite, σ is also infinite. It follows that there are $2^{\aleph_0}$ different types (over the empty set) in an e.c. $\mathcal{C}$-group of type π. For each of these types there is a countable e.c. $\mathcal{C}$-group of type π, in which the type is realized. On the other side each countable e.c. $\mathcal{C}$-group can of course only realize countably many types. q.e.d.

If π' is finite, one can also show that there are $2^{\aleph_0}$ non-isomorphic countable e.c. $\mathcal{C}$-groups of type π. We have to introduce some kind of dimension. Let $G = A \rtimes Z$ be a $\mathcal{C}$-group of type π. Let $(z_n)_{n \in \omega}$ be a sequence of elements in Z such that z_n has order k_n and $z_m^{\frac{k_m}{k_n}} = z_n$ holds for $n < m$. The tree $\mathcal{T}$ constructed as in Definition 24 has only finitely many branches. Thus one can decompose the p-subgroup $A^{(p)}$ of A in the following manner. Let $Z_i := (g_n^{(i)})_{n \in \omega}$ be a branch in $\mathcal{T}$. Set $A^{(i)} := \{a \in A^{(p)}; \forall n \in \omega : g_n^{(i)}(z_n) * a = 0\}$. One gets a decomposition

$$A^{(p)} = A^{(1)} \oplus \ldots \oplus A^{(m)}.$$

We now have to consider how "large" these blocks are as $\mathbb{Z}_p[Z]$-modules.

To this aim we consider $U_i := \{a \in A^{(i)}; p \cdot a = 0\}$. Let $K = \mathbb{F}_p(E)$ the closure of $\mathbb{F}_p$ under the k-th cyclotomic polynomials for all π'-numbers k. We claim that U_i is a K-vector space. It is clear that U_i is a module over the group ring $R := \mathbb{F}_p Z$. We can define the ideal

$$I_i := \{r \in R; \ \forall u \in U_i : r * u = 0\}$$

in R. The ideal I_i is a maximal ideal in R. To show that, let $s \in R - I_i$. As Z is locally cyclic, one can find $n \in \omega$, $z \in Z$ and λ_l for $l \leq n$ such that $s = \lambda_n z^n + \ldots + \lambda_1 z + \lambda_0$. Here z can be chosen such that $z = z_m$ for some $m \in \omega$. Let $f(x)$ be the polynomial of smallest degree in $\mathbb{F}_p[x]$ such that $f(z) * u = 0$ for all $u \in U_i$. According to the assumption f is irreducible and therefore has no common divisor with $\lambda_n x^n + \ldots + \lambda_1 x + \lambda_0$. Thus there exist polynomials $u, v \in \mathbb{F}_p[x]$ with $1 = u(x)f(x) + v(x)(\lambda_n x^n + \ldots + \lambda_1 x + \lambda_0)$. It follows that 1 is contained in the ideal generated by I_i and s. Hence $F_i := R/I_i$ is a field, and U_i is in a natural manner a F_i-vector space. Moreover F_i is isomorphic to the field K we described above.

The dimension of U_i over K can be finite or infinite in e.c. $\mathcal{C}$-groups, as one can easily verify by Theorem 22. As a conclusion we get the following theorem.

Theorem 30. *Let $\pi \neq \emptyset$ be a set of primes, which does not contain all primes. Then there are $2^{\aleph_0}$ non-isomorphic countable e.c. $\mathcal{C}$-groups of type π.*

Proof. For infinite π' the claim was proven in Theorem 29. Now let π' be finite. Then π is infinite. Like above one gets for each $p \in \pi$ a decomposition of the p-subgroup in finitely many direct summands. We can attach to each summand a dimension, which may be finite or countable. For any $\sigma \subseteq \pi$ there exists a e.c. $\mathcal{C}$-group G_σ of type π such that for $p \in \sigma$ the dimension of each direct summand in the decomposition is 1, and for $p \in \pi - \sigma$ the dimension is ω. The so constructed $2^{\aleph_0}$ groups are obviously pairwise non-isomorphic.　　　　　　　　　　　　　　　　　　　　　　　　　　　　　　q.e.d.

9 An Example

We want to give an example for the decomposition. Suppose you want to know how many finite solvable CA-groups of order $147 = 3 \cdot 7^2$ there are. Of course there are the abelian groups $C_3 \oplus C_{49}$ and $C_3 \oplus C_7 \oplus C_7$. Now let us turn to the non-abelian groups. It is clear that the Fitting subgroup must contain the elements of order 7. Thus $|F(G)| = 49$. A complement Z to $F(G)$ is isomorphic to C_3. We consider the cyclotomic polynomial $F_3(x) = x^2 + x + 1$ over $\mathbb{F}_7$. We get $F_3(x) = (x+5) \cdot (x+3) \pmod{7}$. Set $g_1(x) := x + 5$ and $g_2(x) := x + 3$. Let g_i' be the corresponding divisors of F_3 in $\mathbb{Z}_p[x]$ (i.e., $g_i'(x) \equiv g_i(x) \pmod 7$). We get two cases.

Case 1. $F(G) \cong C_7 \oplus C_7$.

Choose a generating element $z \in Z$ and decompose $F(G)$ according to Proposition 13.

Subcase 1. $A_1 \cong A_2 \cong C_7$.

Subcase 2. $A_1 = F(G)$ or $A_2 = F(G)$. These cases are isomorphic (choose z^2 instead of z).

Case 2. $F(G) \cong C_{49}$.

Then $A_1 = F(G)$ or $A_2 = F(G)$.

Thus we get three non-abelian groups.

Now let $\pi = \{7\}$. Let us take a look how $\mathcal{T}_\pi^{(7)}$ is constructed. We will take here the irreducible divisors in $\mathbb{F}_7[x]$. Suppose $q_0 = 3$, $q_1 = q_2 = q_3 = 2$ in the enumeration of the primes of π'. We have

$$\begin{aligned}
F_3(x) &= x^2 + x + 1 \equiv (x+5) \cdot (x+3) \pmod{7}, \\
F_6(x) &= x^2 - x + 1 \equiv (x+4) \cdot (x+2) \pmod{7}, \\
F_{12}(x) &= x^4 - x^2 + 1 \equiv (x^2+4) \cdot (x^2+2) \pmod{7}, \\
F_{24}(x) &= x^8 - x^4 + 1 \equiv (x^2+x+4) \cdot (x^2+2x+2) \\
&\quad \cdot (x^2+5x+2) \cdot (x^2+6x+4) \pmod{7}.
\end{aligned}$$

The tree looks as follows.

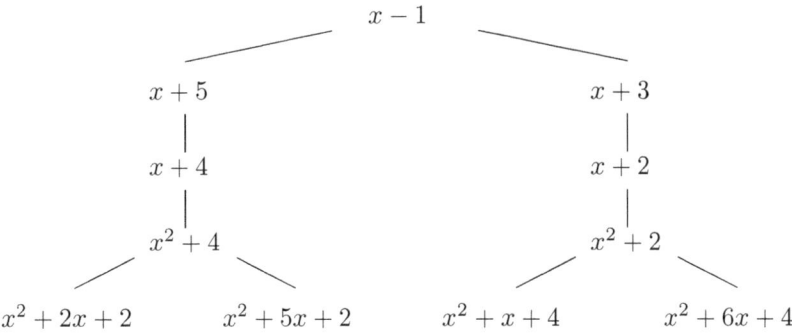

We can now examine how many non-abelian CA-groups exist that are isomorphic to $(C_7 \oplus C_7) \rtimes C_{24}$. We decompose $C_7 \oplus C_7 = A_1 \oplus A_2 \oplus A_3 \oplus A_4$ according to the irreducible divisors of F_{24} and a generating element $z \in C_{24}$.

Case 1. $A_1 \cong C_7 \oplus C_7$.

Isomorphic are the cases $A_j \cong C_7 \oplus C_7$ for $j = 2, 3, 4$.

Case 2. $A_1 \cong A_2 \cong C_7$.

Isomorphic is the case $A_3 \cong A_4 \cong C_7$.

Case 3. $A_1 \cong A_3 \cong C_7$.

Isomorphic is the case $A_2 \cong A_4 \cong C_7$.

Case 4. $A_1 \cong A_4 \cong C_7$.

Isomorphic is the case $A_2 \cong A_3 \cong C_7$.

Thus there are four non-isomorphic non-abelian groups of this type.

10 Some Remarks

There are still some questions left about the model theory of e.c. $\mathcal{C}$-groups. For example one could ask whether two e.c. $\mathcal{C}$-groups of type π are elementarily equivalent (i.e., they fulfill he same first-order sentences). If π is the set of all primes, then this is obviously true. But there are cases where this is not true, as we were able to show in [E99]. Of course two e.c. $\mathcal{C}$-groups of type π must fulfill the same existential sentences, as any finite solvable CA-group of type π is embeddable.

Another question concerns the stability of the e.c. $\mathcal{C}$-groups (in the sense of the stability theory of Morley and Shelah). If π' is infinite, then the decomposition from Section 4 supplies a descending chain of definable subgroups. According to the definable chain conditions in stable groups (see for example [$B_0$88, p. 92]), these groups cannot be superstable. In the case that π is finite we found something like a dimension to count the number of models. For the number of e.c. $\mathcal{C}$-groups of type π in the cardinality $\aleph_\alpha$ one gets $|\alpha + \omega|^{\aleph_0}$. This could be seen as a hint that these groups are superstable or even ω-stable. However, if we take the first-order theory T of an e.c. $\mathcal{C}$-group of type π, there are of course models of T, in which the Fitting subgroup, as well as any complement to the Fitting subgroup, contains elements of infinite order. Thus there are much more models.

References.

[$B_0$88] J. T. BALDWIN: *Fundamentals of Stability Theory.* Perspectives in Mathematical Logic, Springer Verlag, Berlin, 1988.

[B_1SW58] R. BRAUER, M. SUZUKI, G. E. WALL: *A characterisation of the one-dimensional unimodular projective groups over finite fields.* Illinois Journal of Mathematics, **2**, (1958), 718–745.

[E99] A. ECKER: *Zur Modelltheorie der CA-Gruppen.* Dissertation, University of Tübingen, 1999.

[F70]   L. FUCHS: *Infinite Abelian Groups*. Academic Press, New York, 1970.

[GK$_0$M95]   D. GILDENHUYS, O. KHARLAMPOVICH, A. MYASNIKOV: *CSA-groups and separated free constructions*. Bulletin of the Australian Mathematical Society, **52**, (1995), 63–84.

[H$_0$08]   K. HENSEL: *Theorie der algebraischen Zahlen*. Teubner, Leipzig-Berlin, 1908.

[H$_1$93]   W. HODGES: *Model Theory*, vol. 42 of *Encyclopedia of mathematics and its applications*. Cambridge University Press, Cambridge, 1993.

[JH$_2$04]   E. JALIGOT, A. O. HOUCINE: *Existentially closed CSA-groups*. Journal of Algebra, **280**, (2004), 772–796.

[K$_1$91]   P. KISSEL: *Beiträge zur Modelltheorie der linearen und der symplektischen Gruppen*. Dissertation, University of Tübingen, 1991.

[MR96]   A. G. MYASNIKOV, V. N. REMESLENNIKOV: *Exponential groups II: Extensions of centralizers and tensor completion of CSA-groups*. International Journal of Algebra and Computation, **6**, (1996), 687–711.

[S57]   M. SUZUKI: *The nonexistence of a certain type of simple group of odd order*. Proceedings of the American Mathematical Society, **8**, (1957), 686–695.

[W$_0$25]   L. WEISNER: *Groups in which the normaliser of every element except identity is abelian*. Bulletin of the American Mathematical Society, **31**, (1925), 413–416.

[W$_1$98]   Y.-F. WU: *Groups in which commutativity is a transitive relation*. Journal of Algebra, **207**, (1998), 165–181.

Received: October 3, 2005;
In revised version: May 22, 2006;
Accepted by the editors: December 6, 2006.

On Stable Groups in Some Soluble Group Classes

CLAUS GRÜNENWALD
FRIEDER HAUG[*]

[*] Mathematisches Institut
Eberhard-Karls-Universität Tübingen
Auf der Morgenstelle 10
72076 Tübingen, Germany
E-mail: frieder.haug@uni-tuebingen.de

> ABSTRACT. We show that soluble minimax groups and finitely generated soluble groups are unstable, unless they are abelian-by-finite. On the other hand we give an example of a superstable soluble group of finite rank which is not abelian-by-finite.

1 Introduction

The purpose of this paper is to study stable groups in several classes of soluble groups. Recently it was shown that stable pseudofinite groups are finite extensions of soluble groups [M_0T06]. Many examples of stable soluble groups are nilpotent-by-abelian-by-finite. We are therefore interested in classes of groups defined by finiteness conditions which are related to this class. A group G is *soluble* (of derived length n) if it has an increasing series

$$(*) \qquad 1 = A_0 \trianglelefteq A_1 \trianglelefteq \cdots \trianglelefteq A_n = G$$

of normal subgroups such that all factors A_{i+1}/A_i are abelian. If $n = 2$ then G is called *metabelian*. The group G is a *soluble minimax* group if there exists such a series $(*)$ in G in which all the abelian factors satisfy the maximal condition for subgroups (max) or the minimal condition for subgroups (min). Equivalently there exists a finite series $(*)$ of G such that the factors are cyclic or quasicyclic. We say that a soluble group G has *finite rank* if in such a series $(*)$ of G all the abelian factors do have finite total rank. Here the total rank of an abelian group is the sum of the torsion-free rank and all p-ranks for primes p (see [$F_1$70, §16] for this notation). A soluble group G is called *polycyclic*, if in such a series $(*)$ of G all the factors are cyclic. Equivalently, G satisfies max. Finally we say that a group G satisfies the maximal condition for normal subgroups (max-n) if G does not have an infinite increasing sequence of normal subgroups. These classes are connected in the following way (for proofs see [R₀72] or [R₀82]). Polycyclic groups satisfy max-n and they are finitely generated of finite rank. Soluble groups with max-n are finitely generated and finitely generated metabelian groups satisfy max-n. Finitely generated soluble groups of finite rank are minimax groups and soluble minimax groups do have finite rank. Soluble groups of finite rank are nilpotent-by-abelian-by-finite. We show the following three theorems:

Theorem A. *If G is a stable soluble minimax group, then G is abelian-by-finite.*

Theorem B. *If G is a stable finitely generated soluble group, then G is abelian-by-finite.*

Theorem C. *There exists a soluble group of finite rank which is superstable but not abelian-by-finite.*

Theorem B follows already from a result of Noskov [N84] who proved that the theory of a finitely generated soluble group which is not abelian-by-finite is undecidable, by interpreting arithmetic in such a group. But we hope that our proof of just unstability of these groups is simpler then the proof of this more general result of Noskov (which is based on [R₁80, E₀72]).

We would like to thank O. V. Belegradek for the hint on [N84], J. Wilson for the idea to generalize Corollary 12 to Theorem B and B. Zilber for the manuscript [Z93], which is the basis for the proof of Theorem C. The Sections 2 and 3 are essentially due to the first author, Sections 4 and 5 to the second author.

2 Stable actions.

In this section we prepare the proof of Theorem A by showing that, roughly speaking, the action of an element of infinite order on a torsion-free abelian minimax group is unstable. We will give a more precise statement of this fact later. It is easy to see that an abelian group A is a minimax group if and only if there is a finitely generated subgroup $X \leq A$ such that A/X is a direct product of finitely many quasicyclic groups. The finite set of primes which divide the orders of elements of A/X is called the *spectrum* $\mathrm{Sp}(A)$ of A (this and some of the following facts can be found in [R₀72, Chapter 10.3]). If A is in addition a torsion-free abelian minimax group then A has finite (torsion-free) rank, i.e., A is isomorphic to a subgroup of $\mathbb{Q}^r$ for some $r \in \mathbb{N}$. The smallest such r is called the *rank* $\mathrm{rk}(A)$ *of* A. If π is a set of primes then define the subring $\mathbb{Q}_\pi = \{\frac{m}{n} \in \mathbb{Q};\ n \text{ is a } \pi\text{-number}\}$ of $\mathbb{Q}$. Here $n \in \mathbb{N}$ is called a π-*number*, if all its prime divisors are in π. If A is a torsion-free abelian minimax group with rank r and $\mathrm{Sp}(A) = \pi$, then A is embeddable into $\mathbb{Q}_\pi^r$ by the mapping $A \to \bar{A} = A \otimes_\mathbb{Z} \mathbb{Q}_\pi \cong \mathbb{Q}_\pi^r, a \mapsto a \otimes 1$.

We assert that if A is a torsion-free abelian minimax group and $\theta \in \text{Aut}(A)$ has infinite order, then

$$A \geq (\theta^2 - 1)A \geq (\theta^4 - 1)A \geq \ldots \geq (\theta^{2^n} - 1)A \geq \ldots$$

is an infinitely descending chain of subgroups of A (here 1 denotes the identity mapping and $(\theta^n - 1)A = \{\theta^n(a) - a;\ a \in A\}$). From this descending chain we can deduce the instability of the action of θ on A. Let us consider some examples:

(1) Let $A \cong \mathbb{Z} \times \mathbb{Z}$ and $\theta \in \text{Aut}(A)$ is given by the matrix $\theta = \begin{pmatrix} 1 & 2 \\ 1 & 1 \end{pmatrix}$ with respect to a suitable basis of A. Then the semidirect product $A \rtimes \langle \theta \rangle$ of A with $\langle \theta \rangle$ is polycyclic. If we embed $A \rtimes \langle \theta \rangle$ into $K^+ \rtimes K^*$, where $K = \mathbb{Q}(\sqrt{2})$ by mapping A onto the additive group of the ring of integers of K we see that θ acts on A like multiplication with $1 + \sqrt{2}$. Now it is easy to see that the above chain $(\theta^{2^n} - 1)A$, $n \in \mathbb{N}$, is infinitely descending.

(2) Let A be the additive group $\mathbb{Q}_2 = \{m \cdot 2^n;\ m, n \in \mathbb{Z}\}$ and let θ act on A by multiplication with 2. The group $A \rtimes \langle \theta \rangle$ is finitely generated, so there is a bound on the length of chains of centralizers. Here it is even more obvious that the chain $(\theta^{2^n} - 1)A$, $n \in \mathbb{N}$, is infinitely descending.

Similar to these examples we are looking for an embedding of $A \rtimes \langle \theta \rangle$ into $K^+ \rtimes K^*$ for some field K such that via this embedding we can show the descent of the chain $(\theta^{2^n} - 1)A$, $n \in \mathbb{N}$. Therefore we consider the subring E_θ of $\text{End}(A)$ generated by $\{1, \theta\}$ and show that E_θ can be embedded in a ring of S-integers of some algebraic number field K. To this purpose we introduce some notation.

Let K be an algebraic number field. A *valuation* on K is a function $v : K \to \mathbb{R}$ with the properties

A1. $v(x) \geq 0$ for all $x \in K$ and $v(x) = 0$ iff $x = 0$;

A2. $v(x \cdot y) = v(x) \cdot v(y)$;

A3. $v(x + y) \leq v(x) + v(y)$.

If v satisfies even $v(x + y) \leq \max\{v(x), v(y)\}$ (instead of A3), then v is *non-archimedian*, otherwise it is *archimedian*. Two valuations on K are called equivalent, if they induce the same topology on K. An equivalence class of non-trivial valuations on K is called a *prime* on K. The primes containing non-archimedian valuations are called *finite primes*, the others are *infinite primes*. Usually we denote the finite prime on $\mathbb{Q}$ containing the p-adic valuation by the letter $\mathcal{P}$. The set of all primes on K is denoted by $M(K)$. Let L be a finite extension field of the field K and v_1, v_2 primes of K and L respectively. If all valuations in v_2 are continuations of valuations in v_1, then we say that v_2 lies above v_1 and denote this by $v_2 \mid v_1$. Let S be a finite set of primes on K containing all infinite primes. We define

$$\mathcal{O}_S(K) = \{\alpha \in K \,;\, v(\alpha) \leq 1 \quad \text{for all} \quad v \in M(K) \setminus S\};$$

$$U_S(K) = \{\alpha \in K \,;\, v(\alpha) = 1 \quad \text{for all} \quad v \in M(K) \setminus S\}.$$

Then $\mathcal{O}_S(K)$ is the ring of S-integers and $U_S(K)$ is the (multiplicative) group of S-units of K. We need the following proposition.

Proposition 1. Let K be an algebraic number field of degree m, let λ, μ be nonzero elements of K and let S be a finite set of primes on K of cardinality s containing the infinite primes. Then the equation

$$\lambda x + \mu y = 1$$

has at most $3 \cdot 7^{m+2s}$ solutions in S-units $x, y \in U_S(K)$. [E$_1$84, Theorem 1]

Lemma 2. Let K be an algebraic number field and let S be a finite set of primes on K containing all infinite primes. Let $f(x) = a_n x^n + a_{n-1} x^{n-1} + \cdots + a_1 x + a_0$ be a polynomial with coefficients $a_i \in \mathcal{O}_S(K)$ and let $\gamma \in K$ be a zero of f. If $a_n \in U_S(K)$ then $\gamma \in \mathcal{O}_S(K)$ and if $a_n, a_0 \in U_S(K)$ then $\gamma \in U_S(K)$.

Proof. From $f(\gamma) = 0$ it follows

$$a_n = -\left(\frac{a_{n-1}}{\gamma} + \frac{a_{n-2}}{\gamma^2} + \cdots + \frac{a_0}{\gamma^n}\right).$$

Let v be a prime on K such that $v \notin S$ and suppose $a_n \in U_S(K)$. Since v is non-archimedean we get

$$1 = v(a_n) \leq \max\left\{\frac{v(a_{n-1})}{v(\gamma)}, \frac{v(a_{n-2})}{v(\gamma)^2}, \ldots, \frac{v(a_0)}{v(\gamma)^n}\right\}.$$

Now $v(a_i) \leq 1$ for $0 \leq i \leq n-1$, thus it follows that $v(\gamma) \leq 1$ and $\gamma \in \mathcal{O}_S(K)$. If $a_n, a_0 \in U_S(K)$ then $\gamma \in \mathcal{O}_S(K)$. But γ^{-1} is a zero of the polynomial $a_0 x^n + a_1 x^{n-1} + \cdots + a_n$, hence $\gamma^{-1} \in \mathcal{O}_S(K)$. Therefore γ is invertible in $\mathcal{O}_S(K)$, thus $\gamma \in U_S(K)$. q.e.d.

We return to the problem which we stated at the beginning of this section: we want to show that the chain $(\theta^{2^n} - 1)A$, $n \in \mathbb{N}$, is infinitely descending if θ is an automorphism of infinite order of the torsion-free abelian minimax group A. To do this we have to compute the indices $|A : (\theta^{2^n} - 1)A|$. We shall do this in the following lemma. If π is a set of primes and $r \in \mathbb{N}$ then we write $\pi(r)$ for the biggest π-number which divides r and $\pi'(r) = r/\pi(r)$. Remember the remarks on the embeddability of A into $\mathbb{Q}_\pi^r$ for $\pi = \mathrm{Sp}(A)$ and $r = \mathrm{rk}(A)$ which we made at the beginning of this section. In particular A can be embedded into the vector space $\mathbb{Q}^r$. Each endomorphism τ of A has a unique continuation to an endomorphism of $\mathbb{Q}^r$. With this in mind we can define the characteristical polynomial χ_τ, the minimal polynomial m_τ and the determinant $\det \tau$ for any $\tau \in \mathrm{End}(A)$ as the corresponding ones of the continuation of τ to $\mathbb{Q}^r$. If $\pi = \mathrm{Sp}(A)$ then $\det \tau \in \mathbb{Q}_\pi$ and $\chi_\tau, m_\tau \in \mathbb{Q}_\pi[x]$.

Lemma 3. Suppose A is a torsion-free abelian minimax group. If $\tau \in \mathrm{End}(A)$ is injective, then $\infty > |A : \tau(A)| \geq \pi'(\det \tau)$.

Proof. Since A does have finite rank, $|A : \tau(A)|$ is finite (see [R$_0$82, 15.2.3]). Let $\bar{A} = A \otimes_{\mathbb{Z}} \mathbb{Q}_\pi \cong \mathbb{Q}_\pi^r$, where $\pi = \mathrm{Sp}(A)$ and $r = \mathrm{rk}(A)$. Suppose $\bar{\tau}$ is the unique continuation of τ to $\bar{A}$. Then $\bar{\tau}$ is injective and $|A : \tau(A)| \geq |A : A \cap \bar{\tau}(\bar{A})| = |A + \bar{\tau}(\bar{A}) : \bar{\tau}(\bar{A})|$. Furthermore $\bar{A} = A + \bar{\tau}(\bar{A})$ since $\bar{A}/\bar{\tau}(\bar{A})$ is a π'-group and $\bar{A}/A$ is a π-group. Therefore $|A : \tau(A)| \geq |\bar{A} : \bar{\tau}(\bar{A})|$. The exact value of the index $|\bar{A} : \bar{\tau}(\bar{A})|$ follows from the so called "*Elementarteilersatz*" (see for example [v67, §85] or the proof of [F$_1$70, Ex. 15.22] with the help of [F$_1$70, Lemma 15.4]). It says that there are automorphisms $\sigma \in \mathrm{Aut}(\bar{A})$, $\rho \in \mathrm{Aut}(\bar{\tau}(\bar{A}))$ and an isomorphism ϕ from $\bar{A}$ to $\bar{\tau}(\bar{A})$ such that $\bar{\tau} = \rho \phi \sigma$ and ϕ is diagonalizable with respect to a suitable chosen basis of $\bar{A}$. Since $\pi'(\det \tau) = \pi'(\det \bar{\tau}) = \pi'(\det \phi)$ it follows $|\bar{A} : \bar{\tau}(\bar{A})| = \pi'(\det \tau)$ and the lemma is proved. q.e.d.

Let K be an algebraic number field and let π be a finite set of prime numbers. Then $S = S(\pi)$ denotes the set of all infinite primes on K in addition with all primes on K which lie above some prime p on $\mathbb{Q}$ with $p \in \pi$.

Proposition 4. Let A be a torsion-free abelian minimax group. Suppose $\theta \in \mathrm{Aut}(A)$ has infinite order and m_θ is irreducible (over $\mathbb{Q}$). Define the subgroups $B_n = (\theta^{2^n} - 1)A = \{\theta^{2^n}(a) - a;\ a \in A\}$ $(n \in \mathbb{N})$. Then $A \geq B_0 \geq B_1 \geq \cdots \geq B_n \geq \cdots$ and this chain has an infinite strictly descending subchain.

Proof. If $\theta \in \mathrm{Aut}(A)$ has infinite order and if m_θ is irreducible, then

$(m_\theta, q_n) = 1$, where $q_n(x) = x^n - 1$, for all $n \in \mathbb{N}$. By the Lemma of Bézout, it follows that $\theta^{2^n} - 1$ is a monomorphism of A. Now $\theta^{2^{n+1}} - 1 = (\theta^{2^n} + 1)(\theta^{2^n} - 1)$, thus $B_{n+1} = (\theta^{2^n} + 1)B_n \leq B_n$ and $\theta^{2^n} + 1$ is also injective. We will show that $|B_n : B_{n+1}| > 1$ for almost all $n \in \mathbb{N}$. By Lemma 3, we have

$$|B_n : B_{n+1}| \geq \pi'(\det(\theta^{2^n} + 1)).$$

Let $\alpha \in \mathbb{C}$ be a zero of m_θ and put $K = \mathbb{Q}(\alpha)$, $\pi = \mathrm{Sp}(A)$, $S = S(\pi)$ and $r = \mathrm{rk}(A)$. Let

$$E_\theta = \{\sum_{i=0}^{k} z_i \theta^i \ ; \ k \in \mathbb{N}, \ z_i \in \mathbb{Z} \text{ for } 0 \leq i \leq k\} = \{f(\theta) \ ; \ f \in \mathbb{Z}[x]\},$$

be the subring of $\mathrm{End}(A)$ which is generated by θ. Since m_θ is irreducible, the map $\varphi : E_\theta \to K$, $f(\theta) \mapsto f(\alpha)$ is a ring monomorphism of E_θ into K. If $\pi'(\det(\theta^{2^n} + 1)) = 1$, then $\det(\theta^{2^n} + 1)$ is a unit of $\mathbb{Q}_\pi \subseteq \mathcal{O}_S(K)$. Now $\theta^{2^n} + 1$ is a zero of its characteristical polynomial $x^r + \cdots + (-1)^r \det(\theta^{2^n} + 1)$. By Lemma 2, $\varphi(\theta^{2^n} + 1) \in U_S(K)$. It follows that $\varphi(\theta^{2^n} + 1)$, $\varphi(\theta^{2^n})$ are S-unit solutions of the equation $x - y = 1$. By Proposition 1, there are only finitely many solutions of this equation. Since θ has infinite order, we would have infinitely many solutions if $\pi'(\det(\theta^{2^n} + 1)) = 1$ for infinitely many n. Therefore $|B_n : B_{n+1}| = 1$ only for finitely many n and the proposition is proved. q.e.d.

Corollary 5. Let G be a group and let A be a torsion-free abelian minimax group definable in G. Suppose $\theta \in G$ acts on A by conjugation as an automorphism of infinite order. Then G is unstable.

Proof. For $a \in A$ we write θa instead of $a^\theta = \theta^{-1} a \theta$ and we define for $n \in \mathbb{N}$ the subgroups $B_n = (\theta^{2^n} - 1)A$ as in Proposition 4. Since θ has infinite order, the minimal polynomial m_θ has an irreducible factor p such that for all $q_n(x) = x^n - 1$ we have $(p, q_n) = 1$ and the subgroup $\{x \in A; \ p(\theta)x = 0\}$ is a definable torsion-free abelian minimax group. So we may assume that m_θ itself is irreducible (over $\mathbb{Q}$). From Proposition 4 it follows that the chain $A \geq B_0 \geq B_1 \geq \cdots \geq B_n \geq \cdots$ is infinitely descending. But the groups $(h-1)A = \{a^h - a; \ a \in A\}$ ($h \in G$) are uniformly definable with parameter $h \in G$. But a stable group cannot have an infinitely descending chain of subgroups of this form (see for example [F$_0$78, Lemma 4]). q.e.d.

3 Soluble minimax groups.

First we investigate stable nilpotent minimax groups. Together with the results of the previous section this will prove Theorem A. We fix some standard notation. Let G be a group. Then $Z(G)$, $Z_i(G)$, G', $\Gamma_i(G)$ denote the *center*, the i-th term of the *upper central series*, the *commutator subgroup*, and the i-th term of the *lower central series* of G, respectively. We write $C_G(x)$ for the *centralizer* of $x \in G$ in G and $C_G(X)$ for the centralizer of the subset $X \subseteq G$. Furthermore Fit(G) is the *Fitting subgroup* of G. This is is the subgroup generated by all nilpotent normal subgroups of G. The descending chain condition on centralizers will be denoted by min-c. In a group G with min-c there is a unique (and definable) *centralizer-connected* subgroup G^{cc} of finite index, i.e., G^{cc} has no centralizer of finite index > 1. G^{cc} is called the centralizer-connected component of G.

It is known that in any group G the finiteness of $G/Z(G)$ implies that G' is finite. Here, we need the converse.

Lemma 6. Let G be a soluble minimax group. If G' is finite, then $G/Z(G)$ is finite.

Proof. If G' is finite, then G is an FC-group. This means that all sets of conjugacy-classes in G are finite. By [R_082, 14.5.6], $G/Z(G)$ is a residually finite torsion group. Since G is a soluble minimax group, $G/Z(G)$ has a finite abelian series $1 = A_0 \trianglelefteq A_1 \trianglelefteq \cdots \trianglelefteq A_n = G/Z(G)$ such that all factors are cyclic or quasicyclic. If $G/Z(G)$ is a residually finite torsion group, then all these factors have to be finite cyclic and $G/Z(G)$ itself is finite. q.e.d.

Lemma 7. Let G be a nilpotent group of class 2 such that $G/Z(G)$ is torsion-free. If there are $x, y \in G \setminus Z(G)$ such that $Z(C_G(x)) \cap C_G(y) = Z(G)$ and $xZ(G)$ is not p-divisible in $G/Z(G)$ for some prime p, then G is unstable.

Proof. This is exactly the second part of the proof of [GH93, Theorem 4.1]. q.e.d.

Proposition 8. Let G be a stable nilpotent minimax group. Then G' and $G/Z(G)$ are finite.

Proof. Let us suppose for a moment that the proposition holds for all stable nilpotent minimax groups of class 2. Then we argue by induction on the nilpotency class n of G: let $n > 2$. By assumption, $G/Z_{n-1}(G)$ is finite, thus $\Gamma_n(G)$ is finite [$R_0 82$, 14.5.1]. Now $G/\Gamma_n(G)$ has nilpotency class $n-1$ and the induction hypothesis shows that $(G/\Gamma_n(G))' = G'/\Gamma_n(G)$ is finite. From this and Lemma 6 it follows that G' and $G/Z(G)$ are finite. Now we have to show that the proposition holds in the case $n = 2$. Let G be a nilpotent minimax group of class 2 with min-c such that $G/Z(G)$ is infinite. Then $G/Z(G)$ is not periodic (since G is minimax) and we can find $x, y \in G \setminus Z(G)$ with the following properties: for all $k \in \mathbb{N}$ it is $C_G(x) = C_G(x^k)$ and $C_G(y) = C_G(y^k)$ and $[x, y]$ has infinite order. It is easy to see that $x, y \in G^{\infty}$. Now put $H = G^{\infty}$ and $Z = \mathbb{Z}(C_H(x)) \cap C_H(y)$. Obviously we have $\mathbb{Z}(H) \leq Z$. Since G has min-c we may assume that $\mathbb{Z}(H) = Z$ (otherwise we can repeat the procedure by setting $G_1 = C_H(Z) < G$; this must terminate after a finite number of steps). Now H, x, y satisfy the conditions of Lemma 7. Since $H/\mathbb{Z}(H)$ is a torsion-free abelian minimax group there must be a prime p such that $x\mathbb{Z}(H)$ is not p-divisible. This shows that H is unstable. Obviously H is definable. It follows that if G is stable, then $G/Z(G)$ has to be finite. q.e.d.

We shall prove Theorem A by showing that in a stable soluble minimax group G the center of the Fitting-subgroup has finite index in G. In order to do this we need some further facts.

Fact 1. The Fitting subgroup of a soluble minimax group is nilpotent (see [$R_0 67$]).

Fact 2. Let N be a nilpotent subgroup of the stable group G. Then N is contained in a definable nilpotent subgroup H of G of the same nilpotency class as N. If N is normal in G, then H is normal in G (see [P87, 3.17]).

From Fact 2 we conclude that if $\mathrm{Fit}(G)$ is nilpotent in a stable group G, then $\mathrm{Fit}(G)$ is definable. Thus it follows from Fact 1 and Proposition 8 that if G is a stable soluble minimiax group, then $(\mathrm{Fit}(G))'$ is finite. This is the reason why we study in the Lemma 9 minimax groups with abelian Fitting subgroup.

Lemma 9. Let G be a stable soluble minimax group such that $F = \mathrm{Fit}(G)$ is abelian. Then G/F is periodic.

Proof. Suppose that G/F is not periodic. We have $F = C_G(F)$. Therefore there is an element $g \in G$ such that $\langle g \rangle \cap F = 1$ and such that for all $k \in \mathbb{N}$ we have $C_G(g^k) = C_G(g)$ (because G has min-c and min-c is equivalent to max-c, we may choose g such that $C_G(g)$ is maximal). Let $T = T(F)$ be the torsion subgroup of F. Then $T \triangleleft G$ and $G/C_G(T)$ is finite: because G is minimax and has min-c we can find a finitely generated and therefore finite normal subgroup E of T with $C_G(E) = C_G(T)$. But $G/C_G(E)$ is isomorphic to a subgroup of the finite group Aut(E). It follows in particular from the choice of g that $g \in C_G(T)$, i.e., $T \leq C_G(g)$. Furthermore $F/C_F(g)$ is torsion-free: if $f \in F$ such that $f^n \in C_F(g)$ for some $n \in \mathbb{N}$, then $1 = [f^n, g] = [f, g]^n$ since F is abelian. Thus $[f, g] \in T \leq C_F(g)$. By induction it follows that $[f, g^k] = [f, g]^k$ for all $k \in \mathbb{N}$. So we conclude that $f \in C_F(g^n) = C_F(g)$.

The action of g^k on the torsion-free abelian minimax group $F/C_F(g)$ by conjugation induces an automorphism of $F/C_F(g)$ for all $k \in \mathbb{N}$. We have to consider two cases.

If g^n acts trivially on $F/C_F(g)$ for some $n \in \mathbb{N}$, then $H = \langle g^n, F \rangle$ is a nilpotent subgroup of G of class two. If G were stable, then by Fact 2 and Proposition 8 we conclude that H' is finite. This implies that $g^l \in C_G(F) = F$ for some $n \leq l \in \mathbb{N}$ — a contradiction to the choice of g.

If g induces an automorphism of infinite order on $F/C_F(g)$, we can apply Corollary 5 which shows that G is unstable. Thus the lemma is proved.

q.e.d.

Proof of Theorem A. If G is a stable soluble minimax group and $F = \mathrm{Fit}(G)$, then F' is finite as we remarked before Lemma 9. From Hall's criterion for nilpotence (see for example [R$_0$82, 5.2.10]) it follows that $\mathrm{Fit}(G/F') = F/F'$. Now we apply Lemma 9 and conclude that G/F is periodic. Thus G/F is a Cernikov-group. This means G/F is a finite extension of a finite direct sum of quasicyclic groups [R$_0$82, 5.4.23]. But by [R$_0$68, Theorem B iii], G/F is residually finite. This implies that G/F is finite. Since $F/Z(F)$ is finite (Lemma 6), the theorem is proved.

q.e.d. (Theorem A)

4 Finitely generated soluble groups.

In this section we shall prove Theorem B. To do this we restrict ourselves first to the case of finitely generated *metabelian* groups. Then the general result follows by induction. Remember that finitely generated metabelian groups satisfy the maximal condition for normal subgroups by a result of P. Hall (see [R$_0$82, 15.3.1]).

So we have to study a group G satisfying max-n which is an extension of an abelian group A by a finitely generated abelian group Q. The proof splits essentially into three cases: either A is torsion-free of *finite rank* (i.e., $A \leq \mathbb{Q}^r$ for some $r \in \mathbb{N}$); in this case G is a soluble minimax group and Theorem B follows from Theorem A. Or A is torsion-free of *infinite* rank; these groups are studied in Lemma 10. Or A is an elementary abelian p-group (see Lemma 11).

Lemma 10. Let G be an extension of a torsion-free abelian group of infinite rank by a torsion-free abelian group $Q \neq 1$. If G satisfies max-n, then G is unstable.

Proof. We write A additively, Q and G multiplicatively. Furthermore, if $b \in A$ and $h \in Q$, then we write hb instead of $b^h = h^{-1}bh$. If $x = h_1 + \cdots + h_n$ is an element of the group ring $\mathbb{Z}Q$ of Q ($h_1, \ldots, h_n \in Q$), then we write xb for $b^{h_1} + \cdots + b^{h_n} = h_1 b + \cdots + h_n b$. We may assume that A is a definable subgroup of G, otherwise we could replace A by the centralizer $C_G(C_G(A))$ of $C_G(A)$ which is a torsion-free abelian subgroup containing A and which is definable if G is stable. Since G satisfies max-n, we find $a \in A$ such that the normal closure $\langle a^Q \rangle$ of a in G does have infinite rank. Since G satisfies max-n, G is finitely generated and Q is a finitely generated torsion-free abelian group. Therefore there exists $g \in Q$ such that the normal closure $\langle a^{\langle g \rangle} \rangle$ of a in $\langle A, g \rangle$ does have infinite rank. Hence we have $xa \neq 0$ for all $0 \neq x \in \mathbb{Z}\langle g \rangle$. Since Q is a finitely generated torsion-free abelian group the group ring $\mathbb{Z}Q$ is noetherian by Hilbert's Basis Theorem (or see [R$_0$82, 15.3.3]). Therefore we may assume that the *annulator* $\mathrm{ann}_{\mathbb{Z}Q}(a) = \{x \in \mathbb{Z}Q; xa = 0\}$ is maximal $\neq \mathbb{Z}Q$ among the ideals of $\mathbb{Z}Q$ which occur as annulators of elements of $\langle a^Q \rangle$. Then $\mathrm{ann}_{\mathbb{Z}Q}(a)$ is a prime ideal and $\mathbb{Z}Q/\mathrm{ann}_{\mathbb{Z}Q}(a)$ is an integral domain. We choose a basis $\{g_1, \ldots, g_n\}$ of the torsion-free finitely generated abelian group Q such that $g = g_1$ and for $g_1, \ldots, g_r$ we have

$$\mathbb{Z}\langle g_1, \ldots, g_r \rangle \cap \mathrm{ann}_{\mathbb{Z}Q}(a) = 0,$$

while for $j > r$ there exists $0 \neq x \in \mathbb{Z}\langle g_1, \ldots, g_r, g_j \rangle$ with $xa = 0$. Put $Q_0 = 1$, $Q_i = \langle g_1, \ldots, g_i \rangle$, $I_i = \mathrm{ann}_{\mathbb{Z}Q_i}(a)$ ($1 \leq i \leq n$).

Inductively we define for $r \leq i \leq n$ a non-empty subset $P_i \subseteq \mathbb{Z}^r$ and for $\bar{b} \in P_i$ we define a homomorphism $\psi_{\bar{b},i}$ of $\mathbb{Z}Q_i/I_i$ to an algebraic extension $K_{\bar{b},i}$ of $\mathbb{Q}$ such that $\psi_{\bar{b},i}(g + I_i)$ is not a root of unity in $K_{\bar{b},i}$ and in addition with the following property:

If $c \in \mathbb{Z}Q_i$, $c \notin I_i$ then there exists $\bar{b} \in P_i$ with $\psi_{\bar{b},i}(c + I_i) \neq 0$. (†)

If $i = r$, then $I_r = 0$. We denote the polynomial ring

$$\mathbb{Z}[g_1, \ldots, g_r, g_1^{-1}, \ldots, g_r^{-1}]$$

by $\mathbb{Z}Q_r$. Put $P_r = (\mathbb{Z} \setminus \{0, 1, -1\})^r$. Define for $\bar{b} = (b_1, \ldots, b_r) \in P_r$ the homomorphism $\psi_{\bar{b},0}$ from $\mathbb{Z}Q_r$ to $K_{\bar{b},0} = \mathbb{Q}$ via the map $g_j \mapsto b_j$ ($1 \leq j \leq r$). By definition, $\psi_{\bar{b},0}(g)$ is not a root of unity and (†) holds since polynomials have only finitely many zeros.

Suppose we have defined for $r \le i < n$ the set P_i and the homomorphism $\psi_{\bar{b},i} : \mathbb{Z}Q_i/I_i \to K_{\bar{b},i}$ for each $\bar{b} \in P_i$ such that (†) holds.

Now $R = \mathbb{Z}Q_i/I_i$ is an integral domain. Let K be its quotient field. The ring R operates on $\langle a^{Q_i} \rangle$ in a natural way. Since

$$\mathbb{Z}\langle g_1, \ldots, g_{i+1} \rangle \cap \mathrm{ann}_{\mathbb{Z}Q}(a) = \mathrm{ann}_{\mathbb{Z}Q_{i+1}}(a) \neq 0,$$

we can find a polynomial $p(x) \in R[x]$ of minimal degree $\neq 0$ such that $p(g_{i+1})a = 0$. We extend $\psi_{\bar{b},i} : R \to K_{\bar{b},i}$ ($\bar{b} \in P_i$) in a canonical way to a homomorphism $\bar{\psi}_{\bar{b},i} : R[x, x^{-1}] \to K_{\bar{b},i}[x, x^{-1}]$. For $\bar{I}_{i+1} = \{q \in R[x, x^{-1}];\ q(g_{i+1})a = 0\}$ we have an isomorphism $\sigma : \mathbb{Z}Q_{i+1}/I_{i+1} \to R[x, x^{-1}]/\bar{I}_{i+1}$ which replaces g_{i+1} by x. By the Lemma of Gauß, $\bar{I}_{i+1}$ is contained in the ideal $K[x, x^{-1}] \cdot p(x)$ of $K[x, x^{-1}]$ which is generated by $p(x)$. But since $\bar{I}_{i+1}$ is finitely generated, we can find $0 \neq b \in R$ such that $\bar{I}_{i+1} \subseteq \frac{1}{b}(R[x, x^{-1}] \cdot p(x))$. Let β be the highest coefficient $\neq 0$ of p. Put

$$P_{i+1} = \{\bar{b} \in P_i;\ \psi_{\bar{b},i}(b \cdot \beta) \neq 0\}.$$

By induction hypothesis (†), we have $P_{i+1} \neq \varnothing$. Suppose $\bar{b} \in P_{i+1}$. We choose an irreducible factor $f(x)$ of $\bar{\psi}_{\bar{b},i}(p(x)) \in K_{\bar{b},i}[x]$ with

$$K_{\bar{b},i}[x] \cap \bar{\psi}_{\bar{b},i}(\bar{I}_{i+1}) \subseteq K_{\bar{b},i}[x] \cdot f(x).$$

By definition of P_{i+1}, the ideal $K_{\bar{b},i}[x] \cap \bar{\psi}_{\bar{b},i}(\bar{I}_{i+1})$ does not contain a constant polynomial $\neq 0$, in particular f does have degree > 1. Therefore $K_{\bar{b},i+1} = K_{\bar{b},i}[x]/(K_{\bar{b},i}[x] \cdot f(x))$ is an algebraic extension of $K_{\bar{b},i}$ and the map

$$\psi_{\bar{b},i+1} : \mathbb{Z}Q_{i+1}/I_{i+1} \to K_{\bar{b},i+1}, c + I_{i+1} \mapsto \bar{\psi}_{\bar{b},i}(\bar{c})(x) + \left(K_{\bar{b},i}[x] \cdot f(x)\right)$$

(with $\sigma(c + I_{i+1}) = \bar{c}(x) + \bar{I}_{i+1}$, $\bar{c}(x) \in R[x, x^{-1}]$) is a homomorphism of $\mathbb{Z}Q_{i+1}/I_{i+1}$ to $K_{\bar{b},i+1}$. By induction hypothesis, $\psi_{\bar{b},i+1}(g + I_{i+1})$ is not a root of unity. Furthermore we check that $\psi_{\bar{b},i+1}$ does have property (†): if $c \in \mathbb{Z}Q_{i+1} \setminus I_{i+1}$, define as above $\bar{c}(x) \in R[x, x^{-1}]$ with $\sigma(c + I_{i+1}) = \bar{c}(x) + \bar{I}_{i+1}$. Multiplying c with an appropriate power of g_{i+1} (which is mapped by $\psi_{\bar{b},i+1}$ to an element $\neq 0$ of $K_{\bar{b},i+1}$) we may suppose $\bar{c}(x) \in R[x]$. Since $c \notin I_{i+1}$, $\bar{c}(x)$ is coprime to $p(x)$ and we can find in the integral domain R a $0 \neq \alpha \in R$ and polynomials $u(x), v(x) \in R[x]$ with $u(x)p(x) + v(x)\bar{c}(x) = \alpha$. By induction hypothesis (†), we can find $\bar{b} \in P_i$ with $\psi_{\bar{b},i}(\alpha \cdot b \cdot \beta) \neq 0$. In particular $\bar{b} \in P_{i+1}$ and $\bar{\psi}_{\bar{b},i}(\bar{c})(x)$ and $f(x)$ have no common zero (in $\mathbb{C}$). Therefore

$$\psi_{\bar{b},i+1}(c + I_{i+1}) = \bar{\psi}_{\bar{b},i}(\bar{c})(x) + \left(K_{\bar{b},i}[x] \cdot f(x)\right) \neq 0.$$

This proves (†) and the inductive construction is completed.

We fix now $\bar{b} \in P_n$ and define $\psi = \psi_{\bar{b},n}$, $K = K_{\bar{b},n}$, $R = \psi(\mathbb{Z}Q/I_n) \subseteq K$. Since K is algebraic over $\mathbb{Q}$, we can find a finite set π of primes such that the algebraic numbers $\psi(g_1+I_n), \ldots, \psi(g_n+I_n), \psi(g_1^{-1}+I_n), \ldots, \psi(g_n^{-1}+I_n) \in K$ are zeros of polynomials over $\mathbb{Z}$ and the highest coefficients of these polynomials are π-numbers. Put $S = S(\pi)$ (see Section 2 before Lemma 3 for the notations which are used here). Then $\mathbb{Q}_\pi \subseteq \mathcal{O}_S(K)$ and by Lemma 2 we have $R \subseteq \mathcal{O}_S(K)$.

Put $\alpha = \psi(g + I_n) = \psi(g_1 + I_n) \in K$. Then $\alpha \in U_S(K)$ is not a root of unity. Furthermore $\bigcap_{n \in \mathbb{N}} (\alpha^{2^n} - 1)\mathcal{O}_S(K) = 0$: if $0 \neq \beta$ were in this intersection, then we can find a finite set of primes $\pi' \supseteq \pi$ such that for $S' = S(\pi') \supseteq S = S(\pi)$ we have $\beta^{-1} \in U_{S'}(K)$. For $n \in \mathbb{N}$ fix $x_n \in \mathcal{O}_{S'}(K)$ with $(\alpha^{2^n} - 1)x_n = \beta$. Then $\alpha^{2^n}(x_n \beta^{-1})$, $x_n \beta^{-1}$ are solutions of $x - y = 1$ in $U_{S'}(K)$. Since α does have infinite order, we have $x_n \neq x_m$ for $n \neq m$ and hence we have infinitely many solutions of this equation in S'-units. This is a contradiction to the S-unit-theorem (Proposition 1).

Put $A_0 = \langle a^Q \rangle$, the normal closure of a in G. From (†) and the last paragraph it follows that $\bigcap_{n \in \mathbb{N}} (g^{2^n} - 1)A_0 = 0$. Since $xa \neq 0$ for all $0 \neq x \in \mathbb{Z}\langle g \rangle$ (see the beginning of the proof) we have $(g^{2^n} - 1)rA_0 \neq 0$ for all $0 \neq r \in \mathbb{Z}\langle g \rangle$, $n \in \mathbb{N}$. We use max-n to find a group $A_1 \supseteq A_0$ and numbers $m_0, \ldots, m_k \in \mathbb{N}$ with

$$(g^{2^n} - 1)A_1 = (g^{2^n} - 1)A \cap A_1$$

for all $n \in \mathbb{N}$ and $rA_1 \subseteq A_0$ for $r = (g^{2^{m_k}}-1) \cdots (g^{2^{m_0}}-1) \in \mathbb{Z}\langle g \rangle$. Then also the intersection $\bigcap_{n \in \mathbb{N}} (g^{2^n}-1)rA_1$ is 0, but $(g^{2^n}-1)rA_1 \supseteq (g^{2^n}-1)rA_0 \neq 0$ for all $n \in \mathbb{N}$. Thus the chain

$$rA_1 \supset (g-1)rA_1 \supseteq \cdots \supseteq (g^{2^n}-1)rA_1 \supseteq \cdots$$

is infinitely descending. From $(g^{2^n} - 1)A \cap A_1 = (g^{2^n} - 1)A_1$ it follows that $(g^{2^n} - 1)A$, $n \in \mathbb{N}$, is an infinitely descending chain of subgroups of G. But since we assumed that A is definable in G (see the beginning of the proof), this chain is a uniformly definable chain of subgroups of G (as in the proof of Corollary 5). Therefore G is unstable. q.e.d.

Lemma 11. Let G be an extension of an abelian group A, which contains an infinite elementary abelian p-group for some prime p, by a torsion-free abelian group $Q \neq 1$. If G satisfies max-n, then G is unstable.

Proof. As in the proof of Lemma 10 we may suppose that A is definable in G and that there exists an $a \in A$ of order p such that $A_0 = \langle a^Q \rangle$ is infinite

and $\operatorname{ann}_{\mathbb{Z}Q}(a)$ is a prime ideal of $\mathbb{Z}Q$. In particular $\mathbb{Z}Q/\operatorname{ann}_{\mathbb{Z}Q}(a)$ is an integer domain and multiplication by $v \in \mathbb{Z}Q$, $v \notin \operatorname{ann}_{\mathbb{Z}Q}(a)$ is injective on A_0. Suppose multiplication by a $v \in \mathbb{Z}Q \setminus \operatorname{ann}_{\mathbb{Z}Q}(a)$ would be *not* surjective on A_0. Then
$$A_0 \supseteq vA_0 \supseteq \cdots \supseteq v^{p^n} A_0 \supseteq \cdots$$
is an infinitely descending chain of subgroups of A_0. For $i \in \mathbb{N}$ we define the groups $A_{0,i} = \{x \in A;\ v^i x \in A_0 \text{ and } px = 0\}$. Then $A_0 \subseteq A_{0,0} \subseteq \cdots \subseteq A_{0,i} \subseteq \cdots$ is an ascending chain of normal subgroups of G. Since G satisfies max-n, we can find $m \in \mathbb{N}$ such that $\bigcup_{i \in \mathbb{N}} A_{0,i} = A_{0,m} = A_1$. Then $v^m A_1 \subseteq A_0$. For $n \in \mathbb{N}$ choose $k_n \in \mathbb{N}$ with $p^{k_n} \geq p^n + m$. Then
$$v^{p^{k_n}} A_0 \subseteq v^{p^n}(v^m A_1) \subseteq v^{p^n} A_0,$$
and therefore also the chain $v^{p^n}(v^m A_1)$, $n \in \mathbb{N}$, is an infinitely descending chain of subgroups of A_1. Put $A[p] = \{x \in A;\ px = 0\}$. From $(v^{p^n} A[p]) \cap A_1 = v^{p^n} A_1$ it follows that $v^{p^n}(v^m A[p])$, $n \in \mathbb{N}$, is an infinitely descending chain of subgroups of A. These subgroups are definable, since A is definable. But the groups in this chain are even uniformely definable subgroups of G: if $\bar{h} = (h_1, \ldots, h_k) \in Q^k$ then $D_{\bar{h}}(A) = \{(h_1 + \cdots + h_k)x;\ x \in A\}$ is definable with parameters $\bar{h}$ in G. If $v = h_1 + \cdots + h_k$ for $h_i \in Q$ and $\bar{h}_n = (h_1^{p^n}, \ldots, h_k^{p^n})$ for $n \in \mathbb{N}$ then $v^{p^n} A = D_{\bar{h}_n}(A)$. Therefore $v^{p^n}(v^m A[p]) = D_{\bar{h}_n}(v^m A[p])$ is uniformely definable in G and G is unstable (see [F$_0$78, Lemma 4]). Suppose now that multiplication with every $v \in \mathbb{Z}Q \setminus \operatorname{ann}_{\mathbb{Z}Q}(a)$ is surjective and injective on $A_0 = \langle a^Q \rangle$. Then $\mathbb{Z}Q/\operatorname{ann}_{\mathbb{Z}Q}(a)$ is a field. This field is finite since it is finitely generated with prime field $GF(p)$. This is a contradiction to the assumption that $\langle a^Q \rangle \cong \mathbb{Z}Q/\operatorname{ann}_{\mathbb{Z}Q}(a)$ is infinite. q.e.d.

Corollary 12. Let G be a metabelian, stable group which satisfies max-n. Then G is abelian-by-finite.

Proof. Let A be a maximal abelian subgroup containing G'. Then $C_G(A) = A$ and A is definable in G. Since every soluble max-n group is finitely generated, G/A is a finitely generated abelian group. Furthermore G contains a definable subgroup $\bar{G} \geq A$ of finite index such that $\bar{G}/A$ is torsion-free. Therefore we may suppose that $Q = G/A$ is torsion-free. If G is not abelian-by-finite, then $Q \neq 1$.

Case 1. The torsion subgroup $T(A)$ of A is finite. Then $T(A)$ is definable and $G/T(A)$ is a an extension of the torsion-free group $A/T(A)$ by Q which satisfies max-n. By Lemma 10, $G/T(A)$ and hence G are unstable if $A/T(A)$ does have infinite rank. But if $A/T(A)$ does have finite rank, then G is a minimax group and the claim follows from Theorem A.

Case 2. $T(A)$ is infinite. By max-n, we find a prime p such that A contains an infinite elementary abelian p-group. By Lemma 11, G is unstable.

q.e.d.

Proof of Theorem B. The proof is by induction on the derived length of G. If G is finitely generated metabelian (i.e., of derived length at most 2), then G satisfies max-n (see [$R_0$82, 15.3.1]) and G is abelian-by-finite by Corollary 12. If the theorem is proved for groups of derived length n, then let G be a finitely generated stable group of derived length $n+1$. Let A be an abelian subgroup of G such that G/A has derived length n. Then the centralizer $C = C_G(C_G(A))$ is abelian, it contains A and it is definable in G. By induction hypothesis, G/C is abelian-by-finite. Therefore G/C contains a definable normal finitely generated abelian subgroup B/C of finite index. Then B is finitely generated metabelian and satisfies max-n. By Corollary 12, B and G are abelian-by-finite.

q.e.d. (Theorem B)

Obviously Theorem B is correct also for finite extensions of finitely generated soluble groups.

5 Soluble groups of finite rank.

In this section we refer mainly to a result of B. Zilber [Z93]. In [Z93], it was shown that the field of complex numbers with a predicate distinguishing the roots of unity is ω-stable. This proof is based on the following result of H. Mann [$M_1$65].

Theorem. For any linear equation $c_1 x_1 + \cdots + c_n x_n = 0$ with integer coefficients $\neq 0$, there are only finitely many minimal solutions modulo congruence in the complex roots of unity.

Here a solution $\langle u_1, \ldots, u_n \rangle$ of $c_1 x_1 + \cdots + c_n x_n = 0$ is called *minimal*, if for every equation $d_1 u_1 + \cdots + d_n u_n = 0$ ($d_i \in \mathbb{Z}$) with $|d_i| \leq |c_i|$ we have either $d_1 = \cdots = d_n = 0$ or $|d_i| = |c_i|$ for all $1 \leq i \leq n$. Furthermore two minimal solutions $\langle v_1, \ldots, v_n \rangle, \langle u_1, \ldots, u_n \rangle$ in a subgroup U of the multiplicative group $\mathbb{C}^*$ of $\mathbb{C}$ are called *congruent*, if there is a $\zeta \in U$ with $u_i = \zeta v_i$ for all $1 \leq i \leq n$. It is easy to see that a similar theorem is true for other subgroups of $\mathbb{C}^*$.

Lemma 13. Suppose U is the subgroup of $\mathbb{C}^*$ which consists of all powers of a prime p, i.e., $U = \{p^n; n \in \mathbb{Z}\}$. Then every linear equation $a_1 x_1 + \cdots + a_n x_n = 0$ with integer coefficients $\neq 0$ does have only finitely many minimal solutions in U modulo congruence.

Proof. Suppose $\langle u_1, \ldots, u_n \rangle$ is a solution of the linear equation $a_1 x_1 + \cdots + a_n x_n = 0$ in U. Since we work modulo congruence, we may suppose $u_1 \geq \cdots \geq u_n = 1$, after a permutation of the a_i. But then u_j divides u_i for $i \leq j$. For $1 \leq j < n$ we have

$$a_1 \frac{u_1}{u_j} + \cdots + a_{j-1} \frac{u_{j-1}}{u_j} = -\frac{a_n + a_{n-1} u_{n-1} + \cdots + a_{j+1} u_{j+1}}{u_j} - a_j,$$

therefore u_j divides $a_{j+1} u_{j+1} + \cdots + a_{n-1} u_{n-1} + a_n$. If for some $1 \leq j < n$ it would be $a_{j+1} u_{j+1} + \cdots + a_{n-1} u_{n-1} + a_n = 0$, then $\langle u_1, \ldots, u_n \rangle$ would be not minimal. Therefore u_j is bounded by the number $a_{j+1} u_{j+1} + \cdots + a_{n-1} u_{n-1} + a_n$ and we have only finitely many minimal solutions modulo congruence in U. q.e.d.

Corollary 14. *Let $U = \{p^n;\ n \in \mathbb{Z}\} \subseteq \mathbb{C}$. Then the field of complex numbers with a predicate distinguishing U is superstable.*

Proof. This follows from [Z93] and Lemma 13. q.e.d.

Corollary 15. *Let U be the set of p-powers of some prime p. Suppose U acts on the additive group $\mathbb{Q}^+$ of rational numbers by multiplication. Put $G = \mathbb{Q}^+ \rtimes U \leq \mathbb{Q}^+ \rtimes \mathbb{Q}^*$. Then G is a superstable soluble group of finite rank.*

Proof. Let $\bar{\mathbb{Q}}$ be an elementary extension of $\mathbb{Q}^+$ of cardinality $2^{\aleph_0}$ and let U operate on $\bar{\mathbb{Q}}$ by multiplication.

Then $\mathbb{Q}^+ \rtimes U$ is an elementary substructure of $\bar{\mathbb{Q}} \rtimes U$. This can be shown by induction on the length of formulas, since the action of an $u \in U$ on $\mathbb{Q}^+$ is definable already in $\mathbb{Q}^+$. An isomorphism $\bar{\mathbb{Q}} \to \mathbb{C}^+$ lifts to an isomorphism $\bar{\mathbb{Q}} \rtimes U \to \mathbb{C}^+ \rtimes U$. Hence $\mathbb{Q}^+ \rtimes U$ is an elementary substructure of $\mathbb{C}^+ \rtimes U$. But the theory of the group $\mathbb{C}^+ \rtimes U$ is interpretable in the theory of the field of complex numbers with a predicate distinguishing U. Therefore the claim follows from Corollary 14. q.e.d.

Corollary 15 is a more precise version of Theorem C. As H.-P. Schlickewei remarks in the introduction of [S90], Lemma 13 is true for any finitely generated subgroup U of $\mathbb{C}^*$. Therefore also in this case the structure $(\mathbb{C}, +, \cdot, U)$ is superstable and if $U \leq \mathbb{Q}^*$ then $\mathbb{Q} \rtimes U$ is superstable.

References.

[E₀72] Ju. L. Eršhov, *Elementary group theories*. Dokl. Akad. Nauk. SSSR **203** (1972), 1240-1243; Engl. transl. in Soviet Math. Dokl. **13** (1972), 528-532.

[E₁84] J.-H. Evertse, *On equations in S-units and the Thue-Mahler equation*. Invent. Math. **75** (1984), 561 – 584.

[F₀78] U. Felgner, $\aleph_0$-*categorical stable groups*. Math. Zeitschr. **160** (1978), 27-49.

[F₁70] L. Fuchs, *Infinite abelian groups, 2 vols*. Academic Press, New York-London 1970.

[GH93] C. Grünenwald, F. Haug, *On stable torsion-free nilpotent groups*. Arch. Math. Logic, **32** (1993), 451-462.

[M₀T06] D. Macpherson, K. Tent, *Stable pseudofinite groups*. J. of Algebra 2006, to appear.

[M₁65] H. Mann, *On linear relations between roots of unity*. Mathematika **12** (1965), 107-117.

[N84] G. A. Noskov, *On the elementary theory of finitely generated almost soluble groups*. Math.USSR Izvestiya **22** (1984), 465-482.

[P87] B. Poizat, *Groupes Stables*. Nur al-Mantiq wal-Ma'rifah, Villeurbanne 1987.

[R₀67] D. J. S. Robinson, *On soluble minimax groups*. Math. Zeitschr. **101** (1967), 13-40.

[R₀68] D. J. S. Robinson, *Residual properties of some classes of infinite soluble groups*. Proc. London Math. Soc. (3) **18** (1968), 495-520.

[R₀72] D. J. S. Robinson, *On Finiteness Conditions and Generalized Soluble groups, 2 vols*. Springer Verlag, Berlin-Heidelberg-New York 1972.

[R₀82] D. J. S. Robinson, *A Course in the Theory of Groups*. Springer Verlag, Berlin-Heidelberg-New York 1982.

[R₁80] N.S.Romanovskiĭ, *On the elementary theory of an almost polycyclic group*. Mat.Sb. **111(153)** (1980), 135-143; Engl. transl. in Math. USSR Sb. **39** (1981), 125-132.

[S90] H.-P. Schlickewei, *An explicit bound for the number of solutions of S-unit-equations*. J. f. d. reine u. angewandte Mathematik **406** (1990), 109-120.

[v67] B. L. van der Waerden, *Algebra, 2 vols*. Springer Verlag, Berlin-Heidelberg-New York 1967.

[Z93] B. Zil'ber, *ω-stability of the field of complex numbers with a predicate distinguishing the roots of unity*. manuscript, Kemerovo 1993, http://www.maths.ox.ac.uk/ zilber

Received: September 28, 2005;
In revised version: June 18, 2006;
Accepted by the editors: September 1, 2006.

Formal Methods in Concurrent Systems Engineering: Survey and Examples

BRUNO MÜLLER-CLOSTERMANN

University of Duisburg-Essen
Institute for Computer Science & Business Information Systems (ICB)
Schützenbahn 70, D-45117 Essen, Germany
E-mail: bmc@icb.uni-due.de

> ABSTRACT. This contribution introduces modelling techniques for problems in concurrent systems. Formal description techniques like Petri nets, coupled automata systems and networks of timed automata are briefly sketched by means of classic examples. Basic analysis techniques like invariants, state space exploration and model checking are also addressed. The objective of this overview is to emphasize the increasing importance of formal methods for the design, analysis and development of concurrent distributed computer and communication systems. To this end some practical applications and tool systems which are based on formal methods are also considered.

1 Introduction

Many natural and artificial systems are concurrent, in particular all substantial future computer and communication systems will be concurrent and distributed systems. Specification, design, validation and verification, dependability and performance analysis, implementation, testing, tuning, optimization, deployment, operation and maintenance of such systems may be summarized under the term "concurrent systems engineering".

Sequential "classic" programming is based on the paradigm that a program is executed on a single processor, and that under identical start conditions the same function will be performed. All program steps are executed always in the same order and identical results will be achieved. Sequential programs are usually deterministic and determined. This classic paradigm of sequential, deterministic and determined programming is closely related to the John-von-Neumann computer architecture and conventional programming languages have been oriented towards this architecture.

There are several reasons why sequential programming is no longer sufficient. Firstly, the sheer increase of processor speed which has doubled approximately every 18 months from 1970 to 2003 (according to Moore's law) is not sufficient and unsatisfactory from a methodological viewpoint. The processor performance growth has now stopped between 3 and 4 GHz, and even if the increase of processor speed will restart, it will hit the next wall. As a consequence non-concurrent applications will no longer exploit the continued performance growth in modern processors; hence the writing of concurrent software, and more general the mastering of techniques for concurrent systems is necessary. Essentially concurrent programming is based on the renouncement of sequential execution and determinism. Statements can be executed as arbitrary as possible, of course in a sequence restricted by dependencies with respect to shared data, synchronization or necessary communication. Designed, coded and compiled appropriately, many applications can be executed on multi processor systems with an increased efficiency.

A second even more important reason for the "concurrency revolution" is the manifold of multiprocess systems and applications, where possibly a large number of instances do interact with each other to fulfill a certain function, or in newspeak to "provide a service". Examples include multi-processor applications, computer networks, communication protocols and distributed applications.

The design and implementation of concurrent systems is a very complex task. In practice many projects are delayed or even fail completely due to the many problems arising from lack of sufficient requirements, largeness, distribution, concurrency, and other inherent system properties. The record of delayed and failed projects is long. We list just a few costly projects which were reported to have been suffered from severe design problems. The concurrent software for a computer-controlled radiation therapy machine Therac-25 contained a simple but severe software error; two people have been killed and others have been hurt. Other famous examples for failed or delayed projects are the explosion of the (unmanned) european space rocket Ariane 5 in 1996, the computerized baggage handling system at Denver Airport and the German Toll Collect System ("LKW-Maut") in 2004.

To illustrate the inherently high and permanently growing complexity of concurrent distributed systems a citation due to Leslie Lamport, one of the pioneers in concurrent system research, is often used: "A distributed system is one in which the failure of a computer you didn't even know existed can render your own computer unusable."

In the given context formal methods refer to mathematically rigorous techniques and tools for the specification, design and verification of software and hardware systems. The value of formal methods is that they provide techniques to investigate a system design with respect to correctness properties and functional behaviour. Reachability, liveness and safety can be evaluated by state space exploration, model checking, calculation of invariants and other algorithms. However, formal methods are rarely used in practice today (except in the field of safety critical systems) because of the enormous complexity of real systems.

Additionally for most systems the so-called non-functional behaviour (essentially defined by performance and availability) is of outstanding importance. Examples for non-functional "quality-of-service" parameters which are desired include response time (e.g., < 2 seconds), throughput (e.g., 1000 tasks/hour) and availability (e.g., 99.9% over one year). Hence the integration of the concepts time and resources with formal methods has become an active research field. The goal is a unified modelling and evaluation methodology for the investigation of functional and non-functional properties during design and before implementation of a system. Early results are related to stochastic Petri nets and timed automata. MSC and SDL based performance evaluation and stochastic process algebras are more recent approaches to integrate formal methods with performance evaluation.

In the rest of the paper we give an overview of description and modelling techniques for concurrent systems. We start with two variants of Petri nets, namely place/transition nets and timed Petri nets, describe basic results and give some examples (§2). Coupled (or communicating) automata and an example from protocol modelling do follow (§3). Timed automata and their application in combination with the tool UPPAAL give some idea of the value of model checking techniques (§4). As an example for an industrially used tool system that is based on a formal method we sketch the specification and description language SDL; finally we give a short look into an SDL based performance evaluation tool and some applications (§5).

2 Petri Nets

Petri nets have been introduced in the dissertation of Carl Adam Petri in 1969. They belong to the most popular mathematical representations of concurrent systems [R98, GV02]. Besides the formal definition Petri nets have an equivalent graphical representation as bipartite graphs. A Petri net consists of transitions and places that are connected by directed arcs. The state of a Petri net is called a marking which is defined by a distribution of tokens over the places. There are various types of Petri nets, like nets with coloured tokens, hierarchical nets, condition/event nets, (Turing equivalent) Petri nets with inhibitor arcs, and others. Here we introduce the basic Place/Transition Nets and Timed Petri Nets.

2.1 Definitions

Definition 1. A *marked Place/Transition Net (basic Petri Net)* is a 5-tuple (P, T, C^-, C^+, M_0), where

(1) P is a set of n places,

(2) T is a set of m transitions,

(3) C^- is the backward connectivity $n \times m$-matrix, $C_{ij}^- \in \mathbb{N}_0$, where C_{ij}^- describes the multiplicity of arcs from place i to transition j.

(4) C^+ is the forward connectivity $n \times m$-matrix, $C_{ij}^+ \in \mathbb{N}_0$, where C_{ij}^+ describes the multiplicity of arcs from transition j to place i.

(5) M_0 is the initial marking, an n-tuple, where element i describes the number of tokens on place i.

Transitions describe activities and consume and produce tokens; places describe state information by the number of tokens they carry. Arcs connect places and transitions; arc weights are positive integers and describe the firing behaviour of the transitions. If all input places of a transition carry enough tokens the transition is called enabled. Enabled transitions may fire by removing tokens from their input places and putting tokens on all their output places. The number of tokens that vanish from the input places and are put in the output places is defined by the associated arc weights.

Definition 2.

(1) A transition t_j is *enabled* in marking M, iff $M \geq C^- \cdot e_j$.

(2) An enabled transition t_j can *fire* (or *switch*) and the successor marking M' is given as $M' = M + C \cdot e_j$, where $C = C^+ - C^-$.

The terms $C^- \cdot e_j$ and $C \cdot e_j$ determine the j-th row of matrix C^- and C, respectively. The relation $\geq$ in (1) is defined per element, i.e., the relation $(x_1, x_2, ..., x_n) \geq (y_1, y_2, ..., y_n)$ holds, if $x_i \geq y_i$, for all i.

It is possible that more than one transition is enabled; in this case the decision which one of the conflicting and/or concurrent transitions will switch is non-deterministic. Due to this non-determinism Petri nets are well suited for the modelling of concurrent systems.

2.2 Place- and Transition Invariants

Place/Transition Nets can be analysed with respect to certain properties. In particular one is usually interested in liveness, boundedness and reachability of certain markings. Nice properties which can efficiently be evaluated are expressed by P-invariants and T-invariants.

Definition 3.

(1) An n-tuple $v \in \mathbb{Z}^n, v \neq 0$ is a *P-invariant* (Place invariant) if and only if $v \cdot C = 0$.

(2) An m-tuple $w \in \mathbb{Z}^m, w \neq 0$ is a *T-invariant* (Transition invariant) if and only if $C \cdot w = 0$.

We see that P- and T-invariants are defined as solutions of a system of linear equations with coefficient matrix C; i.e., invariants form a subspace in $\mathbb{Z}^n$ or $\mathbb{Z}^m$, respectively.

P-invariants express properties which hold for all markings of the net. T-invariants are candidates to express the reproducibility of markings; in this case the m-tuple w is a frequency vector, where w_j is the number of firings of transition t_j within a non-empty firing sequence $f \subset \{(t_1|t_2|\ldots|t_m)^*\}$ leading from a marking M back to itself. Such a marking M is called reproducible. The following theorem summarizes some of the properties of P- and T-invariants.

Theorem 4.

(1) If v is a P-invariant, then for all reachable markings holds the invariant property $v \cdot M = v \cdot M_0 =$ constant.

(2) If a marking is reproducible, then a T-invariant w does exist.

(3) If a P/T-net has a T-invariant w and there is a feasible firing sequence f with frequency vector w, then all markings are reproducible.

We illustrate these properties by an example.

2.3 Example: Readers/Writers-Problem

We consider a P/T-net that models the Readers/Writers-Problem where up to R readers may simultaneously enter a critical region to access some data; alternatively no reader but at most one writer may enter the critical region. The initial marking is $M_0 = (R, 0, n, W, 0)$ with the interpretation that we have R reader processes, W writer processes, n available "admission tickets" and no process is within the critical region given by the places p_2 and p_5.

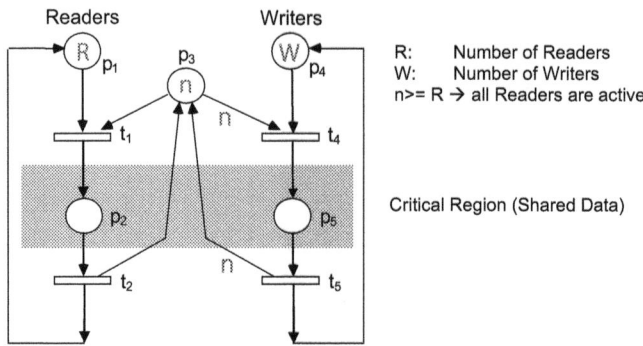

Figure 1. Place/Transition-Net for the Readers/Writers-Problem.

Does the proposed model satisfy the requirement that a single writer on place p_5 excludes all readers from p_2, and vice versa? Of course for fixed values of R, W and n an exhaustive state space exploration under construction of the reachability tree could answer the question. The calculation of the place invariants is the better choice; it is efficient since the algorithm just solves a small set of linear equations and the results are independent of the initial marking.

The linear independent place invariants for this P/T-net can be computed (or guessed by visual inspection in case of a small example) as $x = (1, 1, 0, 0, 0)$, $y = (0, 0, 0, 1, 1)$ and $z = (0, 1, 1, 0, n)$. The marking where exactly one writer process can access the critical region is $(R, 0, 0, W-1, 1)$ and indeed we have $M \cdot z = n = $ constant for this and all other reachable markings. Since all places are covered the net is bounded, i.e., for each place p_i there is an upper limit for the number of tokens that may occupy p_i simultaneously.

The transition invariants are simply $x = (1, 1, 0, 0)$ and $y = (0, 0, 1, 1)$, and firing of transitions t_1 and t_2 or alternatively t_3 and t_4 will reproduce any given marking.

2.4 Timed (Deterministic and Stochastic) Petri Nets

By introducing stochastic durations for all transitions we arrive at Stochastic Petri Nets (SPN), defined as a 6-tuple $(P, T, C^-, C^+, M_0, \Lambda)$, where $\Lambda = (\lambda_1, ..., \lambda_m)$ defines a vector of exponential transition rates, i.e., the time duration for firing transition j is exponentially distributed with rate λ_j. Concurrently enabled transitions perform a race and the winning transition will fire.

If we allow a mixture of transition types, namely timeless transitions, exponential transitions and deterministic transitions (with constant time duration) we obtain the class of Deterministic and Stochastic Petri Nets (DSPN).

2.5 Example: DSPN Model of Leaky Bucket

As an example we display a model for the leaky bucket mechanism; the leaky bucket is a traffic engineering technique in computer networks to control the network-to-network flow or network-to-user flow. In order to work properly we use multiple arcs to model losses and priorities for the immediate transitions.

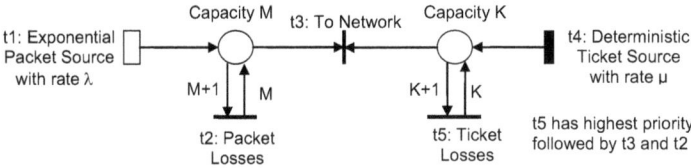

Figure 2. Timed Petri Net for Leaky Bucket Flow Control.

Note that this Petri net model fulfills the basic requirements of a formal method. It is unique, complete, allows visual inspection as well as rigorous proofs to validate the functional behaviour (no starving, no deadlocks, liveness, boundedness). Moreover, non-functional properties may be evaluated by means of Markov or semi-Markov techniques; the most important quantitative measure for this kind of models is the probability of packet loss dependent on the values of λ, μ, M and K. Typically, loss probabilities in networks without "store-and-forward" should not exceed 10^{-3} to 10^{-6} depending on the service to be supported. For voice and video a loss probability of 10^{-3} per packet or frame is fully sufficient, whereas file transfer needs a packet loss probability of approximately 10^{-6} or lower. For quantitative results the Markovian solvers in Petri net tools like TimeNet may be used, see [M$_3$97]. For, e.g., $K = M = 128$ the underlying Markov chain has approx. 16500 states and the stationary distribution may be solved without much computational effort. The packet loss probability is, e.g., given by the stationary probability for marking $(M, 0)$.

3 Communicating Automata and Reachability Analysis

Since the very beginning of computer science various kinds of automata have been a versatile technique [$H_3M_2U_0$01]. Circuit design, definition of formal languages, specification and validation of communication protocols are just a few examples where automata play an important role.

3.1 Coupling of Automata

We introduce the idea of automated protocol validation by means of a pair of coupled finite automata. We assume Mealy automata (with input and output function) which are coupled over unidirectional channels; these channels are perfect, i.e., unbounded, without losses, without errors, and with no time delay. The elements of the alphabets are considered as messages that are exchanged between coupled automata. Sending (output) of a message x is denoted by $x!$, receiving (input) of an element is denoted by $x?$. The channels between the automata store the messages according to the discipline FIFO (First-In-First-Out), the receiving automaton consumes the messages forwarded through the channel asynchronously, i.e., it is not enforced that a message is immediately consumed.

3.2 Automata Based Validation of Communication Protocols

The main application for coupled automata is the specification and validation of communication protocols [$H_2$91]. A protocol is a hierarchically structured software for communication in distributed systems. It is vertically structured as a stack of message exchanging instances which implement a horizontal communication with another stack on a remote computer. In designing such protocols the behaviour of instances is described by automata (in this context also called "finite state machines") and the exchanged messages are the elements of the input and output alphabets. Passing of messages between instances is modelled by channels that buffer messages.

The global state space G of a set of coupled automata is a subset of the Cartesian product over all instances and all channels, $G \subseteq Z_1 \times ... \times Z_n \times K_{1,1} \times ... \times K_{n,n}$, where Z_i and $K_{i,j}$ are the local state spaces of the automata and the channels.

The typical technique to explore the system behaviour is the construction of the global reachability tree (or graph), i.e., a structure that includes all reachable global states G and all executable transitions. Each state can be explored with respect to certain properties. Errors like message loss (in case of finite channels), total deadlock, or non-executable transitions ("dead code") can be found by either exhaustive or partial state space exploration.

A typical error is unspecified reception (unexpected receive); in this case an instance finds a non consumable message in its input channel. Due to the FIFO-property of the channel, the instance is deadlocked because no reaction has been defined for the given pair of state and message. To illustrate the state space exploration we display a simple protocol model that contains an unspecified reception error. The intended protocol function is the establishment of a connection between two instances. Figure 3 displays the two automata and their channels. Sending the message c requests for a connection, sending a d requests for disconnection. The initial global state is $(0, 0, \varepsilon, \varepsilon)$, where the symbol ε indicates an empty channel.

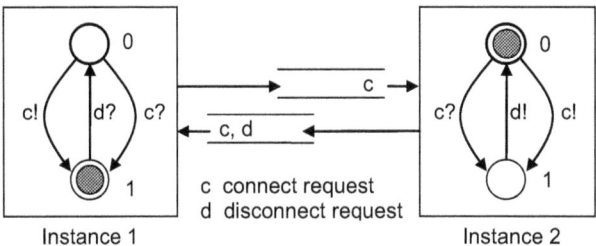

Figure 3. Protocol Model in a Fault State.

The protocol is faulty, since in global state $(1, 0, c, cd)$ the instance 1 is not able to execute a transition. A systematic and automatic state space exploration starts with an initial global state $(0, 0, \varepsilon, \varepsilon)$. Figure 4 shows some states and transitions of the global reachability tree.

Exploration of state 7 reveals that instance 1 is deadlocked; note that there is no global deadlock since instance 2 can forever send messages c and d alternately.

A widely used variation of state space exploration is the bit-state algorithm which is based on hashing without any resolution of collisions. This algorithm performs "only" a partial exploration technique but it is very fast and can explore hundred of million states in reasonable time. An Implementation of the bit-state algorithm has been included in several tools like SPIN [H$_2$91] and QUEST [D$_1$98, D$_1$H$_0$M$_3$96].

4 Synchronous Automata and Model Checking

Here we consider a class of automata with direct coupling which means that each action of an automaton needs a complementary "synchronized" action of another automaton. Again we use "x!" and "x?" to describe action pairs. The symbol x! serves to describe the execution of action x

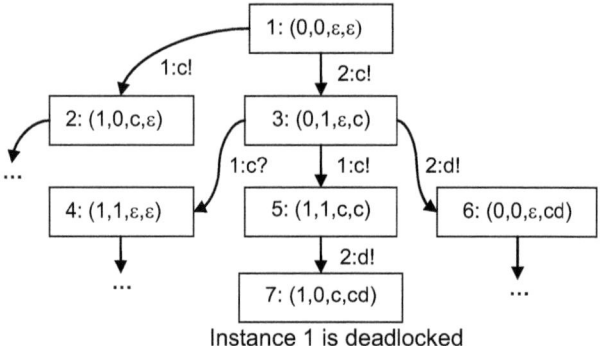

Figure 4. Global (Partial) Reachability Tree.

and the symbol $x?$ the synchronous execution of the co-action. Actions and co-actions may be considered as simultaneous sending and receiving of messages, i.e., there is no queueing or waiting of messages at all. Other more general interpretations are possible.

A very popular class of synchronous automata models that offers additionally much more features are Timed Automata supported by the tool UPPAAL. Introductions to theory and practice of Timed Automata and UPPAAL are given in [$B_0D_0L_0$04, B_1Y04, L_0PY95, $U_1$06]. Here we sketch briefly some of the most important features of Timed Automata and show afterwards two examples.

Timed Automata got their name because of the inclusion of real valued clock variables. The corresponding model checking problem has been proven decidable for a number of timed logics. In UPPAAL, a timed extension of Computational Tree Logic (TCTL) with the operators $\Diamond$ and $\Box$ used in combination with $\forall$ and $\exists$ has been introduced to express checkable model properties. The clocks can be either globally defined or locally within a process; with the help of clock guards, e.g., as "$c < t$", the execution of transitions can be restricted to certain time intervals. In the same way data variables (bounded integers) and data guards are allowed. Also the priorities for the transitions can be controlled by different features.

4.1 Reachabilty, Safety, and Lifeness

Properties of a model are described by means of TCTL (Timed Computational Tree Logic) as state formula or path formula. In case of path formula we distinguish Reachability, Safety, and Liveness; we follow [$B_0D_0L_0$04].

Reachability properties ask whether a given state formula φ can be satisfied by any reachable state, i.e., does there exist a path starting at the initial state, such that φ is eventually satisfied along that path? The path formula $\exists\Diamond\varphi$, in UPPAAL written as E<>p, expresses that there is a path with a state satisfying φ.

Safety properties are of the form: "Something bad will never happen". For instance, in a hard real time system a safety property can be that emergency signals are never queued or the maximum response time is below one second. In UPPAAL, these properties are formulated positively, e.g., something good is invariantly true. If φ is a state formula the path formula $\forall\Box\varphi$ (in UPPAAL written A[]p) expresses that φ should be true in all reachable states. The formula $\exists\Box\varphi$ (in UPPAAL: E[]p) says that at least one path exists, where in all states φ is satisfied.

Liveness properties are of the form: "Something will eventually happen". In the readers/writers-model readers as well as writers should finally get access to the critical region. For a protocol model we require that any message that has been sent should eventually be received. Liveness can also be expressed with the path formula $\forall\Diamond\varphi$ (in UPPAAL: A<>p), meaning that φ is eventually satisfied. The "leads to" or "response" property is written $\varphi \to \psi$ (p $\to$ q) which means that whenever φ is satisfied, then eventually ψ will be satisfied.

We do not comment on the wide field of model checking algorithms ranging from straight-forward state space exploration in case of models without clocks to rather sophisticated techniques for the full model class including data, clocks and guards [P99].

4.2 Petterson's Mutual Exclusion Algorithm

One of the most famous concurrency algorithms is the algorithm of Petterson to solve the mutual exclusion problem. Two (or more) processes share an exclusive critical section; at most one of the processes is allowed to visit the critical section at a given point of time, and no process must be excluded forever, i.e., a solution must show safeness and liveness.

The translation of this concurrent algorithm into a model of coupled automata (cf. Figure 6) is rather straightforward; variable assignments occur as transition labels and they are evaluated when the transition is executed. The composed condition used in the While-loop has been translated into two transitions which are labelled with guards which evaluate to a Boolean expression. The names of the locations are just the comments from the algorithm.

Of course we can now reason about this model or play around to visualize some properties of the model, in particular we can ask whether it is safe,

```
Process P1                              Process P2
Loop                                    Loop
  req1:=1;   // want                      req2:=1;   // want
  turn:=2;   // wait                      turn:=1;   // wait
  While (turn≠1 and req2≠0) wait;         While (turn≠2 and req1≠0) wait;
  // Critical section CS                  // Critical section CS
  Job1();  // is working in CS            Job2();  // is working in CS
  req1:=0;                                req2:=0;
End loop                                End loop
```

Figure 5. Petterson Algorithm for Mutual Exclusion of 2 Processes.

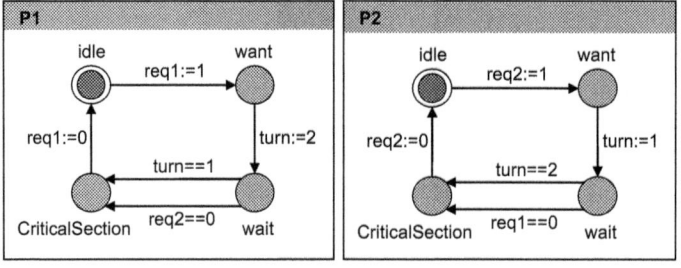

Figure 6. Petterson Algorithm as Coupled Automata.

live, starvation free and has no deadlocks. By means of model checking we may describe appropriate questions and validate them.

The properties to be checked are based on formulas like "P1.CriticalSection" that states that process "P1" is in location "CriticalSection". Table 1 shows some example queries, their interpretation and the result of the check.

Obviously "p = P1.CriticalSection and P2.CriticalSection" expresses that P1 and P2 are both simultaneously in the critical section. The query

"E<>P1.CriticalSection and P2.CriticalSection"

asks for a path with a state where processes P1 and P2 are both in the critical section. This yields indeed the result that this property is not satisfied (=False).

4.3 The Dining Philosophers Problem

The dining philosophers problem is a classic multi-process synchronization problem. A philosopher modelled as a process is either in state eating or

Query	Interpretation	Result
E<>P1.CriticalSection and P2.CriticalSection	P1 and P2 share the critical section	F
A[]not(P1.CriticalSection and P2.CriticalSection)	P1 and P2 do never share the critical section	T
E<>P1.CriticalSection	P1 may enter the critical section	T
A<>P1.CriticalSection	P1 will eventually enter the critical section	F

Table 1. Some Queries for the Petterson Algorithm.

thinking; from time to time he switches between these two states. Now we assume that n philosophers $(n > 1)$ sit at a dining table around a bowl of spaghetti. There are n plates at the table and n forks put between the plates. A philosophers transition from "thinking" to "eating" requires to seize two forks, the left one and the right one. Philosophers are not allowed to communicate (this is another assumption) and have to proceed according to the same internal algorithmic and deterministic behaviour. Depending on their behaviour there may occur deadlock, livelock or starvation. A proper solution is required not to suffer from any of these problems. As an example for deadlock consider the "fair" cycle of actions where each philosopher takes one fork with the intention to take the second fork during the next cycle of actions. This obviously leads to deadlock since we end in a state where each philosophers posses a single fork. The improvement to release the first fork in case the second fork is not available leads to a situation commonly known as livelock: All philosophers will forever repeat endless loops: "take fork, release fork". The usual solution of the dining philosophers problem requires that both forks are taken simultaneously.

Note that the original problem has been posed without using any concept of time. In the context of current systems engineering the consideration of time is of course of outstanding importance. Here we use the philosophers problem to illustrate the usage of time in synchronous automata networks. A timed version of this problem for $n = 3$ will suffice to sketch some of the ideas. A global clock x is introduced and each philosopher has a clock guard "$x > 4$, $x < 5$"; hence seizing the first fork is possible only when the clock is between 4 and 5. After eating the clock is reset by the assignment $x = 0$.

The initial state is (Thinking, Thinking, Thinking, Free, Free, Free); the actions Get_1, Get_2 or Get_3 may be executed not before the clock reaches the value $x = 4$. Note that the philosophers do not interact with each other

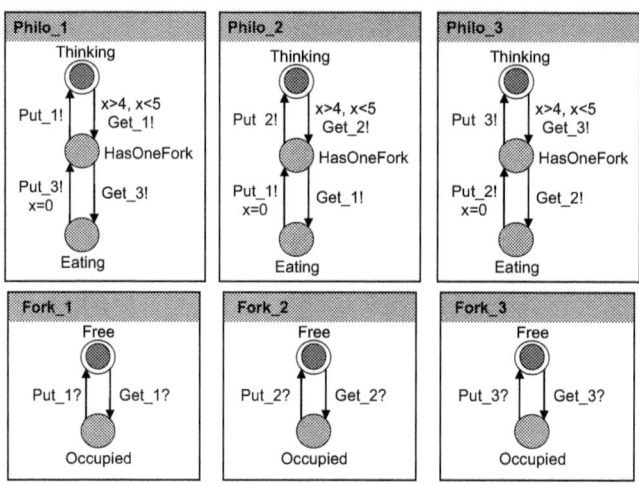

Figure 7. Dining Philosophers as Timed Automata.

directly; the actions put! and get! are synchronized with co-actions of the forks. Since the location "HasOneFork" is a so-called committed location, the philosopher must immediately try to get the second fork. Otherwise it is not excluded that any action will be delayed infinitely. Table 2 shows some example queries denoted in Timed Computational Tree Logic. The results are obtained by model checking techniques

Query	Interpretation	Result
A[]Philo_1.Eating	In all states philosopher 1 is eating	F
E<>Philo_1.Eating	There is a path where philosopher 1 is eating	T
A[]not deadlock	There is no deadlock	F

Table 2. Some Queries for the Dining Philosophers.

The model contains a deadlock, because it is not excluded that all philosophers stay thinking until the clock shows a value ≥ 5.

4.4 Real World Applications

Timed automata and the tool UPPAAL have been applied in a number of projects and case studies. We shortly list some examples from the developers' sources [$U_1$06]. For an audio control protocol which was highly dependent on real-time a known error was not found using conventional testing methods. Using UPPAAL eventually revealed the error. For a field bus communication protocol, a number of imperfections in the protocol logic were found and the error sources were debugged based on abstract models of the protocol. An audio protocol for the exchange of control information between components where the correctness relies on timing delays between signals has been modelled and verified.

5 Performance Evaluation Based on Formal Description Techniques

Automata theory and its extension to coupled automata systems have been the foundation for languages and tools that support developers and engineers to master the complexity of hardware and software design for concurrent systems. Here we give a survey on the SDL method that has been successfully applied over many years and an extension that allows to include additional quantitative aspects [$M_1$01].

5.1 The Specification and Description Language SDL

In industrial practice SDL, the Specification and Description Language, is a very successful tool. It has been used in telecommunication, real-time systems, consumer electronics and in automotive industry. SDL employs a process concept which is based on extended finite state machines; processes exchange signals via signal routes; each process has a common input queue for all incoming signals. Processes can be grouped to blocks; blocks finally build an SDL-system, see Figure 8.

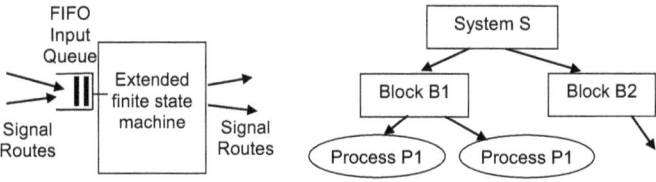

Figure 8. SDL Process and SDL System Diagram.

A system specified with SDL may serve a basis for validation, functional simulation, animation, code generation, prototyping, testing, and more. There is a number of tools which support the SDL methodology, cf. the homepage of the SDL Forum Society http://www.sdl-forum.org/.

5.2 Queueing SDL and the Tool QUEST

In order to support performance analysis, additionally non-functional properties covering time and resource aspects have to be added to support the automatic construction of a performance model [D_1M_398]. To obtain performance measures by model based evaluations implementation related quantitative properties have to be included in the model description. In detail these may cover issues like performance characteristics of hardware devices, concurrency due to exclusive resources, scheduling strategies, processing speeds, bandwidths of communication channels, buffer sizes, timeout values and more [M_101, M_1M_399].

As a response to these requirements the language Queueing SDL (QSDL) and the tool QUEST have been developed [$D_1H_0M_396$, D_198]. QSDL is an extension of SDL that allows the description of performance-related properties like time-consuming actions and competition due to exclusive resources. The extended description includes as main part the functional description in SDL and serves as basis for performance simulation. QUEST is based on the adjunction of time consuming machines that model the congestion of processes due to limited resources. By adding workload models and after defining a mapping of workload to the machines, finally the QSDL description is transformed automatically to an executable simulation program. The execution of the simulation program finally yields performance measures like response time, throughput and utilization. Transient behaviour as well as long-term steady-state measures can be derived from the simulation output.

5.3 Real World Applications

Applications of SDL cover many fields of computer and communication systems [M_101]. There are relatively few reported applications where performance aspects have been integrated directly with the SDL method. Large case studies on the TCP protocol covering performance simulation using QSDL and QUEST and specification driven monitoring have been documented in [$H_0H_1L_1M_1M_301$, H_000].

6 Final remarks

Formal methods help designers and developers of complex software to master the many problems of concurrent systems. In particular if time conditions must be considered the usage of formal methods and associated tools are indispensable. Basic research, development of better techniques and efficient algorithms, tool building and the further improvement of concurrent system engineering are essential topics in computer science. Hence concurrent models are also part of any computer science curriculum. Nowadays there are an increasing number of techniques and tools to support concurrent system engineering that are already used in practice; UPPAAL and SDL are just two examples.

In this survey a brief introduction has been given into some of the typical methods and techniques of concurrent system modelling. The intention of the author was to show how design, development and proper operation of nowadays complex hard- and software systems are finally based on formal methods.

References.

[$B_0D_0L_0$04] G. Behrmann, A. David, K. G. Larsen; A Tutorial on Uppaal, Proceedings of the 4th International School on Formal Methods for the Design of Computer, Communication, and Software Systems (SFM-RT'04), LNCS 3185.

[B_1Y04] J. Bengtsson, W. Yi; Timed Automata: Semantics, Algorithms and Tools, Lecture Notes on Concurrency and Petri Nets; W. Reisig and G. Rozenberg (eds.), LNCS 3098, Springer-Verlag, 2004.

[$D_1$98] M. Diefenbruch; Quantitative Analyse zeit- und ressourcenerweiterter SDL-Systeme mit Hilfe von Zustandsraumexploration und Model-Checking; Universität Essen, Dissertation, Logos Verlag, 1998.

[$D_1M_3$98] M. Diefenbruch, B. Müller-Clostermann; Queueing SDL: A Language for the Functional and Quantitative Specification of Distributed Systems, Workshop on Performance and Time in SDL and MSC; Universität Erlangen-Nürnberg, 1998, Workshop Proceedings.

[$D_1H_0M_3$96] M. Diefenbruch, J. Hintelmann, B. Müller-Clostermann; The QUEST-Approach for the Performance Evaluation of SDL-Systems; Proceedings of FORTE/PSVT '96, Kaiserslautern, Chapman & Hall, 1996.

[GV02] C. Girault, R. Valk; Petri Nets for System Engineering, Springer, 2002.

[$H_0H_1L_1M_1M_3$01] J. Hintelmann, R. Hofmann, F. Lemmen, A. Mitschele-Thiel, B. Müller-Clostermann; Applying techniques and tools for the performance engineering of SDL systems, Computer Networks 35, 2001, pp. 647-665.

[$H_0$00] J. Hintelmann; Entwurfsbegleitende Leistungsanalyse für SDL-basiertes Design multimedialer Internet-Transportsysteme; Dissertation, Universität Essen, 2000.

[$H_2$91] G. Holzmann; Design and Validation of Computer Protocols, Prentice Hall, 1991.

[$H_3M_2U_0$01] J. E. Hopcroft, R. Motwani, J. D. Ullman; Introduction to Automata Theory, Languages and Computation, Addison Wesley, 2001.

[L_0PY95] K. G. Larsen, P. Pettersson, W. Yi; Model-Checking for Real-Time Systems, in: Proceedings of the 10th International Conference on Fundamentals of Computation Theory, H. Reichel (ed.), Dresden, 1995. LNCS 965, pages 62-88.

[M_0K99]   J. Magee, J. Kramer; Concurrency: State Models & Java Programs, Wiley, 1999.

[$M_1$01]   A. Mitschele-Thiel; Systems Engineering with SDL - Developing Performance-Critical Communication Systems, Wiley 2001.

[$M_1M_3$99]   A. Mitschele-Thiel, B. Müller-Clostermann; Performance Engineering of SDL/MSC Systems, Computer Networks 31, 1999, pp. 1801-1815.

[$M_3$97]   B. Müller-Clostermann; Employing Deterministic and Stochastic Petri Nets for the Analysis of Usage Parameter Control in ATM-Networks; In: G. Cooperman, G. Michler and H. Vinck (eds.), Workshop on High Performance Computing and Gigabit Area Networks, University of Essen, 1996, Lecture Notes in Control and Information Sciences, Springer 1997.

[P99]   P. Pettersson; Modelling and Verification of Real-Time Systems Using Timed Automata: Theory and Practice, Ph.D. Thesis, Department of Computer Systems, Uppsala University, 1999.

[R98]   W. Reisig; Elements of Distributed Algorithms — Modeling and Analysis with Petri Nets, Springer, 1998.

[$U_1$06]   UPPAAL 2006: http://www.uppaal.com.

Received: May 9, 2006;
In revised version: August 10, 2006;
Accepted by the editors: December 6, 2006.

On Externally Definable Sets and a Theorem of Shelah

ANAND PILLAY

Department of Mathematics
University of Illinois at Urbana-Champaign
1409 W. Green Street
Urbana IL 61801-2975, United States of America
E-mail: pillay@math.uiuc.edu

In [S∞] Saharon Shelah proves the striking and important result that if T is a complete theory without the independence property, M is a model of T, and M^* is the expansion of M by adjoining predicates for "externally definable" subsets of $M, M \times M, ...$, then $\text{Th}(M^*)$ has quantifier elimination. Shelah's proof is not so long, but arguably rather obscure. The aim of this expository paper is to give a direct and conceptual proof of Shelah's theorem. In fact we give two proofs, the first going through quantifier-free heirs of quantifier-free types and the second through quantifier-free coheirs of quantifier-free types. We would hope that our proofs might suggest various strengthenings of the theorem (namely stronger conclusions). As this paper is expository we feel free to give more details than we normally would.

Special cases of Shelah's result were proved earlier in [B_0P99] (the o-minimal case) and [$B_1$01] (the weakly o-minimal case). Frank Wagner and Victor Verbovskiy ([V05]) also found a somewhat simplified account of Shelah's proof.

Our notation is standard. We work in a large saturated model $\bar{M}$ of a complete theory T in a language L. The symbols x, y range over finite tuples of variables, and $a, b, ...$ over finite tuples of elements of $\bar{M}$.

We will make systematic use of some basic notions of model theory, such as coheirs, heirs, definability of types, etc., but with respect to *complete quantifier-free types*. We will give a brief summary. For a finite tuple a of elements of M, $\text{qftp}(a)$ denotes the quantifier-free type of a. For M a model of T (*viz.* an elementary substructure of $\bar{M}$) $S_{\text{qf}}(M)$ denotes the set of complete quantifier-free types over M. Likewise for $S_{\text{qf}}(A)$, $S_{\text{qf}}(T)$, and so on. We will say that $p(x) \in S_{\text{qf}}(M)$ is *quantifier-free definable* if for each

---

The author held a Marie Curie Chair funded by the European Commission in the EXC programme. He thanks Martin Ziegler for discussions, Udi Hrushovski and Kobi Peterzil for comments on earlier versions, and the referee for a few notational suggestions.

quantifier-free L-formula $\phi(x,y)$ there is a quantifier-free formula $\psi(y)$ over M (namely with parameters from M) such that for each $b \in M$, $\phi(x,b) \in p(x)$ iff $\models \psi(b)$. We call the map taking $\phi(x,y)$ to $\psi(y)$ a (quantifier-free) defining scheme d say, for $p(x)$. If $A \supseteq M$, we can apply the scheme d to A to get $d(A) \in S_{\mathrm{qf}}(A)$, an extension of p. (Namely for $a \in A$, $\phi(x,a) \in d(A)$ iff $\models d(\phi)(a)$.)

Suppose that $p(x) \in S_{\mathrm{qf}}(M)$, $M \subseteq A$ and $q(x) \in S_{\mathrm{qf}}(A)$ is an extension of $p(x)$. In this context we say that $q(x)$ is a *quantifier-free heir* of p if for every quantifier-free formula $\phi(x,y)$ over M such that $\phi(x,b) \in q(x)$ for some $b \in A$ there is $b' \in M$ such that $\phi(x,b') \in p(x)$. Likewise we say that $q(x)$ is a *quantifier-free coheir* of $p(x)$ if q is finitely satisfiable in M, namely every formula in $q(x)$ is satisfied by a tuple from M.

Lemma 1. Suppose that M is a fixed model of T.
(i) If $p(x) \in S_{\mathrm{qf}}(M)$ is quantifier-free definable, with defining schema d, then for any $A \supseteq M$, $d(A)$ is the unique quantifier-free heir of $p(x)$ over A.
(ii) If *every* complete quantifier-free type over M (in any finite number of variables) is quantifier-free definable, then any $p(x) \in S_{\mathrm{qf}}(M)$ has a unique quantifier-free coheir over any $A \supseteq M$.

Proof. (i) is standard. (ii) Suppose a_1 and a_2 realize distinct quantifier-free coheirs of $p(x) \in S_{\mathrm{qf}}(M)$ over Mb. Then $\mathrm{qftp}(b/Ma_i)$ is a quantifier-free heir of $\mathrm{qftp}(b/M)$ for $i = 1, 2$. But for some quantifier-free formula $\phi(x,y)$ over M, we have $\models \phi(a_1,b) \wedge \neg\phi(a_2,b)$, which contradicts (i) as both a_1 and a_2 have the same quantifier-free types over M. q.e.d.

Note that in (ii) there is no reason whatsoever why the unique quantifier-free coheir of p should coincide with its unique quantifier-free heir.

The notion of a *quantifier-free indiscernible sequence* $(a_i : i < \omega)$ over a set A of parameters is clear. The following is standard.

Lemma 2.

(i) Let $p(x) \in S_{\mathrm{qf}}(M)$ be quantifier-free definable, with defining schema d. Let a_0 realize $p(x)$ and for any $n < \omega$, let a_{n+1} realizes $d(Ma_0...a_n)$. Then $(a_i : i < \omega)$ is quantifier-free indiscernible over M (and moreover $\mathrm{qftp}((a_i : i < \omega)/M)$ depends only on p.

(ii) Let $M \prec N$, and $q(x) \in S_{\mathrm{qf}}(N)$ be finitely satisfiable in M. Suppose $a_0 \in N$ realizes $q|M$ and $a_{n+1} \in N$ realizes $q|(Ma_0...a_n)$ then $(a_i : i < \omega)$ is quantifier-free indiscernible over M and again its quantifier-free type over M depends only on q.

Recall that T is said to have *the independence property* if there is a formula $\phi(x,y)$ and $\{b_i : i \in \omega\} \subset \bar{M}$ such that for each $I \subseteq \omega$, $\{\phi(x,b_i) : i \in I\} \cup \{\neg\phi(x,b_j) : j \notin I\}$ is consistent. The following is well-known and easy to prove (see [P00, 12.17]):

Fact 3. A theory T has the independence property if and only if there is an indiscernible sequence $(a_i : i \in \omega)$ in $\bar{M}$ and $\phi(x,c)$ such that $\models \phi(a_n, c)$ if and only if n is even.

We sometimes say that T has the NIP to mean that T does not have the independence property.

We now discuss the structure induced by externally definable sets, and quantifier elimination, as well as setting up notation for the statement and proof of the main theorem.

Let M be a model of T. A subset $X \subseteq M^n$ is said to be *externally definable* if there is a definable (with parameters) subset Y of $\bar{M}^n$ such that $X = Y \cap M^n$. Equivalently, it is said to be externally definable if there is some L-formula $\phi(x,y)$ and $c \in \bar{M}$ such that $X = \{a \in M^n : \models \phi(a,c)\}$. Note that if M_1 is an $|M|^+$-saturated elementary extension of M then every externally definable set in M can be externally defined by a formula with parameters from M_1. Let us fix such M_1. Let L^* be the expansion of L obtained by adding relations for all such externally definable sets X, let M^* be the corresponding expansion of M and $T^* = \text{Th}(M^*)$. Note that to all intents and purposes, every element of M^* is named by a constant, so models of T^* are just elementary extensions of M^*. It is convenient to "label" the new relations of L^* by formulas over M_1 which they come from: so given $\phi(x,y) \in L$ and $c \in M_1$ as above, let $R_{\phi(x,c)}$ be the corresponding new relation symbol in L^*. Note that we have, among other things, added relation symbols for every definable (with parameters) set in M and so:

Remark 4.

(i) $S_{\text{qf}}(T^*)$ can be identified with $S_{\text{qf}}(M^*)$.

(ii) If $p^*(x) \in S_{\text{qf}}(T^*)$ is a complete quantifier-free L^*-type of T^* then there is a unique complete L-type over M which is implied by p^*. Analogously for complete quantifier-free types over arbitrary models of T^*.

As a matter of notation, for a model N^* of T^*, we will typically denote complete quantifier-free L^*-types over N^* by $p^*, q^*, \ldots$, in which case $p, q, \ldots$ will denote the complete L-types over N implied by $p^*, q^*, \ldots$ respectively.

We are interested in the issue of when T^* has quantifier elimination. As with any theory, we have:

Fact 5. The following are equivalent:

(i) T^* has quantifier elimination in L^*.

(ii) Whenever a_1, a_2 are finite tuples in a model of T^* with the same quantifier-free type then they have the same complete type.

(iii) Whenever a_1, a_2 are finite tuples in a model of T^* with the same quantifier-free type then they have the same existential type.

Let us fix a saturated model N^* of T^*. It is convenient to choose N^* such that the L-reduct N of N^* is "well-situated" with respect to M_1.

Lemma 6. Let (N_1, N) be a saturated elementary extension of the pair (M_1, M) (in a language $L_P = L \cup \{P\}$ for P a new unary predicate symbol). Expand N to an L^*-structure N^*, by interpreting $R_{\phi(x,c)}$ as $\{a \in N : N_1 \models \phi(a, c)\}$ (for $\phi(x, y) \in L$ and $c \in M_1$). Then N^* is a saturated elementary extension of M^*.

Proof. Obvious. q.e.d.

We now let N, N^* and N_1 be as in Lemma 6. An elementary but key observation is:

Lemma 7. Any complete quantifier-free (L^*) type $p^*(x) \in S_{\text{qf}}(M^*)$ is quantifier-free-definable.

Proof. This is by the definition of M^*: Let $a \in N^*$, and $R(x, y)$ be a basic relation of L^*, say R is $R_{\phi(x,y,c)}$. Then $\{b \in M^* : N^* \models R(a,b)\}$ $= \{b \in M : \bar{M} \models \phi(a, b, c)\}$ which is an externally definable subset of M, hence quantifier-free L^*-definable. q.e.d.

With the notation above, we will now state and prove the main theorem.

Theorem 8. If T has the NIP, then T^* has quantifier elimination.

Assuming that T^* does not have quantifier elimination we will construct a sequence $(d_i : i < \omega)$ contained in N^*, which is indiscernible over M in the sense of L, but such that for some basic relation $R = R_{\chi(z,c)}(z)$ of L^* we have $N^* \models R(d_i)$ if and only if i is even. Working in $\bar{M}$ this means that $\models \chi(d_i, c)$ if and only if i is even, which, by Fact 3, shows that T has the independence property. The indiscernible sequence can be constructed easily in two (related) ways, so we give both versions below. Version II has a somewhat more explicit character.

So let us begin the proof. Assume that T^* does not have quantifier elimination. By Fact 5 let $p^*(x) \in S_{\text{qf}}(T^*) = S_{\text{qf}}(M^*)$ and let $R(x,y)$ be a basic relation in L^* such that both $p^*(x) \cup \{\exists y R(x,y)\}$ and $p^*(x) \cup \{\neg \exists y (R(x,y))\}$ are consistent with T^*. Assume $R(x,y)$ is $R_{\phi(x,y,c)}(x,y)$. Let $q^*(x,y) \in S_{\text{qf}}(M^*)$ extend $p^*(x) \cup \{\exists y (R(x,y))\}$.

VERSION I. By Lemma 7, let d_{p^*} be the quantifier-free defining schema for $p^*(x)$ and d_{q^*} the one for $q^*(x,y)$. Note that for any $A \subseteq N^*$, we have

(a) the restriction of $d_{q^*}(A)$ to the variable x is precisely $d_{p^*}(A)$, and

(b) $d_{p^*}(A) \cup \{\neg \exists y R(x,y)\}$ is consistent (with the complete diagram of N^*).

[The reason for (b) is as follows. Suppose, by way of contradiction, that $\chi(x,d) \in d_{p^*}(A)$ and $N^* \models \forall x (\chi(x,d) \to \exists y (R(x,y)))$. Let $\delta(z)$ be the defining formula for $\chi(x,z)$. So $N^* \models \delta(d)$. Hence there is $d' \in M^*$ such that $M^* \models \delta(d') \land \forall x (\chi(x,d') \to \exists y (R(x,y)))$. But then $\chi(x,d') \in p^*(x)$ and $p^*(x)$ is inconsistent with $\neg \exists y (R(x,y))$, a contradiction.]

We now build a sequence $\{a_0 b_0, a_1, a_2 b_2, a_3, ...\}$ in N^*, by choosing $a_n b_n$ to realize $d_q(M^* a_0 b_0 .... a_{n-1})$ for even n and choosing a_n to realize

$$d(M^* a_0 b_0 a_1 ... a_{n-1} b_{n-1}) \cup \{\neg \exists y R(x,y)\}$$

for odd n (using (b) above). It follows from the construction, and (a) above, that

(c) the sequence $((a_0 b_0), (a_2 b_2), (a_4 b_4), ...)$ is quantifier-free L^* indiscernible, and

(d) the quantifier-free L^*-type of $(a_0 b_0, a_1, a_2 b_2, a_3, ...)$ equals the quantifier-free L^*-type of $(a_0 b_0, a_2, a_4 b_4, a_6, ...)$.

So by Remark 4 (ii), we conclude from (c), (d) respectively that

(e) the sequence $((a_n b_n) : n = 0, 2, 4, 8, ...)$ is indiscernible over M (in L), and

(f) $(a_0 b_0, a_1, a_2 b_2, a_3, ...)$ and $(a_0 b_0, a_2, a_4 b_4, a_6, ...)$ have the same complete L-types over M.

As N is saturated, we can by (e), (f), and compactness find $b_i \in N$ for i odd, such that $((a_i, b_i) : i < \omega)$ is indiscernible over M (in the sense of L). For i odd we have $N^* \models \neg \exists y (R(a_i, y))$, so $N^* \models \neg R(a_i, b_i)$. Hence $\bar{M} \models \phi(a_i, b_i, c)$ if and only if i is even, implying that T has the independence property. q.e.d. (**Version I**)

VERSION II. By Lemmas 7 and 1, let $q_1^*(x, y) \in S_{\text{qf}}(N^*)$ and $p_1^*(x) \in S_{\text{qf}}(N^*)$ be the (unique) quantifier-free coheirs over N^* of $q^*(x, y), p^*(x)$ respectively. We have

(a) $q_1^* | x = p_1^*$, and

(b) $p_1^*(x) \cup \{\neg \exists y R(x, y)\}$ is consistent (with complete diagram of N^*).

[Both (a) and (b) are by uniqueness: for example we know that $p^*(x) \cup \{\neg \exists y R(x, y)\}$ is consistent, so it extends to a complete type over N^* which is finitely satisfiable in M^*. The quantifier-free part of this type must be $p_1^*(x)$.]

We could now repeat Version I. But we will give a slightly more explicit construction. Let $q_1(x, y)$ be the unique complete L-type over N implied by $q_1^*(x, y)$ (see Remark 4).

Claim. $q_1(x, y) \cup \{\neg R(x, y)\}$ is consistent (with the complete diagram of N^*).

[In some saturated elementary extension of N^*, let a_1 realize $p_1^*(x) \cup \{\neg \exists y R(x, y)\}$, and let (a_2, b_2) realize $q_1^*(x, y)$. Then a_1 and a_2 realize p_1^* hence realize the same complete L-type over N. Hence there is an L-automorphism taking a_2 to a_1. If b_1 is the image of b_2 then (a_1, b_1) realizes $q_1(x, y) \cup \{\neg R(x, y)\}$.]

We now build our sequence $(a_n b_n : n < \omega)$ in N^* by letting $a_n b_n$ realize

$$q_1^*(x, y) | (M^* a_0 b_0 ... a_{n-1} b_{n-1})$$

for n even, and realize

$$q_1(x, y) | (M^* a_0 b_0 ... a_{n-1} b_{n-1}) \cup \{\neg R(x, y)\}$$

for n odd. Then $(a_n b_n : n < \omega)$ is indiscernible over M in the sense of L (as a coheir sequence determined by $q_1(x, y)$). But $N^* \models R(a_n, b_n)$ if and only if n is even, hence $\models \phi(a_n, b_n, c)$ if and only if n is even, showing again that T has the independence property. q.e.d. (**Version II**)

References.

[B₀P99] E. Baisalov and B. Poizat, Paires de structures o-minimales, Journal of Symbolic Logic 63 (1999), 570-578.

[B₁01] B. Baizhanov, Expansion of a model of a weakly o-minimal theory by a family of unary predicates, Journal of Symbolic Logic 66(2001), 1382-1414.

[P00] B. Poizat, *A course in Model Theory*, Springer, 2000.

[S∞] S. Shelah, Dependent first order theories, continued. SH783.

[V05] V. Verbovskiy, Dependent theories. Shelah's theorem, *preprint* 2005.

Received: March 17, 2006;
In revised version: March 30, 2006; June 7, 2006;
Accepted by the editors: December 7, 2006.

A Note on the Categoricity of Countable Interval Structures

KLAUS U. SCHULZ

Centrum für Informations- und Sprachverarbeitung
Ludwig-Maximilians-Universität München
Oettingenstraße 67
80538 München, Germany
E-mail: schulz@cis.uni-muenchen.de

ABSTRACT. The rational intervals (2-tuples of different, ordered rational numbers, the endpoints) introduced by J. van Benthem [v83] and studied by P. Ladkin and R. D. Maddux [LM87] offer a natural way to represent interval-based time by means of countable structures. Structures composed of these rational intervals with relations defined in terms of the relative order of their endpoints were shown to have countably categorical theories. Here we prove an abstract model-theoretic lemma which shows how translation of the interval relations into linear order relations is directly responsible for this categoricity phenomenon. With the lemma, the categoricity of the interval structure follows immediately from the categoricity of the dense linear order of the rationals. New categoricity results for more general intervals in arbitrary dimensions can be proved. We discuss the effect of introducing an additional predicate to express relative length. Some interval relations may be combined with this predicate without losing categoricity. But in general we get various non-isomorphic countable models.

Introductory Remark. The following paper represents a historical manuscript from 1988 that originally was only published as in internal report [S88]. Is has been cited in [M89]. Some of the results, which were new in 1988 to the best of our knowledge, can now be found in [H93]. Since then, the author moved to the field of computational linguistics. The author thanks the anonymous referee for his remarks.

1 Background from Model Theory

We first recall some basic definitions. A theory is called *countably categorical* if it has only one countable model up to isomorphism. A consistent theory in a language L is called *complete* if for every sentence σ of L the theory

---

The author would like to thank Micheal Morreau for helpful comments.

entails exactly one of the sentences σ, $\neg\sigma$. Equivalently, we could say that a consistent L-theory is complete if it has the following property: whenever an L-sentence σ is consistent with T, then σ is already a consequence of T. Obviously, the set of all sentences which are true in a fixed structure is always a complete theory.

Definition 1. Let L be a countable first-order language with variables $x_1, x_2, \ldots$ and let $L^{(n)}$ be the set of all formulas of L which contain at most $x_1, \ldots, x_n$ as free variables. Let T be any consistent L-theory. A subset p of $L^{(n)}$ is called an *n-type* of T if and only if

(a) for every σ in $L(n)$, σ is in p or $(\neg\sigma)$ is in p, and

(b) for every finite subset $\{\sigma_1, \sigma_2, \ldots, \sigma_k\}$ of p, the sentence

$$\exists x_1 \ldots \exists x_n \, (\sigma_1 \, \& \, \ldots \, \& \, \sigma_k)$$

is consistent with T.

A type p is called *principal* if p contains a formula σ such that for all formulas δ of p, the formula $\forall x_1 \ldots \forall x_n \, (\sigma \to \delta)$ is a consequence of T.

The definition of a (principal) n-type plays an important role in theoretical investigations of countable categoricity phenomena. This is made clear by the following theorem of Engeler [E59] and Ryll-Nardzewski [R$_1$59], which allows one to decide whether or not a theory is countably categorical without knowing a single model.

Theorem 2. Let T be a complete first-order theory. Then the following are equivalent:

(a) T is countably categorical.

(b) For every $n > 0$: T has only finitely many n-types.

(c) For every $n > 0$: every n-type of T is a principal type.

Definition 3. Let T_1 and T_2 be theories of the languages L_1 and L_2 respectively. A recursive function * from the sentences of L_1 into sentences of L_2 is called an *interpretation* (of T_1 in T_2) iff for all L_1-sentences σ:

$$T_1 \text{ entails } \sigma \text{ iff } T_2 \text{ entails } \sigma^*.$$

Definition 4. Let * be an interpretation. Suppose σ^* is also defined for all formulas σ and that every variable x_i occurring in a formula σ of L_1 corresponds via * to a finite sequence of variables $y_{i,1}, \ldots, y_{i,m}$ of σ^*, for fixed m. We say $*$ is a *typical* interpretation iff $*$ is homomorphic with respect to $\rightarrow, \&, \neg$ and if $(\exists x_i \sigma)^*$ is $\exists y_{i,1} \ldots \exists y_{i,m} \sigma^*$, for all formulas σ.

In fact most interpretations are typical interpretations. But there is a better reason to call them typical:

Lemma 5. Let T_1, T_2, L_1, L_2, and $*$ as in Definition 3. Suppose $*$ is a typical interpretation. Then for every n-type p of T_1 the set $p^* := \{\sigma^*; \sigma \in p\}$ can be extended to an mn-type of T_2. Furthermore, if p_1 and p_2 are different n-types of T_1 and $p_1^{*\prime}$, $p_2^{*\prime}$ are mn-types of T_2 extending p_1^* and p_2^* respectively, then $p_1^{*\prime}$ and $p_2^{*\prime}$ are different.

Proof. A trivial application of Zorn's Lemma shows that every subset of $L_2^{(n)}$ satisfying Condition 2 of Definition 1 can be extended to a maximal set with this property, which then satisfies also Part 1 of the definition. Hence, to prove the first part of the lemma it suffices to show that p^* fulfills Condition 2. Suppose $\sigma_1^*, \ldots, \sigma_k^*$ is a finite sequence of formulas of p^*. Then $\sigma_1, \ldots, \sigma_k$ belong to p and the sentence $\sigma = \exists x_1 \ldots \exists x_n \, (\sigma_1 \, \& \, \ldots \, \& \, \sigma_k)$ is consistent with T_1. If σ_{false} is any absurd sentence of L_1, then the sentence $\sigma \rightarrow \sigma_{\text{false}}$ is not a consequence of T_1. Hence $(\sigma \rightarrow \sigma_{\text{false}})^*$ is not a consequence of T_2. Since $*$ is a typical interpretation $(\sigma \rightarrow \sigma_{\text{false}})^*$ is the sentence $(\exists y_{1,1} \ldots \exists y_{n,m} \sigma_1^* \, \& \, \ldots \, \& \, \sigma_k^*) \rightarrow \sigma_{\text{false}}^*$ where σ_{false}^* is absurd with respect to T_2. Therefore $\exists y_{1,1} \ldots \exists y_{n,m} \sigma_1^* \, \& \, \ldots \, \& \, \sigma_k^*$ is consistent with T_2. This proves the first part.

Now suppose p_1 and p_2 are different n-types of T_1. By Part 1 of Definition 1, there exists at least one formula σ in $L_1^{(n)}$ such that p_1 contains σ while p_2 contains $\neg \sigma$. (Clearly neither of them contains both formulas, by Condition 2.) Since $*$ is homomorphic with respect to $\neg$, p_1^* contains σ^* while p_2^* contains $\neg \sigma^*$. Hence $p_1^{*\prime}$ and $p_2^{*\prime}$ are different. q.e.d.

Our main lemma is an easy consequence of Lemma 5 and Theorem 2:

Lemma 6. Let T_1 and T_2 be countable theories. Suppose T_2 is countably categorical. If there is a typical interpretation $*$ of T_1 in T_2, then T_1 is countably categorical, too.

Proof. Clearly T_2 is complete. Since we have an interpretation of T_1 in T_2 the former theory is complete, too. Hence, for every $n > 0$ theory T_2 has only a finite number of n-types, by Theorem 2. If T_1 would have infinitely many (different) n-types, then the construction of Lemma 5 would give infinitely many mn-types of T_2, which is impossible. Hence, for every $n > 0$ also T_1 has only a finite number of n-types. By Theorem 2, T_1 is countably categorical. q.e.d.

Let M be a structure and $(i_1, ..., i_n)$ be an n-tuple of different elements. The *n-orbit* of $(i_1, \ldots, i_n)$ is the set consisting of all n-tuples $(j_1, \ldots, j_n)$ such that there exists an automorphism of M which maps i_k onto j_k (for $1 \le k \le n$). A subset O of M^n is called an *n-orbit* of M if and only if there is an n-tuple $(i_1, \ldots, i_n)$ such that O is the n-orbit of $(i_1, \ldots, i_n)$. The next (well-known) lemma gives an algebraic characterization of countably categorical structures:

Lemma 7. Let M be a countable structure. If for all $n > 0$ the number of n-orbits of M is finite, then the theory T of M is countably categorical.

Proof. n-tuples of elements which have the same orbit define the same n-type. (The type defined by an n-tuple consists of all formulas with at most n free variables $x_1, \ldots, x_n$ which are satisfied by the n-tuple in M.) Hence, only finitely many n-types $p_1, \ldots, p_k$ are defined by n-tuples of elements of M. Suppose T has an additional n-type p. Then there is a formula σ_i of p_i such that $\neg \sigma_i \in p$ ($1 \le i \le k$). Since p is a type, $\exists x_1 \ldots \exists x_n \neg \sigma_1 \& \ldots \& \neg \sigma_k$ is consistent with T. But T is complete, therefore $\exists x_1 \ldots \exists x_n \neg \sigma_1 \& \ldots \& \neg \sigma_k$ is a consequence of T. This is a contradiction, since every n-tuple of elements of M satisfies at least one formula σ_i. The lemma follows by Theorem 2.

q.e.d.

2 Applications

2.1 The rational line

Our first aim is to show that we always get a countably categorical structure if we take the set of all m-tuples of different, ordered rationals as our domain and use only relations which are definable in terms of linear order relations on the elements of the tuples. To be precise we need the following definition:

Definition 8. Denote by $(Q, \le)$ the linear order of the rationals and by $\text{Int}^{(m)}(Q)$ the set of all m-tuples of different, ordered rationals. An n-ary relation R on $\text{Int}^{(m)}(Q)$ is called *order-definable* if there is a formula σ in the language of linear orders which contains at most $y_{1,1}, \ldots, y_{n,m}$ as its free variables and which has the following property: for every sequence $i_1, \ldots, i_n$ of elements of $\text{Int}^{(m)}(Q)$:

$$(i_1, \ldots, i_n) \in R \text{ iff } (Q, \le) \text{ satisfies } \sigma[i_{1,1}, \ldots, i_{n,m}]$$

(where $i_{j,k}$ denotes the kth component of i_j).

Remark 9. (a) Since the theory of $(Q, \le)$ allows elimination of quantifiers (see, for example, [CK77]) we may assume without loss of generality that

the formula σ occurring in Definition 8 is quantifier-free. But the results of the first chapter are also applicable to countably categorical structures which do not allow elimination of quantifiers, hence we did not formally restrict the definition to the quantifier-free case.
(b) In what follows we assume that $(Q, \leq)$ satisfies $\sigma[i_{1,1}, \ldots, i_{n,m}]$ only if all m-tuples $(i_{k,1}, \ldots, i_{k,m})$ are elements of $\text{Int}^{(m)}(Q)$ $(1 \leq k \leq n)$. There is no loss of generality: if necessary we may add the formulas $y_{k,1} < \ldots < y_{k,m}$ $(1 \leq k \leq n)$ to σ.

Lemma 10. Any structure S with $\text{Int}^{(m)}(Q)$ as its domain and only order-definable relations has a countably categorical theory.

Proof. For any order-definable relation R of S let σ_R be the formula of the language of linear orders which defines R in the sense of Definition 8. Now we define a typical interpretation * as follows:

- if σ is an atomic formula $R(x_1, \ldots, x_n)$, then σ^* is $\sigma_R(y_{1,1}, \ldots, y_{n,m})$,

- if σ is $\sigma_1 j \sigma_2$ where $j \in \{\&, \rightarrow\}$, then σ^* is $\sigma_1^* j \sigma_2^*$,

- if σ is $\neg \sigma_1$ then σ^* is $\neg \sigma_1^*$, and

- if σ is $\exists x_i \sigma_1$, then σ^* is $\exists y_{i,1} \ldots \exists y_{i,m} \sigma_1^*$.

It is trivial to show that * is in fact a typical interpretation of the theory of S in the theory of $(Q, \leq)$. It is well known that the latter theory is countably categorical (see, for example, [CK77]). By Lemma 6, the theory of S is countably categorical. q.e.d.

Let us give a number of examples: $\text{Int}^{(2)}(Q)$ is just the set of rational intervals which was used by van Benthem [v83] and Ladkin-Maddux [LM87] to represent temporal intervals. Van Benthem dealt with two relations *precedes* and *contained-in*, which are defined as follows:

$$i_1 = (i_{1,1}, i_{1,2}) \text{ precedes } i_2 = (i_{2,1}, i_{2,2}) \quad \text{iff} \quad i_{1,2} < i_{2,1},$$
$$i_1 \text{ contained-in } i_2 \quad \text{iff} \quad i_{1,1} \geq i_{2,1} \ \& \ i_{1,2} \leq i_{2,2}.$$

The relations of Ladkin-Maddux are the thirteen relations introduced by Allen [A83], which represent all possible ways of ordering the endpoints (where endpoints of different intervals are allowed to coincide). As a matter of fact all these relations are order-definable. Hence, the situations discussed in [v83] and [LM87] are special cases of Lemma 10 for $m = 2$.

2.2 The rational plane and higher dimensions

Rational intervals are natural candidates for representing temporal intervals. Extended plane locations or spatial locations are closely related objects which have their own interest: reasoning on such locations is of central importance to various fields of robotics (see, for example, B. R. Donald [D87]). From a logical point of view, the question arises as to whether higher dimensional structures remain categorical. The answer is yes but the suitable notions of an interval and an order-definable relation are more restrictive. Since all results can easily be generalized we will restrict ourselves to two dimensions. We denote by Q^2 the set of all points of the 2-dimensional euclidean space which have two rational coordinates.

Definition 11. An *interval* of Q^2 is a pair (p,q) of elements of Q^2 such that both coordinates of p are strictly greater than the corresponding coordinates of q. The points p and q are called the *components* of the interval.

Graphically, an interval is an rectangle whose sides are parallel to the axes of the coordinate system. The rectangle is represented using two endpoints of one diagonal.

Definition 12. We denote by $\text{Int}(Q^2)$ the set of all intervals of Q^2. An n-ary relation R on $\text{Int}(Q^2)$ is called *order-definable* iff there exists a formula σ of the language of linear orders with at most $4n$ free variables such that for all sequences $i_1, \ldots, i_n$ of intervals of Q^2 we have: $(i_l, \ldots, i_n) \in R$ iff $(Q, \leq)$ satisfies $\sigma[i_{1,1}, \ldots, i_{n,4}]$ (where i_k is the pair with the components $(i_{k,1}, i_{k,2})$ and $(i_{k,3}, i_{k,4})$). We assume without loss of generality that a sequence $i_{1,1}, \ldots, i_{n,4}$ of rational numbers satisfies σ in $(Q, \leq)$ only if all pairs $((i_{k,1}, i_{k,2}), (i_{k,3}, i_{k,4}))$ are elements of $\text{Int}(Q^2)$ ($1 \leq k \leq n$).

Lemma 13. Any structure with $\text{Int}(Q^2)$ as its domain and only order-definable relations has a countably categorical theory.

Proof. As above. q.e.d.

We give some examples: To have a graphical description of some order-definable relations on $\text{Int}(Q^2)$ we make use of the rectangles corresponding to the intervals. The relation *contained-in* which expresses that the first rectangle is a subset of the second is order-definable. The relation that the intersection of the rectangles is a rectangle (is empty, is a single point) is order-definable. The relation *touches* which expresses that the intersection of the rectangles is a nonempty subset of a line is order-definable. In all cases it is not difficult to find a defining formula in the sense of Definition 12. Furthermore it would not be difficult to introduce an Allen-style system [A83] of mutually excluding order-definable relations which is complete in the sense that every pair of two-dimensional intervals fulfills exactly one of the relations.

The above notion of an interval might be thought too restrictive. Why, for example, do we not allow arbitrary convex polygons to be intervals? In fact we could, but then the order-definable relations would not be very interesting. Even *contained-in* would not be order-definable. To see this, take all vertices of a polygon P and draw both parallels to the axes. These parallels define a finite system of rectangles. Moreover, if one side of P is not parallel to an axis, some rectangle is decomposed by the side into two parts of which the first is a subset of the polygon while the second lies outside. We can find two other (small) convex polygons, one contained in the first part, the other in the second. By construction, both polygons will fulfill the same order-definable relations with respect to P. The argument shows that we must have the sides of the intervals parallel to the axes if we wish to have reasonable order-definable relations. But there is no need to use only intervals which correspond to rectangles. We may use unions of such basic intervals as elements. The sequence of the pairs of diagonal endpoints can be used as a formal representation. (If desired, order-definable constraints may exclude disconnected unions.) Arbitrary figures with finite extension can be approximated by such generalized intervals.

3 Limits of Categoricity Results

Let us return to the rational intervals. The relations introduced by van Benthem, Allen and Ladkin-Maddux are order-definable. They only express statements on the relative order of the endpoints. But temporal reasoning in general has to deal with more expressive relations. For example, it would be interesting to assign a length to the intervals. Clearly, if we want to maintain categoricity we should be as cautious as possible. As a first step we discuss what happens if we add the relation *congruent*, which expresses that two intervals have the same length. We restrict the discussion to the thirteen interval relations introduced by Allen [A83] and we assume the reader is familiar with these relations and their definitions. Ladkin-Maddux [LM87] have shown that Allen's transitivity table (see [A83]) defines a relational algebra in the sense of Jonsson-Tarski [JT52]. This relational algebra is generated by each of the relations *meets*, *precedes*, *overlaps*, and their inverses ([LM87]). This entails that all thirteen relations are definable in terms of such a generating relation. It is interesting to note that exactly these generating relations destroy the countable categoricity in connection with *congruent*:

Lemma 14. Let $I = \text{Int}^{(2)}(Q)$ be the set of rational intervals. Let R be any one of Allen's interval relations. Denote by C the binary interval relation which is true for two rational intervals if and only if these intervals have the same length. If R is one of the relations *meets*, *precedes* or *overlaps*, or the inverse of one of these relations, then the theory of (I, R, C) is not countably categorical. If R is one of the seven other relations *during*, *starts*, *finishes*, their inverses and *equals*, then the theory of (I, R, C) remains countably categorical.

Proof. We first show that the theory of (I, R, C) is not countably categorical in the first case. Since R generates the relational algebra we may assume without loss of generality that R is *meets*. For every $n > 0$ consider the set p_n of all formulas $\sigma(x_1, x_2)$ of the language of (I, R, C) with at most x_1 and x_2 as free variables such that the intervals $(-1, 0)$ and $(n, n+1)$ satisfy σ in (I, R, C). Clearly each set p_n is a 2-type of the theory of (I, R, C). The set p_n contains the formula "there exist m intervals $x_{2+1}, \ldots, x_{2+m}$ which all have the same length as x_1 such that x_1 meets x_3 and x_3 meets x_4 and ... and x_{2+m} meets x_2" if and only if $m = n$. Hence, for $m \neq n$ the types p_m and p_n are different. By Theorem 2, the theory of (I, R, C) is not countably categorical.

We now give a sketch for the remaining part of the proof. As an example, we treat the relation *starts*. Call two n-tuples $(i_1, \ldots, i_n)$ and $(j_1, \ldots, j_n)$ *similar* iff for all $k \neq l \in \{1, \ldots, n\}$ the following conditions hold: $length(i_k) < length(i_l)$ iff $length(j_k) < length(j_l)$ and $starts(i_k, i_l)$ iff $starts(j_k, j_l)$. Clearly, "similar" defines an equivalence relation. Moreover the number of equivalence classes is obviously finite. Next we show that for similar n-tuples $(i_1, \ldots, i_n)$ and $(j_1, \ldots, j_n)$ there always exists an automorphism Φ of (I, R, C) such that $\Phi(i_k) = j_k$ ($1 \leq k \leq n$). Then the result follows by Lemma 7. To find an automorphism for similar n-tuples the following observation is fundamental: every bijective map of I onto I which satisfies the following conditions (a) and (b) is an automorphism of (I, R, C). Let f be any bijective and monotonic function from the positive rationals into themselves:

(a) All rational intervals of length q are mapped on rational intervals of length $f(q)$.

(b) Intervals with the same starting point are mapped onto intervals with the same starting point.

Let $(i_1, \ldots, i_n)$ and $(j_1, \ldots, j_n)$ be similar n-tuples. Since the rationals represent a dense linear order, the function f_0 mapping $length(i_k)$ to $length(j_k)$ ($1 \leq k \leq n$) can be extended to a bijective and monotonic function f as described above. It then follows easily that a bijective map of I onto I satisfying (a) and (b) and mapping i_k to j_k ($1 \leq k \leq n$) exists. The other relations *finishes*, *during*, their inverses and *equals* are treated in a similar way. q.e.d.

It should be noted that it is not difficult to construct a model of the theory of $(I, meets, C)$ which is not isomorphic to $(I, meets, C)$. The basic methods of Non-Standard Analysis (see, for example, [R$_0$66] or [K76]) can be used to construct countable interval structures which have infinitesimal small (but non-point-) intervals. Within these structures there exist infinite sequences of meeting intervals of the same length which are not cofinal in the sense that every interval of the structure precedes one interval of the sequence. Hence, the archimedean order of the standard model excludes any isomorphism. Non-standard models obviously exist also for the theories which are countably categorical. This shows that the categoricity results should be interpreted in the sense that the standard model allows isomorphisms to structures which are –in a naïve graphical way– quite different.

References.

[A83] J. F. Allen. Maintaining Knowledge about Temporal Intervals. *Communications ACM*, 26(11):832–843, 1983.

[CK77] C. C. Chang and H. J. Keisler. *Model Theory*. North-Holland, Amsterdam, 1977.

[D87] B. R. Donald. A search algorithm for motion planning with six degrees of freedom. *Artificial Intelligence*, 31(3):295–353, 1987.

[E59] E. Engeler. A characterization of theories with isomorphic enumerable models. *Notices Am. Math. Soc.*, 6:161, 1959.

[H93] W. Hodges. *Model Theory*. Cambridge University Press, 1993.

[JT52] B. Jónsson and A. Tarski. Boolean algebras with operators II. *Am. J. Math.*, 74:127–162, 1952.

[K76] H. J. Keisler. *Elementary calculus*. Prindle, Weber & Schmidt, Boston, 1976.

[LM87] P. Ladkin and R. D. Maddux. The algebra of convex time intervals. Technical Report Research Report KES.U.87.2, Kestrel Institute, 1987.

[M89] R. D. Maddux. Some algebras and algorithms for reasoning about time and space. Presented on April 5, 1989 to the Applied Mathematics-Engineering Colloquium, Iowa State University, 1989.
http://citeseer.ist.psu.edu/maddux90some.html.

[R$_0$66] A. Robinson. *Non-Standard Analysis*. North-Holland, Amsterdam, 1966.

[R$_1$59] C. Ryll-Nardzewski. On the categoricity in power $\aleph_0$. *Bull. Acad. Polon. Sci. Ser. Sci. Math, Astron. Phys.*, 7:545–548, 1959.

[S88] K. Schulz. On the categoricity of countable interval structures. Technical Report SNS-Report 88-34, Seminar für natürlich-sprachliche Systeme, University of Tübingen, 1988.

[v83] J. van Benthem. *The Logic of Time*. Reidel, Dordrecht, 1983.

Received: September 30, 2005;
In revised version: April 27, 2006;
Accepted by the editors: December 5, 2006.

Countable Homogeneous and Partially Homogeneous Ordered Structures

JOHN K. TRUSS

Department of Pure Mathematics
University of Leeds
Leeds LS2 9JT, United Kingdom
E-mail: pmtjkt@leeds.ac.uk

> ABSTRACT. I survey classification results for countable homogeneous or 'partially homogeneous' ordered structures. This includes some account of Schmerl's classification of the countable homogeneous partial orders, outlining an extension of this to the coloured case, and also treating results on linear orders, and their generalizations, trees and cycle-free partial orders.

1 Introduction

In this paper I shall describe a number of classifications of countable homogeneous or partially homogeneous structures, concentrating on partial and linear orders and coloured versions of these. The starting points are therefore [S79] and [D85]. By saying that a structure is 'homogeneous' we mean that any isomorphism between finite substructures extends to an automorphism. (In some papers this is called 'ultrahomogeneous' to distinguish it from other uses of the word 'homogeneous'.) We may regard it as an interesting exercise to try to classify at least some of the structures in a certain class, in the hope that this will enable us to understand the whole class better; in addition, since homogeneity is a restriction on the automorphism group, such classifications can furnish interesting examples of permutation groups having a rich structure, which may in some cases enable us to 'recognize' the structure from its group.

It was Schmerl who succeeded in [S79] in giving a complete list of all the countable homogeneous partial orders. This list is relatively restricted. In particular it only contains countably many structures, and the corresponding automorphism groups have been studied in a number of papers, with most work being done on the 'generic' partial order. Simplicity results were for instance presented in $[G_1M_1R_193]$. If one wishes to consider larger classes, various weakenings of the notion of homogeneity are available. The first is to consider 'partial homogeneity'. We say, for instance, that a structure is *k-homogeneous* if any isomorphism between substructures of size k extends to an automorphism. We also say that it is *k-transitive* if *some* isomorphism between isomorphic substructures of size k (not necessarily the originally given one) extends to an automorphism. The case of 1-transitive linear orders was studied by Morel [$M_2$65], and in his thesis [D85], Droste obtained results about k-homogeneous partial orders for various k. He concentrated on the case of trees, the most important classification being of all the countable 2-transitive trees. This was extended to 'weakly 2-transitive trees' in [$DH_1M_0$89], and to 1-transitive trees in [$C_2$04]. Certain non-trees, called 'cycle-free partial orders' were classified in [C_3T_1W99, $T_1$99, W97].

Another direction is to consider 'coloured' versions of the original structures. A *coloured partial order* is a structure of the form $(P, <, F)$ where $(P, <)$ is a partial order and F is a function from P onto a set C, thought of as the 'colour set'. Here isomorphisms are required to preserve colours as well as the order, and we immediately have notions of homogeneity and partial homogeneity for such structures. Colours actually arise naturally anyway, even in the monochromatic context, since any structure can automatically be made 1-homogeneous, say, by giving different colours to each of the orbits of its points under the action of its automorphism group. In addition, colours arose naturally in [$DH_1M_0$89] and [W97] as certain cuts in chains, which meant that it was necessary to classify particular coloured chains, which arose in the general description of these other more involved structures. This led directly to the work done by Campero-Arena [$C_0$02], generalizing Morel's work on 1-transitive chains to the coloured case. For finite sets of colours this amounted to a modification of Morel's work, but allowing infinitely many colours greatly added to the complexity of the problem. The results for finite colour sets were related to work of Rosenstein [$R_0$82]. Instead of homogeneity or partial homogeneity as a hypothesis he considered $\aleph_0$-categoricity, and gave a complete classification of all $\aleph_0$-categorical linear orders.

In yet another direction, we may consider structures which are not even partial orders at all, namely graphs or directed graphs, and I only mention these in passing to compare with the partial orders results. In fact classifications in these cases were one of the main initial motivations for a study of various homogeneous orders. Lachlan and Woodrow [LW80] extended Gardiner's classification [$G_0$76] of finite homogeneous graphs to the countable case, and in later work, Lachlan [L84] and Cherlin [$C_1$98] classified the countable homogenous tournaments, and the countable homogeneous digraphs respectively. We only mention the following points about these results. Gardiner's list, and the extension by Lachlan and Woodrow, each comprises just countably many structures (and there are only five countable homogeneous tournaments). However, Cherlin's classification contains $2^{\aleph_0}$ structures (the majority constructed by Henson in [$H_0 72$]). Before Cherlin's proof, it was thought that because there were known to be uncountably many examples, that in itself would rule out any 'classification'. Despite this, Cherlin's list is reasonably explicit, and most of the uncountably many structures are described in terms of the choice of a set of integers (a 'real number'). In the classifications I describe in this paper, there are just countably many in these cases: homogeneous partial orders, $\aleph_0$-categorical linear orders, 2-transitive trees, there are $2^{\aleph_0}$ for homogeneous coloured partial orders, 1-transitive coloured linear orders when the colour set is infinite, weakly 2-transitive trees, and 1-transitive trees. There is a third possibility, namely $\aleph_1$ (which of course may be different from $2^{\aleph_0}$) which applies to these cases: 1-transitive linear orders, 1-transitive coloured linear orders where the colour set is finite, and 3-CS-transitive cycle-free partial orders.

What we ultimately seek in each case is a 'classifier'. In the easiest and simplest cases this may just be a natural number, or pair of natural numbers, as for instance in the classification of finite fields (we may just use the cardinality of the field, or else the pair consisting of characteristic and dimension). The idea however is that we should be able to read off from a classifier what the structure is, and that it should give us some insight into how a general structure in the class being classified is built up, or what it looks like, and therefore it is often preferable for the classifier itself to carry some structure. Now Rosenstein described the $\aleph_0$-categorical countable linear orders by saying that they are formed from singletons by closing up under two operations, sums (concatenations) and 'shuffle' (see later for the details). If we want to describe these more explicitly, then a natural method is to represent them by a (finite) tree describing which operations were used, and at what stage. This is precisely the route taken in [$C_0 T_1 04a$], and these 'coding trees' provide quite a neat and visual method for describing the countable 1-transitive coloured linear orders with finite colour set. In the extension to infinitely many colours, it was found that the use of coding trees, previously merely cosmetic and organizational, is really essential, since the class of structures being classified is so much more involved. In the other instances mentioned above the classifiers comprise, for homogeneous partial orders and for 2-transitive trees, at most two cardinality parameters from 1 to $\aleph_0$; for homogeneous directed graphs, a real (subset of ω) together with cardinalities in some cases; likewise for countable weakly 2-transitive trees; for countable 1-transitive linear orders, essentially just a countable ordinal; and for homogeneous countable coloured partial orders, what Torrezão calls a 'skeleton'. This is really just an analogue of 'coding tree' tailored for the type of structure under consideration.

Now describing the countable homogeneous structures of a particular relational similarity type is equivalent to determining all the amalgamation classes of finite structures in that class, as was shown by Fraïssé. Specifically, if $\mathcal{A}$ is a countable homogeneous structure in a countable relational language, then the *age* of $\mathcal{A}$, which is defined to be the class of all structures isomorphic to finite substructures of $\mathcal{A}$, has the following properties: it is closed under isomorphisms and substructures, it has at most countably many members up to isomorphism, and it has the joint embedding and amalgamation properties. Conversely, for any class of finite structures fulfilling these properties (an *amalgamation class*), there is a countable homogeneous structure unique up to isomorphism, whose age is the class that we started with. Often the homogeneous structure arising from a given amalgamation class in this way is referred to as the corresponding *Fraïssé generic* or *Fraïssé limit*. So classifying a class of countable homogeneous structures is equivalent to classifying the corresponding amalgamation classes, and we shall pass freely between the two as the occasion arises.

The structures $\mathbb{Q}_C$ and P_C

A Fraïssé generic structure which will feature frequently in the paper is the 'C-coloured version of the rationals' $\mathbb{Q}_C$ where C is a finite non-empty or countable set. This has domain (isomorphic to) $\mathbb{Q}$, and is 'interdense', meaning that between any two points, there are points of all possible colours. It exists and is unique up to isomorphism, and it arises as the Fraïssé limit of the class of all finite C-coloured linear orders, so is homogeneous. In a similar way we may see that there is a countable generic coloured partial order P_C, which is the Fraïssé limit of the class of all finite C-coloured partial orders. It has to be verified that this is indeed an amalgamation class, which is not completely obvious and requires a little work.

The paper is organized as follows. In §2 we describe Schmerl's classification of the countable homogeneous partial orders, and its modification by Torrezão in the coloured case. In §3 we move on to consider 1-transitive linear orders as classified by Morel, and the corresponding coloured versions, also touching on $\aleph_0$-categorical linear orders. Then in §4 we look at trees, considering three main possibilities, the 2-transitive, weakly 2-transitive, and 1-transitive cases. Finally in §5 we outline Warren's work on cycle-free partial orders.

We conclude the introduction by listing a few further definitions and notations. We usually take a partial order as being given by a 'strict' relation, that is $<$ rather than $\leq$, though occasionally we may use $\leq$, and $\parallel$ stands for the relation of incomparability. Set notation such as $A < B$ or $A \parallel B$ means that every element of A is less than (incomparable with respectively) every member of B. A *chain* is a linearly ordered set (or a subset of a partially ordered set which is linearly ordered by the induced relation). An *antichain* is a partial order in which every two distinct elements are incomparable. A 3-element partial order $\{x, y, z\}$ with $x > y, z$ and $y \parallel z$ is a Λ-*shape* (with colouring unspecified) and a 3-element partial order $\{x, y, z\}$ with $x < y, z$ and $y \parallel z$ is a V-*shape*. We write $\mathcal{P}_|$ for a 2-element chain, $\mathcal{P}_-$ for a 2-element antichain, $\mathcal{P}_{|\cdot}$ for a 3-element partial order containing a 2-element chain both of whose elements are incomparable with the third point, $\mathcal{P}_\Lambda$ for a Λ-shape, and $\mathcal{P}_V$ for a V-shape. As usual we write $[x, y] = \{z : x \leq z \leq y\}$ and $(x, y) = \{z : x < z < y\}$ (which are not necessarily linear orders).

2 Homogeneous partial orders and the coloured versions

Schmerl's classification of the countable homogeneous partial orders.

In this section, I outline Schmerl's classification, and the way it is modified in the coloured case. He showed that every countable homogeneous partial order is of one of the following types:

- a dense *chain of antichains*, which is a partial order $(P, <)$ obtained from a chain $(X, <)$ by replacing each point x by an antichain A_x of some fixed size $\leq \aleph_0$, and decreeing that if $x < y$ then all points of A_x lie below all points of A_y, with no other relations,

- an *antichain of chains*, which is a partial order which can be written as the disjoint union of chains, so that points in distinct chains are incomparable, and which are all singletons, or all isomorphic to $\mathbb{Q}$,

- and the (Fraïssé) *generic partial order* $(P, <)$.

There are one or two 'degenerate' cases. A single chain may be viewed either as a chain of antichains or an antichain of chains, and we take it to be the former, if non-trivial, though a singleton is viewed as an antichain. In other words, by an *antichain of (non-trivial) chains* we understand that there are at least two chains involved (though an antichain may have size 1). Most of the time we can treat antichains and antichains of chains as parallel cases, but on one or two occasions their behaviours differ, in which case we refer explicitly to an 'antichain of non-trivial chains'. The same conventions will be followed in the coloured case.

The fact that each of these structures is homogeneous is easy to verify. Conversely one supposes that $(P, <)$ is a countable homogeneous partial order, and shows that it must be of one of the above types. This is done by considering which of $\mathcal{P}_|$, $\mathcal{P}_-$, $\mathcal{P}_{|\cdot}$, $\mathcal{P}_\Lambda$, $\mathcal{P}_V$ are embeddable in P.

Since the first stage of the classification of countable homogeneous coloured partial orders (the 'interdense' case) is more-or-less the same as Schmerl's, we give the outline for these, and the monochromatic classification will be an immediate consequence. The definition is that $(X, <, F)$ is *interdense* if for any $x < y$ in X and $c \in C$, there is z such that $x < z < y$ and $F(z) = c$. This notion is therefore dependent on precisely which colour set C we are taking, and as usual to make sense of the notion, it is assumed that C is the set of all colours actually appearing.

Most of the interesting new structures in the coloured case arise when interdensity fails. After outlining the interdense classification, we show how a general coloured homogeneous partial order may be written as a union of interdense components. The major part of the classification consists of describing how the components can fit together, which we sketch at the end of this section. Notice that if we think of trying to cut a general coloured partial order into interdense pieces, then each piece will only be interdense with respect to *its own* colour set.

In terms of P_C and $\mathbb{Q}_C$, given in the introduction, we can now list the following partial orders, which are all easily seen to be countable homogeneous interdensely coloured partial orders, and will form the members of our classification in the interdense case.

Any antichain at all, even coloured by many colours, is interdense, (since $x < y$ is always false), and it is also clearly homogeneous, since it is essentially a (multi-)set. (In practice we prefer in this case to reduce to the monochromatic subsets as 'components'.)

Each $\mathbb{Q}_C$ is interdense and homogeneous, and any non-trivial countable homogeneous interdense chain is of this form.

Building on these two cases, we have an *antichain of chains*, which is a union of a finite or countable set of copies of some $\mathbb{Q}_C$, with elements in distinct copies incomparable, and we also have a *chain of antichains*, which is obtained from some $\mathbb{Q}_C$ by replacing all points coloured by the same colour by a finite or countable coloured antichain, where points of the same colour must be replaced by isomorphic antichains, and the colour sets of antichains replacing differently coloured points of $\mathbb{Q}_C$ must be disjoint. The ordering is given by $x < y$ if for some $q < r$ in $\mathbb{Q}_C$, x and y lie in the antichains replacing q and r respectively.

Finally we have the generics P_C.

Following Schmerl's method, we may characterize these by which of $\mathcal{P}_|$, $\mathcal{P}_-$, $\mathcal{P}_{|\cdot}$, $\mathcal{P}_\Lambda$, $\mathcal{P}_V$ embed (for appropriate colours on their points). We remark that any (coloured) partial order not embedding $\mathcal{P}_|$ is an antichain, and any partial order not embedding $\mathcal{P}_-$ (under any colouring) is a chain.

Lemma 1. Let $\mathcal{P}$ be a countable homogeneous coloured partial order. If $\mathcal{P}_|$ and $\mathcal{P}_\Lambda$ embed in $\mathcal{P}$ for every possible colouring of their points, then $\mathcal{P}$ is isomorphic to the generic coloured partial order. Similarly if we are given that $\mathcal{P}_|$ and $\mathcal{P}_V$ embed in $\mathcal{P}$ for every possible colouring of their points.

Proof. Schmerl shows in [S79] that if (the monochromatic versions of) $\mathcal{P}_|$ and $\mathcal{P}_\Lambda$ both embed in a countable homogeneous partial order, then it is isomorphic to the (monochromatic) generic. This proof is readily adapted to the situation described in the lemma. q.e.d.

Lemma 2. Let $\mathcal{P}$ be a countable homogeneous coloured partial order. If $\mathcal{P}_|$ embeds in $\mathcal{P}$ for all possible colourings of its points and $\mathcal{P}_\Lambda$ embeds in $\mathcal{P}$ for some colouring of its points then $\mathcal{P}_\Lambda$ embeds in $\mathcal{P}$ for all possible colourings of its points, with a similar statement for $\mathcal{P}_|$ and $\mathcal{P}_V$. In either case, $\mathcal{P}$ is then isomorphic to the generic coloured partial order.

Proof. Starting from an instance of $\mathcal{P}_\Lambda$, we may add points above and below using the hypothesis on copies of $\mathcal{P}_|$ to find another instance of $\mathcal{P}_\Lambda$ correctly coloured. The proof for $\mathcal{P}_V$ is the dual, and the final remark follows from Lemma 1. q.e.d.

In the interdense case we have the following rather stronger implication.

Lemma 3. Let $\mathcal{P}$ be a homogeneous interdensely coloured partial order. Then $\mathcal{P}_\Lambda$ embeds in $\mathcal{P}$ for some colouring of its points if and only if $\mathcal{P}_V$ embeds in $\mathcal{P}$ for some colouring of its points.

Proof. We just do one direction, and suppose for a contradiction that $\mathcal{P}_V$ does not embed for any colouring of its points but that $\mathcal{P}_\Lambda$ does. Let $x, y < z$, with $x \parallel y$ be a Λ-shape in $\mathcal{P}$. One uses homogeneity and interdensity to find $t, u < x$, $v, w < y$ such that $t \parallel u$, $v \parallel w$ and $F(u) = F(v)$. Since $\mathcal{P}_V$ does not embed, $t, u \parallel y$ and $v, w \parallel x$. By homogeneity there is an automorphism g fixing t and w and taking u to v. Since $\mathcal{P}_V$ does not embed, x and gx are comparable. If $gx \le x$ then $v = gu < gx \le x$, contrary to $v \parallel x$. Hence $x < gx$. Since $gu = v < y$, $u < g^{-1}y$, x so again as $\mathcal{P}_V$ does not embed, x and $g^{-1}y$ are comparable. We cannot have $g^{-1}y \le x$, as this would give $w = g^{-1}w < g^{-1}y \le x$, contradiction, and so $x < g^{-1}y$. Hence $x < gx < y$, which again is impossible, completing the proof. q.e.d.

Lemma 4. Let $\mathcal{P}$ be a countable homogeneous interdensely coloured partial order with colour set C. If $\mathcal{P}_\Lambda$ does not embed for any colouring of its points then $\mathcal{P}$ is an antichain, or an antichain of chains isomorphic to $\mathbb{Q}_C$.

Note that by the previous lemma, we could replace $\mathcal{P}_\Lambda$ by $\mathcal{P}_V$ in this.

Proof. Since $\mathcal{P}_\Lambda$ does not embed, nor does $\mathcal{P}_V$ by Lemma 3, so $\mathcal{P}$ is the disjoint union of its maximal chains. Since the colours are interdense, each chain must be a singleton, or isomorphic to $\mathbb{Q}_C$. q.e.d.

Lemma 5. Let $\mathcal{P}$ be a countable homogeneous coloured partial order with interdense colours, which does not embed $\mathcal{P}_|$ for any colouring of its points and which is not an antichain. Then $\mathcal{P}$ is a chain of antichains. More specifically, C may be written as the disjoint union of sets $D_{c'}$ for $c' \in C'$, and $\mathcal{P}$ is obtained from $\mathbb{Q}_{C'}$ by replacing each element having colour $c' \in C'$ by an antichain coloured by $D_{c'}$, such that if y, z have the same colour then $A_y \cong A_z$, and if y, z have different colours, then A_y and A_z have disjoint colour sets, and for $p_1 \in A_y$ and $p_2 \in A_z$, $p_1 < p_2 \Leftrightarrow y < z$ in $\mathbb{Q}_{C'}$.

For a full description of a chain of antichains, it is not enough to give the colour sets $D_{c'}$; we also have to specify for each c', how many elements of the antichain are coloured by each element of $D_{c'}$. The fact that this is the same for each point corresponding to c' is ensured by requiring that they are isomorphic. Once this choice has been made, then this description does uniquely determine $\mathcal{P}$ up to isomorphism. We call this partition $\{D_{c'} : c' \in C'\}$ of the colour set of a chain of antichains $\mathcal{P}$, its *colour-structure partition*.

Proof. 'Incomparability or equality' is an equivalence relation $\sim$ on $\mathcal{P}$ (since $\mathcal{P}_|$ does not embed), so that $\mathcal{P}$ can be partitioned into maximal antichains. By homogeneity, any two maximal antichains sharing a colour are isomorphic. Let Y be the set of these antichains and for each $y \in Y$ let D_y equal the set of colours in C occurring in that antichain. Let $C' = \{D_y : y \in Y\}$ and we view C' as a set of colours, and colour the points of Y accordingly. We order Y by letting $y_1 < y_2$ if some member of y_1 is below some member of y_2. It follows by homogeneity that this is equivalent to saying that every member of y_1 is below every member of y_2, and $<$ is clearly a partial order of Y. It is linear, since if $y_1, y_2 \in Y$ are incomparable, then they would have to be contained in the same antichain, hence equal. It is then easy to check that $Y \cong \mathbb{Q}_{C'}$ and that $\mathcal{P}$ is obtained from Y in the way described. The other option, that Y is a singleton, is ruled out by the hypothesis that $\mathcal{P}$ is not an antichain. q.e.d.

Lemma 6. Any countable homogeneous interdensely coloured partial order $\mathcal{P}$ containing a Λ-shape, and also points x, y, z, t such that $x < y$, $t < z$, $x \parallel t, z$ and $y \parallel t, z$, is isomorphic to the generic coloured partial order.

Proof. By interdensity, any colour in the colour set C occurs in (x, y) and also in (t, z). Since $x \parallel t, z$ and $y \parallel t, z$, $\mathcal{P}_|$ embeds for any colouring of its points. But we are assuming that $\mathcal{P}_\Lambda$ embeds for some colouring of its points, so by Lemma 2, it embeds for any colouring of its points and hence, by Lemma 1, $\mathcal{P}$ is the generic partial order. q.e.d.

Lemma 7. Let $\mathcal{P}$ be a countable homogeneous interdensely coloured partial order. If $\mathcal{P}_\Lambda$ (or $\mathcal{P}_V$) and $\mathcal{P}_|$ embed in $\mathcal{P}$ for some colouring of their points then $\mathcal{P}$ is isomorphic to the generic coloured partial order.

Proof. By the previous lemma, it is sufficient to show that the partial order $Q = \{x_1, x_2, x_3, x_4\}$ where $x_1 < x_2$, $x_3 < x_4$, $x_1 \parallel x_3, x_4$ and $x_2 \parallel x_3, x_4$, also embeds in $\mathcal{P}$ for some colouring of its points. Suppose otherwise for a contradiction. As $\mathcal{P}_|$ embeds, there are x, y, z with $x < y$ and $x, y \parallel z$, and clearly $s \parallel z$ for all $s \in [x, y]$. From the fact that Q does not embed we deduce that any $t > z$ is strictly above every element of $[x, y]$ (and dually, any $t < z$ is strictly below every element of $(x, y]$).

Now, by interdensity, there is $z' \in (x, y)$ such that $F(z') = F(z)$. By homogeneity, there are x' and y' such that $x' < z < y'$ with $F(x') = F(x)$ and $F(y') = F(y)$. By the above remark, $y' > [x, y)$ and $x' < (x, y]$. Also, $z \parallel z'$. By homogeneity, there is an automorphism f of $\mathcal{P}$ that interchanges z and z'. By interdensity, there is $y'' \in (z', y)$ such that $F(y'') = F(y') = F(y)$. Since $z' < y''$, then $z < fy''$ and therefore, by our previous remark, $y'' < fy''$. But also $z = f^{-1}z' < f^{-1}y''$ and therefore, $y'' < f^{-1}y''$, a contradiction. q.e.d.

We can now prove the result for the interdense case.

Theorem 8. Any countable homogeneous interdensely coloured partial order $\mathcal{P}$ is isomorphic to one of the following:

- an antichain,
- an antichain of chains each isomorphic to $\mathbb{Q}_C$,
- a chain of antichains obtained from $\mathbb{Q}_{C'}$ by replacing each point by a coloured antichain, so that points coloured the same are replaced by isomorphic antichains, and the colour sets of antichains replacing differently coloured points are disjoint,

- the C-coloured generic.

Furthermore, each of the coloured partial orders described is homogeneous.

Proof. If $\mathcal{P}_|$ does not embed, then $\mathcal{P}$ is an antichain, and if $\mathcal{P}_-$ does not embed then it is a chain, so as remarked above is isomorphic to $\mathbb{Q}_C$ (or a singleton). All these cases are homogeneous. Now suppose that both $\mathcal{P}_|$ and $\mathcal{P}_-$ embed (for some colouring of their points). It follows easily that $\mathcal{P}_{|.}$ or $\mathcal{P}_\Lambda$ or both embed for some colouring of their points. Recall also that $\mathcal{P}_\Lambda$ embeds if and only if $\mathcal{P}_V$ embeds.

If $\mathcal{P}_{|.}$ and $\mathcal{P}_\Lambda$ both embed for some colouring of their points then, by Lemma 7, $\mathcal{P}$ is the generic coloured partial order.

If $\mathcal{P}_{|.}$ embeds for some colouring of its points but $\mathcal{P}_\Lambda$ does not embed for any colouring of its points, then, by Lemma 4, $\mathcal{P}$ is an antichain of chains. If $\mathcal{P}_\Lambda$ embeds for some colouring of its points but $\mathcal{P}_{|.}$ does not embed for any colouring of its points, then, by Lemma 5, $\mathcal{P}$ is a chain of antichains. q.e.d.

The following lemma will be quite useful in various places.

Lemma 9. Let $\mathcal{P}$ be a homogeneous coloured partial order with colour set C, and let C' be a non-empty subset of C. Then the restriction $\mathcal{P}'$ of $\mathcal{P}$ to C' is also homogeneous. Furthermore, if $\mathcal{P}$ is the Fraïssé limit of an amalgamation class $\mathcal{K}$, then $\mathcal{P}'$ is the Fraïssé limit of the family $\mathcal{K}'$ of restrictions of members of $\mathcal{K}$ to C'.

Proof. Any finite partial automorphism of $\mathcal{P}'$ is also a partial automorphism of $\mathcal{P}$, so extends to an automorphism of $\mathcal{P}$ whose restriction to C' is the desired automorphism of $\mathcal{P}'$. The final statement follows from Fraïssé's theorem, since the age of $\mathcal{P}'$ is clearly the family of restrictions to C' of members of the age of $\mathcal{P}$. q.e.d.

Now moving on to the general case in which interdensity is not assumed, we introduce an equivalence relation $\sim$ whose equivalence classes are the maximal interdensely coloured substructures of a given homogeneous coloured partial order $\mathcal{P}$ with colour set C (except that antichain components will be monochromatic). Let $\approx$ on C and $\sim$ on $\mathcal{P}$ be given by $c_0 \approx c_1$ if there are x, y and z such that $x \leq y \leq z$ and $F(x) = F(z) = c_0$, $F(y) = c_1$, and $x \sim y$ if $F(x) \approx F(y)$. (This is an adaptation of [$C_0T_1$04a, Lemma 3.2] to the partially ordered case.) We refer to the $\sim$-classes as *components* of $\mathcal{P}$.

Lemma 10. If $\mathcal{P} = (P, <, F)$ is a homogeneous coloured partial order, then $\approx$ and $\sim$ are equivalence relations on C and $\mathcal{P}$ respectively. Each component is convex, homogeneous and interdensely coloured (with respect to its colour set), and the colour sets of distinct components are disjoint.

Proof. The *definition* of $\approx$ is asymmetric, but by homogeneity it easily follows that actually it is symmetric and transitive.

It is immediate that $\sim$ is an equivalence relation on $\mathcal{P}$. Convexity of each component follows from the definition of $\approx$. The fact that each component is itself homogeneous follows from Lemma 9.

To show that each component X has interdense colours for its colour set C', let $x < y$ in X, and $c \in C'$. Let $F(x) = c_1$ and $F(y) = c_2$. We find points $u < v < w$ coloured c_1, c, c_2 respectively, and by taking (u,w) to (x,y) using homogeneity we find a point (the image of v) between x and y coloured c as required. The existence of u, v, w follows from $c_1 \approx c_2 \approx c$ and homogeneity (splitting into the cases $c_1 \neq c$, $c_1 = c$). q.e.d.

Since each component is homogeneous, it is isomorphic to one of the structures described in Theorem 8, namely, a chain of antichains, an antichain of chains or the generic. To completely describe the general case, we need to see how the different components can be 'fitted together'. In the following lemmas we outline what the different cases are, and how they can be handled. In treating a general countable coloured homogeneous partial order, it suffices to consider just the cases where there are finitely many components. This is because of the following 'compactness' type result.

Lemma 11. Let $\mathcal{P}$ be a countable coloured partial order which is expressible as a union of a family $\mathcal{F}$ of subsets which are coloured by pairwise disjoint colour sets and are interdensely coloured. Then $\mathcal{P}$ is homogeneous if and only if the union of every finite subfamily of $\mathcal{F}$ is homogeneous.

We would like to say that $\mathcal{P}$ is homogeneous if and only if every finite union of components is homogeneous, but we cannot because the proof of Lemma 10 requires homogeneity, so we have to refer to the components 'indirectly'.

Proof. By Lemma 9, if $\mathcal{P}$ is homogeneous, the union of any finite subfamily of $\mathcal{F}$ is homogeneous.

Conversely, we use a back-and-forth argument. Let p be a finite partial automorphism of $\mathcal{P}$ and $x \in \mathcal{P}$. We shall show that p can be extended to a finite partial automorphism q having x in its domain. Since p is finite, there is a union $\mathcal{Q}$ of a finite subfamily of $\mathcal{F}$ such that p is a partial automorphism of $\mathcal{Q}$. Since $\mathcal{Q}$ is homogeneous, we can extend p to an automorphism f of

Q, and then $q = p \cup \{(x, f(x))\}$ is the desired finite partial automorphism of $\mathcal{P}$. This shows that any finite partial automorphism can be extended to include any specified element of $\mathcal{P}$ in its domain. A similar argument applies to the range. Since $\mathcal{P}$ is assumed countable, it follows by back and forth that $\mathcal{P}$ is homogeneous. q.e.d.

From now on we may therefore if we wish restrict to just finitely many components. Of course as the number of components increases towards infinity, we shall have $2^{\aleph_0}$ possibilities, but they are essentially 'controlled' by the finite component substructures.

The key steps are working out how two components can be related, and similarly for three components in two possible patterns, a 'V-shape', or a chain of length 3 ('Λ-shapes' are just the dual to V-shapes). Once we know this, the possibilities for the overall structure can be expressed in terms of the components, with natural restrictions on the relations between them.

If P_1 and P_2 are components, let us write $P_1 \prec P_2$ to mean that some element of P_1 lies below some element of P_2. It follows easily from homogeneity, and the fact that the components are convex, that this is a partial order. What will be needed for the classification is to know first of all which kind of component we have at each point, and what the relations are between different components. Let us say that P_1 is *completely below* P_2 if for all $x \in P_1$ and $y \in P_2$, $x < y$, and we write $P_1 < P_2$ (which accords with the earlier notation). This can happen for all possible types of component for P_1 and P_2 (since there is essentially 'no interaction' between the structures of P_1 and P_2). We also say that P_1 is *partially below* P_2 if there are $x_1, x_2 \in P_1$ and $y_1, y_2 \in P_2$ such that $x_1 < y_1$ and $x_2 \parallel y_2$. This can happen in some but not all cases.

Lemma 12 (The 2-chain lemma). Let $\mathcal{P}$ be a homogeneous coloured partial order having two components P_1 and P_2 such that $P_1 \prec P_2$. Then one of the following must hold. Furthermore, each of the possibilities listed can occur, and is uniquely determined up to isomorphism, given the appropriate cardinalities and colour sets arising:

(i) P_1 and P_2 are of any possible type and $P_1 < P_2$,

(ii) P_1 and P_2 are both chains of antichains, and P_1 is partially below P_2 (written $<_{cc}$),

(iii) P_1 and P_2 are both antichains, or antichains of chains, and there is a 1–1 correspondence between the sets of constituent chains such that $x < y$ if and only if x and y lie in chains which correspond; we write $P_1 <_{pm} P_2$ (for 'perfect matching'),

(iv) P_1 and P_2 are both antichains, or antichains of chains, and there is a 1–1 correspondence between the sets of constituent chains such that $x < y$ if and only if x and y lie in chains which do not correspond; we write $P_1 <_{\mathrm{cpm}} P_2$ (for 'complement of perfect matching'),

(v) P_1 and P_2 are both antichains, or antichains of chains, and for any x_1, x_2 in the same chain of P_1 and y_1, y_2 in the same chain of P_2, $x_1 < y_1 \Leftrightarrow x_2 < y_2$, and for any finite disjoint unions U and V of chains of P_1 there is $y \in P_2$ such that y is above all members of U and not above any member of V, and for any finite disjoint unions U and V of chains of P_2 there is $x \in P_1$ such that x is below all members of U and not below any member of V; we write $P_1 <_{\mathrm{g}} P_2$ (for 'generic'),

(vi) P_1 is an infinite antichain and P_2 is generic, and P_1 is partially below P_2 (written $<_{\mathrm{ag}}$), or the same thing with P_1 and P_2 interchanged (written $<_{\mathrm{ga}}$),

(vii) P_1 and P_2 are both generic, and P_1 is partially below P_2 (written $<_{\mathrm{gg}}$).

Proof. (ii) We remark that in this case we can give an explicit description of the structure of $P_1 \cup P_2$. Let $\{D_i(c) : c \in C'_i\}$ for $i = 1, 2$ be the colour structure partition for P_i (together with 'multiplicities', which we do not indicate explicitly). We obtain $P_1 \cup P_2$ from $\mathbb{Q}_{C'_1 \cup C'_2}$ by replacing each point coloured $c \in C'_i$ by an antichain of points coloured by the members of $D_i(c)$, letting P_i be the set of points arising from C'_i, and decreeing that $x < y$ provided that x and y are ordered this way in $\mathbb{Q}_{C'_1 \cup C'_2}$, and either they both lie in the same one of P_1, P_2, or else $x \in P_1$ and $y \in P_2$. The fact that this is homogeneous follows quite easily from its bi-definability with the chain of antichains derived from $\mathbb{Q}_{C'_1 \cup C'_2}$ and the colour structure partition $\{D_1(c) : c \in C'_1\} \cup \{D_2(c) : c \in C'_2\}$.

(iii), (iv), and (v) arise from the classification of countable homogeneous bipartite graphs, which is mentioned in [G$_2$G$_3$K96] for instance. These are of five possible kinds, empty (which doesn't arise here, since we assume $P_1 \prec P_2$), complete (that is $P_1 < P_2$), perfect matching, complement of a perfect matching, and 'generic' (where P_1 and P_2 must both have $\aleph_0$ constituent chains). The main additional point is that for homogeneity, if X_1 and X_2 are maximal chains of P_1 and P_2 respectively, then either $X_1 < X_2$ or $X_1 \parallel X_2$, which relies on there being at least two constituent chains for each (except where one or both are antichains, in which case it is clear anyway).

For (vi) and (vii) the relevant structures are most easily constructed by Fraïssé amalgamation. For (vii) the class $\mathcal{K}$ is taken to be the family of all finite partial orders of the form $X_1 \cup X_2$ where the colours of X_1 and X_2 are in $F(P_1)$, $F(P_2)$ respectively, and no point of X_1 is above any point of X_2, and for (vi) it is the same except that X_1 also has to be an antichain. The fact that these are amalgamation classes, establishing existence, is easy. The main labour is to demonstrate uniqueness, which is quite involved, particularly for (vii).

The other thing needed to complete the proof is to show that for all combinations $P_1 \prec P_2$ not otherwise mentioned, we must have $P_1 < P_2$, for instance if P_1 is a chain of antichains, and P_2 is generic. This is not hard, but varying details are needed in each of the cases. q.e.d.

The following lemmas are useful when considering more than two components.

Lemma 13. Let $\mathcal{P}$ and $\mathcal{Q}$ be homogeneous coloured partial orders with disjoint colour sets. Suppose that $\mathcal{P} \cup \mathcal{Q}$ is a partial ordering extending each of $\mathcal{P}$ and $\mathcal{Q}$, and such that for each component X of $\mathcal{P}$ and Y of $\mathcal{Q}$, $X < Y$, $Y < X$, or $X \parallel Y$. Then $\mathcal{P} \cup \mathcal{Q}$ is also homogeneous.

Proof. Let $p : A \to B$ be a finite partial automorphism of $\mathcal{P} \cup \mathcal{Q}$, and let p_1 and p_2 be its restrictions to $\mathcal{P}$ and $\mathcal{Q}$ respectively. As each of $\mathcal{P}$ and $\mathcal{Q}$ is homogeneous, these may be extended to automorphisms f_1 and f_2 of $\mathcal{P}$, $\mathcal{Q}$. Then $f = f_1 \cup f_2$ is an automorphism of $\mathcal{P} \cup \mathcal{Q}$ extending p. This is because f_1 must preserve components X of $\mathcal{P}$ and f_2 must preserve components Y of $\mathcal{Q}$, and by hypothesis, all points of X are related in the same way to the points of Y. q.e.d.

For countable homogeneous coloured partial orders having three components P_1, P_2, P_3 forming a V-shape with $P_2 \parallel P_3$, since each component can be a chain of antichains, an antichain of chains, or generic, there appear to be 27 cases to consider. In view of the evident symmetry between P_2 and P_3, this at once reduces to 18. Several more are eliminated by Lemma 13, since by the 2-chain lemma 7 the majority of 2-component relations must be complete. In all cases, the next lemma gives the uniqueness of the V-shape structures given the 2-component relations.

Lemma 14. Let $\mathcal{P}$ and $\mathcal{Q}$ be countable homogeneous coloured partial orders having three components P_1, P_2 and P_3 forming a V-shape so that $P_1 \prec P_2, P_3$ and $P_2 \parallel P_3$, and similarly for $\mathcal{Q}$. If the union of any two components of $\mathcal{P}$ is isomorphic to the union of the corresponding two components of $\mathcal{Q}$, then $\mathcal{P}$ and $\mathcal{Q}$ are isomorphic.

Proof. Since $\mathcal{P}$ and $\mathcal{Q}$ are countable and homogeneous, it suffices to show that they have the same age. Let $X_1 \cup X_2 \cup X_3$ be a finite substructure of $\mathcal{P}$ with $X_i \subseteq P_i$. Then by hypothesis, $X_1 \cup X_2$ embeds in $Q_1 \cup Q_2$ and $X_1 \cup X_3$ embeds in $Q_1 \cup Q_3$. By homogeneity of $\mathcal{Q}$ we may assume that the two embeddings agree on X_1. Since $P_2 \parallel P_3$ and $Q_2 \parallel Q_3$, the union of the two embeddings is an embedding of $X_1 \cup X_2 \cup X_3$ into $\mathcal{Q}$. Thus the age of $\mathcal{P}$ is contained in the age of $\mathcal{Q}$. The same argument applies in reverse, so they have the same ages, as required. q.e.d.

Lemma 15 (The V-shape lemma). Let $\mathcal{P}$ be a countable homogeneous coloured partial order having three components P_1, P_2 and P_3, where $P_1 \prec P_2, P_3$ and $P_2 \parallel P_3$. Then one of the following must hold. Furthermore, each of the possibilities listed can occur, and is uniquely determined up to isomorphism, given the appropriate cardinalities and colour sets arising:

(i) $P_1 < P_2$, and P_1, P_3 follow one of the cases listed in the 2-chain lemma 7, and similarly with P_2 and P_3 interchanged,

(ii) P_1, P_2, and P_3 are all chains of antichains, and $P_1 <_{cc} P_2$ and $P_1 <_{cc} P_3$.

(iii) P_1, P_2 and P_3 are all antichains of chains and $P_1 <_g P_2$ and $P_1 <_g P_3$,

(iv) P_1 is an antichain, P_2 is an antichain of chains and P_3 is a generic, and $P_1 <_g P_3$ and $P_1 <_{ag} P_3$ (or the same with P_2 and P_3 interchanged),

(v) P_1 is an antichain, P_2 and P_3 are generics, and $P_1 <_{ag} P_2$ and $P_1 <_{ag} P_3$,

(vi) P_1, P_2 and P_3 are generics and $P_1 <_{gg} P_2$ and $P_1 <_{gg} P_3$,

(vii) P_1 and P_2 are generics, P_3 is an antichain and $P_1 <_{gg} P_2$ and $P_1 <_{ga} P_3$ (or the same with P_2 and P_3 interchanged),

(viii) P_1 is generic, P_2 and P_3 are antichains, and $P_1 <_{ga} P_2$ and $P_1 <_{ga} P_3$.

The next remark is straightforward but useful, and in the 3-chain lemma enables us to discount the cases where either lower or upper link is complete.

Lemma 16. Let $\mathcal{P}$ be a countable homogeneous coloured partial order having three components P_1, P_2 and P_3 such that $P_1 \prec P_2 \prec P_3$. If $P_1 < P_2$ or $P_2 < P_3$ then $P_1 < P_3$. Furthermore, if P_1 and $P_2 \cup P_3$ (similarly for $P_1 \cup P_2$ and P_3) are known to be countable homogeneous coloured partial orders with P_1 interdense, P_2 and P_3 components of $P_2 \cup P_3$ with $P_1 < P_2 \prec$

P_3, (or $P_1 \prec P_2 < P_3$ in the other case), then $P_1 \cup P_2 \cup P_3$ is homogeneous and uniquely determined up to isomorphism.

Lemma 17 (The 3-chain lemma). Let $\mathcal{P}$ be a homogeneous coloured partial order consisting of three components P_1, P_2 and P_3, such that $P_1 \prec P_2 \prec P_3$. Then the relation between P_1 and P_3 is the transitive closure of the other two relations, and the relation between the three components is one of the following:

(i) $P_1 < P_2$ and $P_1 < P_3$ and the relation between P_2 and P_3 is any of the ones allowed by the 2-chain lemma 7,

(ii) $P_1 < P_3$ and $P_2 < P_3$ and the relation between P_1 and P_2 is any of the ones allowed by the 2-chain lemma 7,

(iii) P_1, P_2 and P_3 are all chains of antichains, $P_1 <_{cc} P_2 <_{cc} P_3$ and $P_1 <_{cc} P_3$,

(iv) P_1, P_2 and P_3 are all antichains of chains, $P_1 <_{pm} P_2$ and the relation between P_2 and P_3 is one of $<_{pm}$, $<_{cpm}$, $<_g$, in which case the relation between P_1 and P_3 is the same as that between P_2 and P_3,

(v) P_1, P_2 and P_3 are all antichains of chains, $P_2 <_{pm} P_3$ and the relation between P_1 and P_2 is one of $<_{pm}$, $<_{cpm}$, $<_g$, which is the same as the relation between P_1 and P_3,

(vi) P_1, P_2 and P_3 are all antichains of chains, $P_1 <_g P_2$ and $P_2 <_g P_3$, and the relation between P_1 and P_3 is $<_g$, $<_{cpm}$ or $<$,

(vii) P_1, P_2 and P_3 are all generics, $P_1 <_{gg} P_2 <_{gg} P_3$, and $P_1 <_{gg} P_3$ or $P_1 < P_3$,

(viii) P_1 and P_3 are antichains, P_2 is a generic, $P_1 <_{ag} P_2 <_{ga} P_3$, and the relation between P_1 and P_3 is $<_g$, $<_{cpm}$ or $<$,

(ix) P_1 and P_2 are antichains, P_3 is a generic, $P_1 <_{pm} P_2$ or $P_1 <_g P_2$, and $P_1, P_2 <_{ag} P_3$,

(x) P_1 is a generic, P_2 and P_3 are antichains, $P_1 <_{ga} P_2, P_3$, and $P_2 <_{pm} P_3$ or $P_2 <_g P_3$,

(xi) P_1 is an antichain of chains, P_2 is an antichain, P_3 is a generic, $P_1 <_g P_2 <_{ag} P_3$, and $P_1 < P_3$,

(xii) P_1 is a generic, P_2 is an antichain, P_3 is an antichain of chains, $P_1 <_{ga} P_2 <_g P_3$, and $P_1 < P_3$,

(xiii) P_1 and P_2 are generics, P_3 is an antichain, $P_1 <_{\text{gg}} P_2 <_{\text{ga}} P_3$, and $P_1 < P_3$ or $P_1 <_{\text{ga}} P_3$,

(xiv) P_2 and P_3 are generics, P_1 is an antichain, $P_1 <_{\text{ag}} P_2 <_{\text{gg}} P_3$, and $P_1 < P_3$ or $P_1 <_{\text{ag}} P_3$.

(xv) P_1 and P_3 are generics, P_2 is an antichain, $P_1 <_{\text{ga}} P_2 <_{\text{ag}} P_3$, and $P_1 < P_3$ or $P_1 <_{\text{gg}} P_3$.

Guided by these preliminary lemmas, we can now describe how the final classification of all countable homogeneous coloured partial orders with finitely many components is achieved.

Skeletons.

In this case the classifiers are referred to as 'skeletons'. We start by saying what the skeleton of a homogeneous coloured partial order $\mathcal{P} = (P, <, \mathcal{F})$ is. What then will remain is to characterize abstractly which structures can arise as skeletons, and to show that $\mathcal{P}$ is uniquely determined by its skeleton.

Definition 18. The *skeleton of* $\mathcal{P}$ is the coloured partial order $\mathcal{Q}$ with labels on its elements and pairs of comparable elements given by:

- the domain of $\mathcal{Q}$ is the set of components of $\mathcal{P}$, partially ordered by $\prec$,

- each element X of $\mathcal{Q}$ is labelled CA, A, AC, or Ge according as it is a (non-trivial) chain of antichains, an antichain, an antichain of at least two non-trivial chains, or generic, by the set of colours occurring in X, by its colour structure partition if it is a non-trivial chain of antichains, and by the number of constituent chains if X is an antichain or an antichain of chains,

- pairs (X, Y) of elements of $\mathcal{Q}$ such that $X \prec Y$ are labelled C if $X < Y$, PM, CPM, or G if both X and Y are antichains or antichains of chains, and the relation between them is a perfect matching, its complement, or generic respectively, CC if both X and Y are chains of antichains, AG if X is an antichain and Y generic, GA the dual of this, and GG if both are generic, and in all of the last four cases, X is partially below Y.

It is clear then from the previous lemmas that any countable homogeneous coloured partial order has a skeleton which obeys all these conditions.

We now turn to the 'converse' idea. We have defined skeleton *of* $\mathcal{P}$, and we can now define 'skeleton' abstractly. Although by Lemma 11 we could restrict to finite skeletons, in fact the definition is given for the countable case.

Definition 19. A *skeleton* is a finite or countable partial order $\mathcal{Q}$, with labels on the vertices and pairs of comparable vertices fulfilling the following conditions:

(i) each vertex x is labelled by CA, A, AC, or Ge, and by a set C_x (of colours), so that $x \neq y \Rightarrow C_x \cap C_y = \emptyset$,

(ii) if x is labelled CA then it is also labelled by a colour structure partition $\{D_{c'} : c' \in C'\}$ where $\bigcup C' = C_x$,

(iii) if x is labelled A or AC, then it also has a label $N(x) \in \{1, 2, \ldots, \aleph_0\}$, and if A, then $|C_x| = 1$ (antichain components are monochromatic), where if the label is AC, then $N(x) \neq 1$,

(iv) (conditions required by the 2-chain lemma)   pairs of comparable elements $x_1 < x_2$ are labelled by one of C, CC, PM, CPM, G, AG, GA and GG (and we write $x_1 X x_2$ in place of $(x_1, x_2) \in X$ for such labels), where if the label is CC, then x_1 and x_2 are both labelled CA, if PM, CPM, G then they are labelled A or AC, if AG then they are labelled A and Ge, if GA then they are labelled Ge and A, and if GG then they are both labelled Ge,

(v) (cardinality restrictions)   if $x_1 PM x_2$ or $x_1 CPM x_2$ then $N(x_1) = N(x_2)$, if $x_1 G x_2$ then $N(x_1) = N(x_2) = \aleph_0$, if $x_1 AG x_2$ then $N(x_1) = \aleph_0$, and if $x_1 GA x_2$ then $N(x_2) = \aleph_0$,

(vi) (conditions arising from the V-shape lemma) if $x_1 < x_2, x_3$ and $x_2 \parallel x_3$, then one of the following must apply: $x_1 C x_2$; $x_1 C x_3$; $x_1 CC x_2$ and $x_1 CC x_3$; $x_1 G x_2$ and $x_1 G x_3$; $x_1 G x_2$ and $x_1 AG x_3$; $x_1 AG x_2$ and $x_1 G x_3$; $x_1 AG x_2$ and $x_1 AG x_3$; $x_1 GG x_2$ and $x_1 GG x_3$; $x_1 GG x_2$ and $x_1 GA x_3$; $x_1 GA x_2$ and $x_1 GG x_3$; $x_1 GA x_2$ and $x_1 GA x_3$;

(vii) (conditions arising from the Λ-shape lemma)   the duals of (vi);

(viii) (conditions arising from the 3-chain lemma)   if $x_1 < x_2 < x_3$, then one of the following must apply: $x_1 C x_2$ and $x_1 C x_3$; $x_1 C x_3$ and $x_2 C x_3$; $x_1 CC x_2 CC x_3$ and $x_1 CC x_3$; $x_1 PM x_2 R x_3$ and $x_1 R x_3$ where R is PM, CPM, or G; $x_1 R x_2 PM x_3$ and $x_1 R x_3$ where R is PM, CPM, or G; $x_1 G x_2 G x_3$ and $x_1 R x_3$ where R is G, CPM, or C; $x_1 GG x_2 GG x_3$ and $x_1 GG x_3$ or $x_1 C x_3$; $x_1 AG x_2 GA x_3$ and $x_1 R x_3$ where R is G, CPM, or G;

$x_1\text{PM}x_2$ and $x_1, x_2\text{AG}x_3$; $x_1\text{G}x_2$ and $x_1, x_2\text{AG}x_3$; $x_1\text{GA}x_2\text{PM}x_3$ and $x_1\text{GA}x_3$; $x_1\text{GA}x_2, x_3$ and $x_2\text{G}x_3$; $x_1\text{G}x_2\text{AG}x_3$ and $x_1\text{C}x_3$; $x_1\text{GA}x_2\text{G}x_3$ and $x_1\text{C}x_3$; $x_1\text{GG}x_2\text{GA}x_3$ and $x_1\text{GA}x_3$ or $x_1\text{C}x_3$; $x_1\text{AG}x_2\text{GG}x_3$ and $x_1\text{AG}x_3$ or $x_1\text{C}x_3$; $x_1\text{GA}x_2\text{AG}x_3$ and $x_1\text{GG}x_3$ or $x_1\text{C}x_3$.

Theorem 20. For any skeleton $\mathcal{Q}$ there is a countable homogeneous coloured partial order $\mathcal{P}$ having $\mathcal{Q}$ as its skeleton.

Proof. The structure $\mathcal{P}$ is built by Fraïssé amalgamation from a class $\mathcal{K}$ of finite structures. The definition of $\mathcal{K}$ is obtained by 'interpreting' the instructions enshrined in $\mathcal{Q}$. The ideas are clear, though there are some annoying technical problems caused by the requirement to ensure that $\mathcal{K}$ is closed under formation of substructures. This means for instance, that with regard to perfect matchings, we have to allow substructures which may not actually *be* perfect matchings, but which can still be extended to them. We omit the details. q.e.d.

3 Linear orders and and coloured linear orders

In the remainder of the paper, I shall consider partially homogeneous structures, beginning with linear orders. The particular interest here is that this provides a natural class of structures, namely the countable 1-transitive linear orders, with precisely $\aleph_1$ members (whether or not the continuum hypothesis is true). This might be thought to have some bearing on Vaught's conjecture, but this is certainly not the case as it stands, since the notion of '1-transitivity' is not first order axiomatizable. Conceivably, a clever coding procedure could be used to derive an axiomatizable class from it, but at present, we regard the class of interest just in its own right. First we explain why the only interesting value of k to take here in k-homogeneous is 1.

Lemma 21. Any 2-homogeneous chain is homogeneous (hence k-homogeneous for each k). Similarly for coloured chains.

Proof. Given an isomorphism from $a_1 < \ldots < a_k$ to $b_1 < \ldots < b_k$, by 2-homogeneity there are automorphisms taking $(-\infty, a_1], [a_1, a_2],$ ..., $[a_{k-1}, a_k], [a_k, \infty)$ to $(-\infty, b_1], [b_1, b_2], \ldots, [b_{k-1}, b_k], [b_k, \infty)$ respectively, and we can patch them to give a single automorphism taking a_i to b_i for each i. q.e.d.

We remark that we could just have said that any countable non-trivial 2-homogeneous chain is isomorphic to $\mathbb{Q}$. However, this would not carry across quite so simply to the coloured case, where there are slightly more complicated examples which can arise, namely finite or countable unions of convex subsets, each of which is isomorphic to some $\mathbb{Q}_n$ or a singleton, where the pieces are coloured by pairwise disjoint colour sets.

The most obvious examples of countable 1-transitive chains are $(\mathbb{Z}, \leq)$ and $(\mathbb{Q}, \leq)$. The key difference between these two is therefore that $\mathbb{Z}$ is only 1-transitive, whereas $\mathbb{Q}$ is (fully) homogeneous. We now give Morel's result, which lists all the countable 1-transitive linear orders. The lexicographic product of chains $(X, \leq)$ and $(Y, \leq)$, 'X copies of Y', written $X.Y$, is the cartesian product ordered by $(x_1, y_1) < (x_2, y_2)$ if $x_1 < x_2$ or $(x_1 = x_2 \wedge y_1 < y_2)$, and we then define $\mathbb{Z}^\alpha$, where α is an ordinal, by transfinite induction thus: $\mathbb{Z}^0 = \{0\}$, $\mathbb{Z}^{\alpha+1} = \mathbb{Z}.\mathbb{Z}^\alpha$, and, viewing $\mathbb{Z}^\alpha$ as a subset of $\mathbb{Z}^{\alpha+1}$ via $\{(0, z) : z \in \mathbb{Z}^\alpha\}$, we let $\mathbb{Z}^\lambda = \bigcup_{\alpha < \lambda} \mathbb{Z}^\alpha$, if λ is a limit ordinal.

Morel's classification.

Theorem 22.

(i) If $(X, \leq)$ and $(Y, \leq)$ are 1-transitive linear orders, then so is $(X.Y, \leq)$.

(ii) Suppose that $\{X_\alpha : \alpha \in A\}$ are 1-transitive chains and $\leq$ is a linear ordering of A such that if $\alpha_1 \leq \alpha_2$, then X_{α_1} is a convex subset of X_{α_2}. Then $\bigcup_{\alpha \in A} X_\alpha$ is a 1-transitive chain.

(iii) For every ordinal α, $\mathbb{Z}^\alpha$ and $\mathbb{Q}.\mathbb{Z}^\alpha$ are 1-transitive chains.

(iv) Every countable 1-transitive chain is isomorphic to $\mathbb{Z}^\alpha$ or $\mathbb{Q}.\mathbb{Z}^\alpha$ for some countable ordinal α.

Proof. (i) In $X.Y$ to pass from any point to another, we first use the 1-transitivity of X to move it to the correct copy of Y. Then we use the 1-transitivity of Y to move it to the precisely correct point in this copy of Y (fixing all the other copies).

(ii) If $x, y \in \bigcup_{\alpha \in A} X_\alpha$ are given, there is $\alpha \in A$ such that $x, y \in X_\alpha$. Now move x to y in X_α, fixing all points outside X_α (possible by convexity).

(iii) This follows from (i) and (ii), since $\mathbb{Z}^\alpha$ is convex in $\mathbb{Z}^\beta$ whenever $\alpha \leq \beta$.

(iv) (Sketch) If $(X, \leq)$ is densely ordered, then 1-transitivity implies that it has no endpoints, therefore $(X, \leq) \cong (\mathbb{Q}, \leq)$. If it is not dense (and $|X| > 1$), then each point lies in a copy of $\mathbb{Z}$. Amalgamate them and repeat, transfinitely if necessary. Eventually we reach 1 or $\mathbb{Q}$, and these give the two cases $\mathbb{Z}^\alpha$ and $\mathbb{Q}.\mathbb{Z}^\alpha$. q.e.d.

For uncountable chains, things are much more complicated, and we do not investigate them here. See [$G_1$81] for instance.

We now move on to consider the case of coloured chains, as given in [$C_0T_1$04a, $C_0T_1$04b]. Now for (monochromatic) linear orders, the main building blocks were $\mathbb{Z}$ and $\mathbb{Q}$. In the coloured case, we also need to consider coloured versions $\mathbb{Q}_C$ of the rationals, which were described in the introduction. There is a subdivision into two cases, those in which the colour set is finite, and those in which it is (countably) infinite. The former, treated in [$C_0T_1$04a], is relatively straightforward, amounting to a modification of Morel's list. In the latter some much more complicated ideas are involved. This is mirrored in the fact that for the former there are just $\aleph_1$ examples, in the latter $2^{\aleph_0}$, the $\aleph_1$ arising just because of the part played by Morel's structures.

To analyze a general countable 1-transitive coloured linear order coloured by a set C, we need the following constructions. The first is derived from $\mathbb{Q}_n$ where $2 \leq n \leq \aleph_0$ ($\aleph_0$ only needed when C is infinite). If Y_i for $i < n$ are coloured linear orders, then $\mathbb{Q}_n(Y_0, \ldots, Y_{n-1})$ denotes the result of substituting a copy of Y_i for each point of $\mathbb{Q}_n$ coloured i, called a $\mathbb{Q}_n$-*combination*, or in [$R_0$82] 'shuffle'. (Here, if n is infinite, we just mean $\mathbb{Q}_n(Y_0, \ldots)$, and in future we shall take this kind of minor modification in the notation for granted.) The next is *concatenation*, in general over a countable linear ordering $(\gamma, <)$ (though in practice it suffices to concatenate two sets at a time). If $\{Y_x : x \in \gamma\}$ are coloured linear orders, then the concatenation is obtained from the linear ordering γ by replacing each x by Y_x, ensuring that these sets are disjoint, ordering each Y_x as before, and inducing the ordering between different Y_xs from that of γ. The third is that if Z is a (monochromatic) linear order, and Y is a coloured linear order, then $Z.Y$ is the *lexicographic product*, with colours given by the second coordinates. There are two other constructions which arise when C is infinite, one is lim, which corresponds to taking the union of an infinite nested family, the other, which is more obscure, is select_n, again for $2 \leq n \leq \aleph_0$, which makes a 'selection' among possible points.

Coding trees.

To keep track of what is happening, in [$C_0$02] the notion of 'coding tree' was introduced. For the finite colour set case this is not really necessary, and indeed it is not mentioned at all in Rosenstein's related work on $\aleph_0$-categorical structures, but for an infinite C, some device like this seems essential. It is similar to the idea of a 'skeleton' used in §2. For technical reasons, we require coding trees to be Dedekind–MacNeille complete (a notion explained in the next section). For finite colour sets, the coding trees will be be finite, hence automatically Dedekind–MacNeille complete. For infinite coding trees, Dedekind–MacNeille completeness corresponds to the behaviour of certain subsets of the colour set which we refer to as 'clumps' (meaning that they colour a convex subset of the linear order). The definition then is that a *coding tree* is a Dedekind–MacNeille labelled tree with a root (at the top, since coding tees grow 'downwards', see the next paragraph), at most $\aleph_0$ leaves, every vertex has to be a leaf or above a leaf, all cones at ramification points have greatest elements (see the next section for the definition of these notions), at most countably many vertices have only one child, and the labels are among $\mathbb{Q}_n$, a countable linear order $(\gamma, <)$, select$_n$ for some n, $1 < n \leq \aleph_0$, a countable 1-transitive linear order Z, lim, 1; and these must obey various conditions expressing the intended interpretation, for instance vertices labelled $\mathbb{Q}_n$ or select$_n$ have n children, and a bijection between n and these children is specified, vertices labelled γ have children in bijective correspondence with γ, vertices labelled lim have no children but just one cone below, vertices labelled 1 are leaves and they also have (distinct) colours attached, together with some other conditions.

We remark that 'trees' will be used in more than one sense in this paper. To help distinguish them from the trees whose classification we discuss in the next section, and which have no direct connection with coding trees, we envisage coding trees as growing downwards, and the trees being classified as growing upwards.

Theorem 23.

(i) Any coding tree encodes some countable 1-transitive coloured linear order.

(ii) The countable 1-transitive coloured linear order encoded by any coding tree is unique up to isomorphism.

(iii) Any countable 1-transitive coloured linear order is encoded by some coding tree.

Proof. We remark that it isn't even clear in general what is meant by saying that a coding tree 'encodes' a coloured linear order. For finite coding trees there is no problem, since they are so explicit, and merely a 'book-keeping device'. For instance, in Figure 1, I show the coding tree for the coloured order $\mathbb{Q}_3(\mathbb{Z}^2.(\mathbb{Q}(\text{green})^\wedge\mathbb{Q}(\text{yellow})), \mathbb{Q}_2(\text{red}, \text{blue}), \mathbb{Z}.\mathbb{Q}(\text{white}))$ where $^\wedge$ stands for 'concatenation'. For infinite coding trees, which may not be well-founded or conversely well-founded, it is not so clear. What we do is to introduce an auxiliary notion, called 'expanded coding tree', which is the same as the coding tree, but with each non-leaf vertex having below it the whole

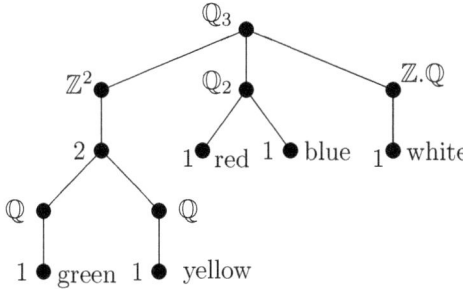

Figure 1. Example of a coding tree.

tree represented at that point, not just the code. Thus for instance, at a vertex labelled $\mathbb{Q}_n$, in the coding tree there are just n children, but in the corresponding expanded coding tree, there are infinitely many, indexed by the members of $\mathbb{Q}_n$, and each child in the coding tree corresponds to all the children in the expanded coding tree having its colour. For concatenation and lim there is no difference between coding tree and the expanded version. For lexicographic product, in the expanded coding tree, instead of 1 child, we have children indexed by the label Z at that point. The most subtle point is the handling of select_n. Here in the coding tree there are n children of the vertex x (indicating the possibilities for what comes below), but in the expanded coding tree there is just one (the choice actually made). So that all possibilities are chosen somewhere, though in different places, we also have to say that in the expanded coding tree, for every vertex y above x, all possible choices are made at some point below y.

This outlines the definition of 'expanded coding tree'. There is then a fairly easy notion of this being *associated with* a coding tree, and we say that a coding tree *encodes* a coloured linear order if it is isomorphic to the coloured linear ordering of the leaves of some expanded coding tree associated with it.

The proofs of the three parts are then accomplished in outline as follows. For (i) we derive an expanded coding tree associated with the given coding tree by taking suitable functions on the branches having 'finite support' (to avoid increasing the cardinality too much, and even to be able to verify the most basic properties). Then (ii) may be proved using a careful application of back-and-forth. Finally for (iii) one starts with a given countable coloured linear ordering, and seeks to recognize inside it the ingredients constituting an expanded coding tree, which can then be 'collapsed' to form a coding tree. As mentioned before, a key technique here is look at 'clumps', which are sets of colours colouring convex subsets, and the coding tree is built up from a chosen maximal tree of clumps. q.e.d.

We also now mention the related class of $\aleph_0$-categorical linear, or coloured, linear orders. These were treated by Rosenstein [$R_0$82], and are rather like the finite colour set 1-transitive coloured linear orders with modifications. The differences are these. In the first place, there is no longer any requirement of 1-transitivity, so that in one sense there are *more* examples. This is replaced instead by deducing from the Ryll–Nardzewski Theorem that there are only finitely many orbits under the action of the automorphism group. On the other hand, we no longer have any of the examples in which $\mathbb{Z}$ appears, since this would give infinitely many 2-types (pairs at greater and greater distances), contrary to the Ryll–Nardzewski Theorem, so in another sense we also have *fewer* examples. This means that when taking lexicographic products, we can only take them over $\mathbb{Q}$ and not over any non-trivial $\mathbb{Z}^\alpha$, and to ease notation, we may regard this as a $\mathbb{Q}_n$-combination with $n = 1$. The conclusion is thus that a countable linear ordering is $\aleph_0$-categorical if and only if it can be built up in finitely many steps from singletons by taking concatenations and $\mathbb{Q}_n$-combinations where $1 \leq n < \aleph_0$. The proof of this result can be carried out by methods which are quite similar to that of the classification of countable 1-transitive coloured linear orders with a finite colour set. Coding trees may be used as a method for describing the stages in the construction.

4 Trees

Trees form an important example intermediate between chains and cycle-free partial orders. In fact, these last were proposed as a generalization of trees by Rubin [$R_1$93]. In his thesis [D85], Droste initiated the study of sufficiently transitive countable trees. The idea is that trees can grow upwards, or downwards—the theory is identical; here we suppose that they grow upwards. A *tree* then is a partially ordered set $(T, \leq)$ in which every two elements have a common lower bound, and for every $x \in T$, $\{y \in T : y \leq x\}$ is linearly ordered. We generally assume that T is not itself linearly ordered, in which case it is called *proper*, thus there are at least two incomparable elements. Trees are also called 'semilinear orders' (that is, linear in one direction).

A key feature required to give a description of a tree is how it 'ramifies'. A *ramification point* is a point of the Dedekind–MacNeille completion which is the infimum of two incomparable points. Merely requiring 1-transitivity of a tree, the weakest sensible hypothesis, guarantees that either all ramification points lie in T, and we then say that it has *positive type*, or none

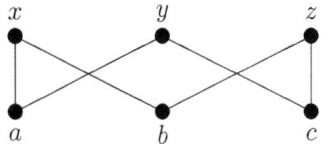

Figure 2. A 6-crown.

do, in which case it has *negative type*. In order to make sense of this, and because it is also used even to give the *definition* of a cycle-free partial order, we now introduce the Dedekind–MacNeille completion X^D of an arbitrary partially ordered set $(X, <)$. Intuitively, this is obtained by adjoining all the Dedekind cuts of maximal chains, and the upper and lower bounds which 'ought' to be present. To make this more precise, we require the following definitions.

Definition 24. $I \subseteq X$ is an *ideal* if $x \leq y \in I \Rightarrow x \in I$. For $A \subseteq X$, let $A^+ = \{x \in X : (\forall a \in A) a \leq x\}$ and $A^- = \{x \in X : (\forall a \in A) x \leq a\}$.

Then A^- is always an ideal. Ideals of the form $\{a\}^- = \{x \in X : x \leq a\}$ are called *principal*. We say that $(X, \leq)$ is *Dedekind–MacNeille complete* (D–M *complete*) if every non-empty subset I of X, bounded above, and such that $I = I^{+-}$, is a principal ideal.

Let us give some examples: In the 6-crown shown in Figure 2, the only suitable sets I are $\{a\}$, $\{b\}$, $\{c\}$, $\{a, b, x\}$, $\{a, c, y\}$, and $\{b, c, z\}$. Here $\{a, b, x\} = \{x\}^-$, $\{a, c, y\} = \{y\}^-$, and $\{b, c, z\} = \{z\}^-$, so these ideals are all principal, and this partially ordered set (clearly not a tree) is D-M-complete.

However the partial order in Figure 3(a) is not D-M-complete, since $I = \{a, b\}$ satisfies $I^{+-} = I$: $I^+ = \{x, y\}$ and $I^{+-} = \{a, b\}$, and I is not principal. The Dedekind–MacNeille completion here is shown in Figure 3(b).

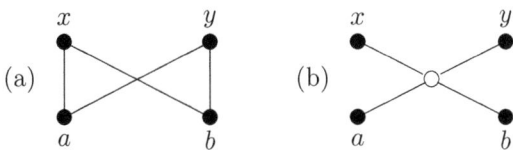

Figure 3. (a) A '4-crown', which is not D-M-complete, and (b) its completion.

Generally, for any partial order $(X, \leq)$ we can form a minimal D-M-complete partially ordered set $(X^D, \leq)$ extending X, called its *D-M-completion*. In fact we can take the elements of X^D to be the non-empty bounded above ideals I satisfying $I^{+-} = I$, partially ordered by $\subseteq$, and with X identified with a subset of X^D via principal ideals.

Lemma 25. If $(X, \leq)$ is D-M-complete, then

(i) every maximal chain C of X is Dedekind-complete (in the usual sense),

(ii) if $x, y \in X$ have a common upper bound (common lower bound) then they have a least upper bound $\sup(x, y)$ (greatest lower bound $\inf(x, y)$ respectively).

Proof. (i) Let $A \subseteq C$ be non-empty bounded above, and let $I = A^{+-}$. Then $A \subseteq I$ and I is bounded above and satisfies $I^{+-} = I$, thus $I = \{a\}^-$ for some a. This is the least upper bound of A in C.

(ii) If $I = \{x, y\}^{+-}$, then $I = \{a\}^-$ for some a, and $a = \sup(x, y)$. If x, y have a common lower bound, let $I = \{x, y\}^{-+-}$. Then if $I = \{b\}^-$, $b = \inf(x, y)$. q.e.d.

If $c = \sup(a, b)$, where a and b are incomparable, then c is called a *lower ramification point*. Similarly, if $c = \inf(a, b)$ where a and b are incomparable, then c is called an *upper ramification point*.

In a tree we have upper ramification points, but not lower ones. In general we may have both.

Definition 26. For every partially ordered set $(X, \leq)$ with Dedekind–MacNeille completion X^D, let $\uparrow\mathrm{Ram}(X), \downarrow\mathrm{Ram}(X)$ be the sets of upper and lower ramification points of X^D, and $\mathrm{Ram}(X) = \uparrow\mathrm{Ram}(X) \cup \downarrow\mathrm{Ram}(X)$. We also write X^+ for $X \cup \mathrm{Ram}(X)$ (not now with the same meaning as before).

We define $\sim$ on $\{y \in T : x < y\}$, where x is a fixed element of T^+, and T is a tree, as follows: let $y_1 \sim y_2$ if $\exists y (x < y \leq y_1, y_2)$. This is clearly reflexive and symmetric. and to see that it is transitive, let $y_1 \sim y_2 \sim y_3$ and $x < y \leq y_1, y_2$ and $x < y' \leq y_2, y_3$. Since $y, y' \leq y_2$ and T is a tree, $y \leq y'$ or $y' \leq y$, suppose the former. Then $x < y \leq y_1, y_3$, so $y_1 \sim y_3$.

The $\sim$-classes are called the (upper) *cones* at x. Thus $x \in \mathrm{Ram}(T) \Leftrightarrow$ there are at least 2 cones, and the number of cones is called the *ramification order of* x, $r.o.(x)$.

In set theory, 'tree' is usually used to mean that for every $x \in T$, $\{y \in T : y \leq x\}$ is *well*-ordered rather than just linearly ordered. In this case there is a unique minimal element, called the *root*, and we have a notion of 'level'

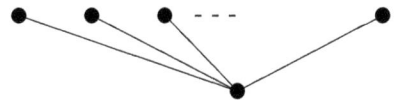

Figure 4. A 2-transitive tree with just 2 levels.

in the tree, enumerated by ordinals. To distinguish these from 'trees' in our sense, we call them *well-founded* trees. They will play a (small) part in what follows. (There is another notion of 'well-founded tree' in descriptive set theory, see [M₃80], namely trees whose dual is well-founded.)

The classification of proper countable k-transitive trees, $k \geq 2$.

Lemma 27. If $(T, \leq)$ is proper countable k-transitive tree where $k \geq 2$, then, either all its maximal chains have length 2, and T has the form shown in Figure 4, or $k \leq 3$, and all its maximal chains are densely ordered without endpoints.

Proof. First suppose that there is a chain $x_0 < x_1 < x_2 < \ldots < x_k$ of length $k+1$. Then p defined by $p(x_i) = x_{i+1}$ for $0 \leq i < k$ is an isomorphism of k-element substructures, so by k-transitivity there is an automorphism f which takes $\{x_0, \ldots, x_{k-1}\}$ to $\{x_1, \ldots, x_k\}$, and which must take x_i to x_{i+1}. Then $\{f^n x_0 : n \in \mathbb{Z}\}$ is an infinite chain. Let C be a maximal chain containing this set. Repeating the argument (with different choices of the x_i) we see that C has no endpoints. To see that C is densely ordered, let $x_0 < x_1$ be any two points of C. As C has no endpoints, we can choose $x_k > x_{k-1} > \ldots > x_2 > x_1$. This time let $p(x_0) = x_0, p(x_i) = x_{i-1}$ for $2 \leq i \leq k$, and let f be an automorphism which extends p. Since $x_0 < x_1 < x_2$, $f(x_0) < f(x_1) < f(x_2)$, so $x_0 < f(x_1) < x_1$.

Now let C' be any other maximal chain. Since T is a tree, $C \cap C' \neq \emptyset$. Let $x \in C \cap C'$. Then $\{y \in C : y \leq x\}$ is infinite. But this is contained in C', so C' is also infinite, Therefore, by the preceding argument, C' is densely ordered without endpoints.

As T is proper, there are incomparable x and y. Since T is a tree, there is $z_k \leq x, y$. Since all the maximal chains have no endpoints, there are $z_1 < z_2 < \ldots < z_k$. Consider p given by $p(z_i) = z_{i+1}$, for $i < k$, $p(z_k) = x$ (or y), and extending to an automorphism, we see that there are incomparable $t, u > x$ and incomparable $v, w > y$, and repeating the argument, incomparable $r, s > t$. This is illustrated in Figure 5. Now $\{r, s, u, y\}$ and $\{t, u, v, w\}$ are 4-element antichains, so

$$\{r, s, u, y, z_1, \ldots, z_{k-4}\} \cong \{t, u, v, w, z_1, \ldots, z_{k-4}\}.$$

However, no automorphism can take $\{r, s, u, y\}$ to $\{t, u, v, w\}$, since the infima of each two of r, s, u, y are comparable, but of t, u and v, w are not. We deduce that $k \leq 3$.

If all chains of T have length $\leq k$, then T is a well-founded rooted tree, with at most k levels. Since T is infinite, the tree shown in Figure 4 embeds, with root a and upper level infinite. If a is not the root of T, then there is $b < a$, and some automorphism takes $\{x_1, \ldots, x_{k-1}, a\}$ to $\{x_1, \ldots, x_{k-1}, b\}$ where $x_1, \ldots, x_{k-1}$ are $k-1$ of the points on the upper level. But this takes a down, which is impossible, since the chains are finite. Similarly, each point above a is maximal, since otherwise we could find an automorphism moving it up. We deduce that all the maximal chains have length 2, and T has the claimed form. q.e.d.

Lemma 28. Let $(T, \leq)$ be a proper infinite k-transitive tree with $k \geq 2$, and with infinite chains. Then all the ramification points of T have the same ramification order, and either all or none of the ramification points are in T.

Proof. Let x, y be ramification points. Then there are $a, b > x$ $c, d > y$, in T, such that $x = \inf(a, b)$ and $y = \inf(c, d)$. Choose $z_1 < z_2 < \ldots < z_{k-2} < x$ and $t_1 < t_2 < \ldots < t_{k-2} < y$ in T. Then $\{z_1, z_2, \ldots, z_{k-2}, a, b\} \cong \{t_1, t_2, \ldots, t_{k-2}, c, d\}$, so there is an automorphism f which takes the first set to the second. This takes $\{a, b\}$ to $\{c, d\}$, hence x to y. We deduce that x and y have equal ramification order.

In addition, $x \in T \Leftrightarrow y \in T$, so either $\mathrm{Ram}(T) \subseteq T$ or $T \cap \mathrm{Ram}(T) = \emptyset$. q.e.d.

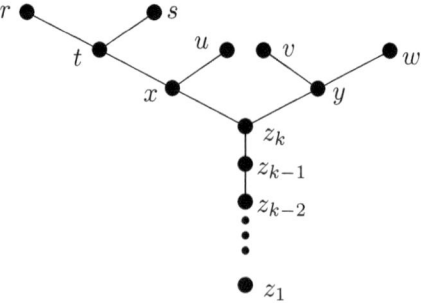

Figure 5. The argument in Lemma 27.

We say that T has *positive type* if $T^+ = T$, (that is $\text{Ram}(T) \subseteq T$), and *negative type* otherwise.

Theorem 29 (Droste). For every cardinal κ, $2 \leq \kappa \leq \aleph_0$, there are countable 2-homogeneous trees T_κ and T'_κ such that all points of T_κ, T'_κ have ramification order κ, T_κ of positive type, T'_κ of negative type. Moreover, any proper countable infinite k-transitive tree for $k \geq 2$ is isomorphic to a tree as in Figure 4 or to some T_κ or T'_κ.

Proof (Sketch). Intuitively, T_κ is constructed by starting with one maximal chain ordered like $\mathbb{Q}$ and allowing all its points to ramify to $\kappa - 1$ other branches, and then repeating infinitely many times on the new branches formed, and taking the union. Formally, the elements of T_κ may be represented by finite sequences of rationals. For T'_κ we modify this by first choosing a countable dense set of irrationals at which the ramification is performed.

The fact that T_κ and T'_κ are 2-homogeneous relies on the constructions' following the same 'recipe' at each stage. Since the only 2-element substructures are a chain and an antichain, it suffices to consider just these cases. Given any two 2-element chains we choose maximal chains added during the construction containing them. Since $\mathbb{Q}$ is 2-homogeneous, we take the first to the second so that the first 2-element chain goes to the second. Now this isomorphism can be extended since the construction starting from any two of these maximal chains is the same. For 2-element antichains the same argument is used, but now using two pairs of maximal chains.

Given any countable k-transitive proper tree with infinite chains, by Lemma 27 we know what κ has to be, and also whether it has positive or negative type. This enables us to show that there is an isomorphism to T_κ (T'_κ) by using back-and-forth, in the style of the previous paragraph, where we map the whole of a maximal chain at a time. q.e.d.

Weak 2-transitivity.

A key point about the above classification, apart from the fact that it is rather explicit, is that it comprises just countably many structures. It was discovered by Droste, Holland, and Macpherson [DH$_1$M$_0$89], that an apparently minor change in the definition, from 2-transitivity, to 'weak 2-transitivity', increases the supply of examples to uncountably many, even though they are still fairly easy to describe. By saying that a partial order is *weakly 2-transitive* is meant that its automorphism group acts transitively on the set of 2-element chains (whereas 2-transitivity requires it also to be transitive on the family of 2-element antichains).

I now describe the general form that a weakly 2-transitive tree can take. First observe that once again, (except in the trivial case of just 2 levels), all maximal chains are ordered like $\mathbb{Q}$. This follows as for the 2-transitive case, since only transitivity on 2-element chains was used for this. We can also deduce by back-and-forth that the automorphism group acts transitively on the set of all maximal chains. It is no longer the case however that the group must act transitively on the ramification points. Letting T be such a tree, we adjoin its ramification points to form T^+, and let C be any maximal chain (by the above remarks, it doesn't matter which) and let C^+ be the union of C and all members of T^+ lying below a member of C.

Now one possibility which can arise is that a cone at some ramification point a has a least element. If so, the ramification point is called *special*, and as one easily shows that no two members of $T^+ - T$ are consecutive, each such least element of a cone at a lies in T. Furthermore, special ramification points clearly form a single orbit under the action of $\text{Aut}(T)$. A notion of *type* is then given for such a tree (which I do not define here precisely), which provides the following information:

(i) whether T has positive or negative type, and in the former case, what the ramification order is at points of T,

(ii) whether there are any special ramification points a, and if so, how many cones there are at a with or without minimal members,

(iii) all other ramification orders.

The main result is then the following.

Theorem 30. Any two countable weakly 2-transitive trees having the same type are isomorphic. Furthermore, any type is the type of some countable weakly 2-transitive tree.

We remark that the second sentence is carried out by starting with a suitable coloured version of $\mathbb{Q}$. If there are special ramification points, then one colour is reserved to stand for pairs of points consisting of a special ramification point and the point of T above it; if there are none, then this colour is just reserved for points of T. The other colours correspond to all other ramification orders specified by the type. Now branches are added in stages. At each stage we add the correct number of branches above each ramification point; for the 'typical' point this is just as for the 2-transitive case. At the lower point of a pair, we have to add the correct number of immediate successors (lying in T), and the correct number without successors. Then one iterates. It follows from this characterization that there are $2^{\aleph_0}$ pairwise non-isomorphic countable weakly 2-transitive trees.

The 1-transitive case.

Finally in this section we consider a further relaxation of transitivity to just 1-transitive, a case which was tackled in [$C_2$04]. Here there are three things which apply even in the weakly 2-transitive case which we can no longer assert. These are that all maximal chains are ordered like $\mathbb{Q}$, are 2-homogeneous linear orders, and are all isomorphic. The argument that they are all isomorphic to $\mathbb{Q}$ clearly requires the stronger hypothesis, so we expect examples now for instance having maximal chains ordered like $\mathbb{Z}$. We might also expect 2-homogeneity of maximal chains to be replaced by 1-transitivity, but even this is false in general, and we can construct examples for instance having maximal chains ordered like $\omega.\mathbb{Z}$ (ω copies of $\mathbb{Z}$). Third, it need not be the case that all maximal chains are isomorphic.

Given these setbacks, it might seem that the quest for any meaningful 'classification' would be hopeless in this generality. However Chicot [$C_2$04] successfully carried this out, though a purist might object that the characterization given doesn't really provide a classification in the strict sense of the word (the classifiers are 'too complicated'). As in the case of countable 1-transitive coloured linear orders, the characterization does though provide a great deal of information about what the possibilities are, and details on how they are constructed, directly generalizing the weakly 2-transitive case.

I now outline the main steps in Chicot's analysis (though the full details are far too complicated to describe in a brief survey). First, some definitions.

Definition 31. A linear order $(X, <)$ is *lower 1-transitive* if for every $a, b \in X$, $(-\infty, a] \cong (-\infty, b]$.

This is a weakening of 1-transitivity. The most obvious example of a lower 1-transitive but not 1-transitive linear order is ω^*, that is, ω backwards. The point is that any branch of a 1-transitive tree is lower 1-transitive, though it need not be 1-transitive, so the first step in classifying all the countable 1-transitive trees is to find all the countable lower 1-transitive linear orders. Even this, it turns out, is a substantial extension of Morel's list. The corresponding natural notion in place of 'isomorphic' is this:

Definition 32. $(X, <)$ and $(Y, <)$ are *lower isomorphic* if for some $x \in X$ and $y \in Y$, $(-\infty, x] \cong (-\infty, y]$ (where these intervals are taken in X, Y respectively), and we write $X \cong_l Y$.

For example $\omega^* \cong_l \mathbb{Z}$.

Lemma 33. : If $(T, \leq)$ is a 1-transitive tree, then all its branches are lower 1-transitive linear orders, and they are all lower isomorphic.

To describe the countable lower 1-transitive linear orders, we again use coding trees. We shall have additional constructions, and corresponding labels of points in the coding trees. We have already mentioned ω^*, and $\dot{\mathbb{Q}}$ and $\dot{\mathbb{Q}}_n$ are used for $\mathbb{Q}$ and $\mathbb{Q}_n$ respectively with an extra point added on the right. The reason for the presence of $\mathbb{Q}_n$ and $\dot{\mathbb{Q}}_n$ is that if $Y_0, Y_1, \ldots, Y_n$ are lower 1-transitive, and all lower isomorphic, then $\mathbb{Q}_n(Y_0, Y_1, \ldots, Y_{n-1})$ and $\dot{\mathbb{Q}}_n(Y_0, Y_1, \ldots, Y_n)$ are also lower 1-transitive, and this represents a typical way of constructing many new examples of lower 1-transitive linear orders. For instance, $\mathbb{Q}_2(\omega^*, \mathbb{Z})$ is a case in point.

Now the coding trees used here differ somewhat from those used earlier, in that they are required to be *levelled*, which means that they can be partitioned into maximal antichains, so that the partition is linearly ordered compatibly with the tree ordering, meaning that for levels L_1 and L_2, $L_1 < L_2$ if and only if there are some $x_i \in L_i$ such that $x_1 < x_2$ in the tree, if and only if for every $x_1 \in L_1$ ($x_2 \in L_2$) there is $x_2 \in L_2$ ($x_1 \in L_1$ respectively) such that $x_1 < x_2$. In what follows we shall use the word 'branch' for a maximal chain of a coding tree of points greater than or equal to some leaf (or for any maximal chain of the trees being classified).

Definition 34. A *coding tree* here is a Dedekind–MacNeille complete levelled tree $(T, \leq)$ with greatest element (root) having countably many leaves such that every element is greater than or equal to some leaf, and the vertices are labelled by $\mathbb{Z}$, ω^*, $\mathbb{Q}$, $\dot{\mathbb{Q}}$, $\mathbb{Q}_n$, $\dot{\mathbb{Q}}_n$ (for $2 \leq n \leq \aleph_0$), 1, or lim, and

(i) x is labelled 1 if and only if it is a leaf,

(ii) two vertices on the same level have equal, or lower-isomorphic, labels,

(iii) the number of children of a vertex is 1, 2, n, $n+1$ if it is labelled $\mathbb{Z}$ or $\mathbb{Q}$, ω^* or $\dot{\mathbb{Q}}$, or $\mathbb{Q}_n$ or $\dot{\mathbb{Q}}_n$ respectively, and if it is labelled lim, then it has no children and just one cone below it, and for ω^*, $\dot{\mathbb{Q}}$, and $\dot{\mathbb{Q}}_n$ one child is designated as the 'right child',

(iv) at each level, the 'left forests' from that level are isomorphic as levelled labelled forests, and

(v) the (labelled levelled) trees with roots at distinct left children of any parent vertex are not isomorphic.

To make sense of the final clauses we say that any child which is not 'right' is *left* (for instance, *all* children of $\mathbb{Q}_n$ vertices) and the *left forest* of a vertex is the forest consisting of its left children and their descendants. For $\mathbb{Z}$, ω^*, $\mathbb{Q}$, and $\dot{\mathbb{Q}}$ the left forest is a single tree, but for $\mathbb{Q}_n$ and $\dot{\mathbb{Q}}_n$ it is a union of n trees. Saying that two forests are isomorphic means that their trees can be put into 1–1 correspondence in such a way that corresponding trees are isomorphic.

As for 1-transitive coloured linear orders we have a notion of *expanded coding tree* which is formed from a coding tree by interpreting the labels. If E is an expanded coding tree associated with a coding tree T, then the set of leaves of E under the natural left-right order is the linear ordering *encoded* by T.

Theorem 35. Any coding tree encodes some countable lower 1-transitive linear order, which is unique up to isomorphism. Conversely, any countable lower 1-transitive linear order is encoded by some coding tree.

The next stage in this particular problem is to analyze certain countable lower 1-transitive coloured linear orders, since by analogy with the case of weakly 2-transitive trees these will describe the possible structures of branches of a 1-transitive tree with information about ramification included. It is (fortunately) not necessary to classify *all* such, as for instance, the points corresponding to the points of the structure, which will be assigned one fixed colour $\bar{c}$, must occur densely. The precise definitions here are motivated by the case of weakly 2-transitive trees, but the extra requirements in the 1-transitive case are considerably more involved. In essence however there is a definition of *colour coding tree* and with respect to this, an analogue of Theorem 35 for the coloured countable lower 1-transitive linear orders needed as branches for the description of all the countable 1-transitive trees.

Moving towards a description in outline of all the countable 1-transitive trees T, let us denote by Υ the class of branches of T^+ viewed as coloured chains, up to isomorphism. We remark that members of Υ will have no maximal elements, and they will form a subset of a lower isomorphism class.

We next need to introduce the notion of 'cone type' of a ramification point. Now in the weakly 2-transitive case we had the notion of a *special* ramification point, and we had to count the numbers of cones at such a point with or without least members. This time, we need to do a similar thing, but corresponding to all possible levels of the coding trees of members of Υ (and a key point is that since all members of Υ are lower isomorphic, their colour coding trees have the same sets of levels, and are closely linked, so may be considered simultaneously). This gives rise to a pair of sequences indexed by the levels of the coding tree of an element of Υ telling us how many special and normal cones there are in the corresponding quotient of that lower 1-transitive coloured linear order at a particular ramification point a, and this is called the *cone type* of a.

We say that a countable tree $(T, <)$ is *structured* if it is proper, and there is a colouring F of T^+ such that the set Υ of branches of T^+ is (up to isomorphism) a subset of some colour lower isomorphism class of lower 1-transitive coloured linear orders without maxima, and having one colour $\bar c$ dense in the others (which precisely colours the elements of T), any two elements having the same colour have the equal cone types, every final segment of a member of Υ occurs infinitely often above every point of T, and whenever $x < y$ are consecutive points of a member of Υ and $F(x) = \bar c$ then also $F(y) = \bar c$.

Finally we have the notion of the *type* $t(T)$ of a tree $(T, <)$, which comprises the set Υ of isomorphism types of branches of T^+, the colour set of T^+ (which may be viewed as the set of orbits of T^+ under the action of $\mathrm{Aut}(T)$), and the family of cone types of ramification points. Conversely one may describe in abstract terms what a 'type' is, based on all the restrictions that one can show are necessary for it to arise in this way. Given all this, the main result is then as follows.

Theorem 36. Any countable tree is structured if and only if it is 1-transitive. Furthermore, any countable 1-transitive tree has a type, two countable 1-transitive trees are isomorphic if and only if they have the same type, and any type is the type of some countable 1-transitive tree.

5 Cycle-free partial orders

M. Rubin [R$_1$93] carried out an exhaustive study of trees (not just countable, and not just 2-transitive ones), and proposed that his work be generalized to what he called 'cycle-free' partial order, CFPOs. He had a clear intuition as to what this should mean, but there were some problems in formulating the correct definition. Roughly speaking, we may say that the idea is that whereas in a tree one can only branch while passing upwards, in a CFPO one can branch going up, or down, and repeatedly, provided one never returns to the starting point.

Peter Neumann [AN98] proposed the following 'desiderata' for such a notion:

(1) subsets of CFPOs should be cycle-free,

(2) trees and their duals should be cycle-free,

(3) the diagram representing a cycle-free partial order should be 'treelike' (in the graph-theoretical sense),

(4) cycle-freeness should be first order expressible,

Figure 6. A substructure of a CFPO need not be a CFPO.

(5) the Dedekind–MacNeille completion of a cycle-free partial order should be cycle-free.

He included (4) and (5) because the definition of a CFPO given by Richard Warren [W97] went via the Dedekind–MacNeille completion, and was ostensibly not first order expressible. The reason why Warren felt it necessary to go by way of X^D is explained by considering (1).

According to Rubin's ideas, the partial ordering in Figure 3(b) definitely should be a CFPO, and when we began considering the notion, we thought that the one in Figure 3(a) should not be, since one can traverse $a - x - b - y - a$ and obtain a cycle. This however already contradicts (1), since the one in Figure 3(a), which (we thought) *has* a cycle, is clearly a subset of that in Figure 3(b), which does *not*. In the light of this example, Warren felt that, to assess whether X was cycle-free, one should really look at X^D, and then (guided by (3) and the graph-theoretical idea) see whether there were still any 'genuine' cycles. His definition was this: $(X, \leq)$ is a cycle-free partial order (CFPO) if between any two points u and v of X, there is in X^D a unique path from u to v. So, in our example, we get over the fact that in X there are distinct paths from a to b, via x or via y, by noting that in X^D these become $a\ t\ c\ t\ b$, $a\ t\ d\ t\ b$ where $t = \sup(a,b)$, neither of which is a 'path', as it 'retraces its steps'. To formalize this, we say that a *connecting set* in a partial order from x to y is a finite sequence $x = x_0, x_1, x_2, \ldots, x_n = y$ such that, for each i, x_i and x_{i+1} are comparable. A *path* is a set of the form $\bigcup_{0 \leq i < n} C_i$ for some connecting set $\{x_i : 0 \leq i \leq n\}$ where C_i is a maximal chain of $[x_i, x_{i+1}] = \{y : x_i \leq y \leq x_{i+1}\}$ if $x_i \leq x_{i+1}$ ($= \{y : x_{i+1} \leq y \leq x_i\}$ if $x_{i+1} \leq x_i$ for simplicity), and such that the C_i only overlap at the endpoints.

Now this definition captured some of what was desired, and it was possi-

Figure 7. The infinite 'alternating chain' ALT.

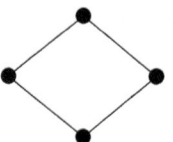

Figure 8. A diamond.

ble to develop the desired theory. However, (4) was still unlikely. In fact, the definition as it stood could not be first order, and even (1) would be false. That is because Warren's definition entailed 'connectedness'—between any two points there is a connecting set. So for instance the partial order shown in Figure 6(a) would be a CFPO, but its substructure in Figure 6(b) would not. In addition, by general model-theoretic considerations, one easily sees that the infinite 'alternating chain' ALT shown in Figure 7, which is connected, and is surely a CFPO, is elementarily equivalent to two incomparable copies of ALT, which is not. To have any chance of axiomatizing the class of CFPOs therefore, we have to relax the connectedness stipulation, and instead say that a CFPO is a partial order $(X, <)$ in which for any $x, y \in X$ there is *at most* one path from x to y. This is equivalent to saying that all its connected components are CFPOs in Warren's sense.

To achieve an axiomatization as given in [$T_1$96] we require the special partial order shown in Figure 8, a *diamond* and $2n$-*crowns* for $n \geq 3$ (a 6-crown is shown in Figure 2).

Theorem 37. $(X, \leq)$ is a CFPO (not necessarily connected) if and only if no diamond or $2n$-crown for $n \geq 3$ embeds in X^+.

To get an axiomatization for CFPOs from this theorem, one has to eliminate the reference to X^+. Since X^+ is interpretable in X, that is, we may express all properties of X^+ just by talking about X, this can be done. In fact there is an apparently slightly weaker equivalent (X^+ embeds no diamond, and X embeds no $2n$-crown for $n \geq 3$).

Theorem 38.

(i) The class of (not necessarily connected) CFPOs is axiomatizable by universal axioms, but is not finitely axiomatizable.

(ii) The class of connected CFPOs is not axiomatizable.

An important and related concept is that of 'being in between'. Any (now connected) CFPO has a natural betweenness relation $B(x; y, z) : x$ is between y and z if x lies on the unique path from y to z. This fulfils various axioms studied by Adeleke and Neumann [AN98], and is occasionally employed in our work.

The CFPOs studied by Warren were just those which are countable, proper (now meaning 'not a tree or its dual'), and k-CS-transitive for some $k \geq 3$. In non-trivial cases we cannot expect any interest in restricting to k-transitive CFPOs, since we could then not even embed the alternating chain with 5 points shown in Figure 6(a), since $\{a, c\} \cong \{a, e\}$, but as a is at different 'distances' 2 and 4 from c and e, there can be no automorphism taking $\{a, c\}$ to $\{a, e\}$. Instead we go for 'k-connected-set-transitive', meaning that for any two isomorphic connected k-element substructures there is an automorphism taking the first to the second. This rules out the above counter-example, as the substructures used are not connected. The classification falls into the following cases, for a CFPO M:

(1) Sporadic [W97]: ALT embeds in M, but all chains of M^+ are finite,

(2) Skeletal [W97]: all the maximal chains of M have length 2, but M^+ has infinite chains (from which it follows that ALT embeds),

(3) Infinite chain [$C_3 T_1$W99]: M has an infinite chain,

(4) ALT does not embed [$T_1$99].

The most surprising feature of this classification one may say is the 'skeletal' case. The word is intended to suggest that the structure M is a mere 'skeleton', consisting of just upper and lower points (maximal chains of length 2); however, on passing to M^+, where all the ramification points are included, suddenly infinite chains appear; it is as if flesh has been put on the bones. This situation contrasts starkly with that for trees, where the finite chain case is essentially trivial. So we have a much richer situation, with

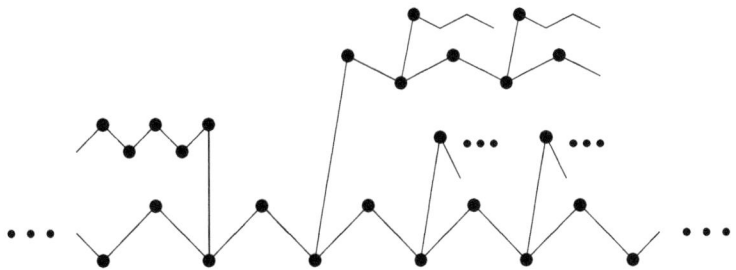

Figure 9. $\mathcal{M}_{3,2}$; maximal chains of M^+ have length 2.

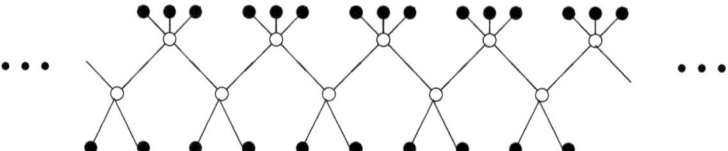

Figure 10. $\mathcal{N}_{\kappa_1,\kappa_2}$; a 'decorated' version of ALT.

many more beautiful and interesting structures, and having complicated automorphism groups.

Lemma 39. Let M be an infinite, proper, k-CS-transitive CFPO, for some $k \geq 2$, all of whose chains are finite. Then all maximal chains of M have length 2.

If $k = 2$, the proof is as for trees. If $k > 2$ one has to argue that, given a chain $\{a, b, c\}$ of length 3, one can find $k-2$ points 'at the side' so that there is a non-trivial order-preserving map on a connected set of size k fixing each of these $k-2$ points, and whose domain and range each intersect $\{a, b, c\}$ in a set of size two. This ensures that one can map strictly upwards within a finite chain, leading to a contradiction.

We now describe in outline the first three cases (1), (2), (3).

Sporadics.
These are CFPOs M in which all the chains of M^+ have finite length, in which case it turns out that this length is ≤ 4, but which nevertheless embeds ALT. There are four kinds:

- $M = \mathcal{M}_{\kappa_1\kappa_2}$. Each minimal point is an upward ramification point of order κ_1, and each maximal point is a downward ramification point

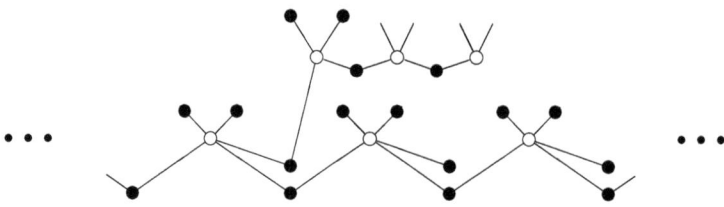

Figure 11. $\mathcal{P}_{\kappa_1,\kappa_2}$.

of order κ_2. See Figure 9.

- $M = \mathcal{N}_{\kappa_1 \kappa_2}$. This is a 'decorated' version of ALT. See Figure 10.
- $M = \mathcal{P}_{\kappa_1 \kappa_2}$. Each minimal point is an upward ramification point of order 2. The maximal points do not ramify. The middle points ramify upwards with order κ_1, and downwards with order κ_2. See Figure 11.
- In addition there is $M = \mathcal{P}'_{\kappa_1 \kappa_2}$, the dual of $\mathcal{P}_{\kappa_1 \kappa_2}$, that is $\mathcal{P}_{\kappa_1 \kappa_2}$ turned upside down. Notice that in these diagrams, we fully shade in points of M. Points of $M^+ - M$ are not shaded in. However, all the points of M^+ are *definable* from points of M.

Skeletals.

These are the countable k-CS-transitive CFPOs M for $k = 3$ or 4 in which all maximal chains of M have length 2, but the maximal chains of M^+ are infinite. For these, as for the sporadics, to specify the whole of M^+ it suffices to give one maximal chain of M^+, (with top and bottom points in M) and instructions about which points ramify up and down and with what ramification order. In these cases, for upward ramification points x and y there is an automorphism taking x to y (similarly for the downward ramification points), since we may find $a, b, c, d, e, f \in M$ such that $x = \inf(b,c)$, $y = \inf(e,f)$, $a < b, c$, and $d < e, f$. In fact this shows that $\mathcal{P}_{\kappa_1 \kappa_2}$ is *not* 3-CS-transitive (though it may be k-CS-transitive for larger k). The maximal chains of M^+ are best thought of as *coloured chains*, that tell us what kinds of points there are. The automorphism group will then act transitively on the points of any fixed colour.

Case 1: $\uparrow \text{Ram}(M) = \downarrow \text{Ram}(M)$ (see Definition 26): $\mathcal{A}^Z_{\kappa \lambda}$. Here Z is an infinite 1-transitive countable linear order (from Morel's list, Theorem 22). The order-type of a maximal chain of M^+ is $1 + Z + 1$; all points of $\text{Ram}(M)$ ramify both upwards and downwards, with orders κ, λ respectively.

Otherwise $\uparrow \text{Ram}(M) \cap \downarrow \text{Ram}(M) = \emptyset$.

Case 2: $\text{Ram}(M)$ densely ordered: $\mathcal{B}_{\kappa \lambda}$. Here the upward and downward ramification points are interdensely ordered.

Case 3: Fully covered CFPOs; every upper cone has a least element, and every lower cone has a greatest element. Here the maximal chain consists of $\mathbb{Q}$ 'pairs' (together with the top and bottom elements): $\mathcal{C}_{\kappa \lambda}$.

Case 4: Partially covered CFPOs: Every lower cone has a greatest element, but $\uparrow \text{r.o.}(M) = 2$ and just one of the two upper cones has a least element (and the dual of this case). The maximal chains are listed:

$\mathcal{D}_\lambda$	$\mathbb{Q}_2(2)$	$\mathcal{E}_\lambda$	$\mathbb{Q}.(\mathbb{Q}.2+1)$
$\mathcal{F}_\lambda^Z$	$Z.(\mathbb{Q}.2+1)$	$Z \not\cong \mathbb{Q}.$	

Case 5: Non-covered CFPOs: $\uparrow$r.o.$(M) = \downarrow$r.o.$(M) = 2$, and at each upward ramification point just one cone has a least element and similarly for downward ramification points: $\mathcal{G}^Z$, $\mathcal{H}^Z$, $\mathcal{I}$, $\mathcal{J}$, $\mathcal{K}$ (with some duals).

Infinite chain CFPOs.

Now we move on to the classification of infinite chain CFPOs. These structures are superficially 'less surprising' in the sense that they are ones one might have suspected existed, whereas the finite chain ones one might have doubted. Their classification is rather more complicated, though it turns out that there is a close connection, which can be made explicit, between this and the finite chain case.

As indicated in the finite chain case, the whole behaviour of any one of the CFPOs M in our list can be described by a maximal chain of M^+ with sufficient information about the ramification points.

Lemma 40. Suppose that M is a proper k-CS-transitive CFPO, $k \geq 2$, with an infinite chain. Then

(i) for every $x \in M$ there are infinite chains A_x, B_x in M such that $A_x < x < B_x$,

(ii) M is dense and 1-transitive,

(iii) for every $x \in M$ there are infinite antichains X_x, Y_x in M such that $X_x < x < Y_x$,

(iv) $\uparrow$Ram(M), $\downarrow$Ram(M) are dense in M.

Proof. As for proofs about trees, the important thing is to establish this for one maximal chain. Let C be an infinite maximal chain of M, and let $x \in C$. Then either $\{y \in C : x < y\}$ or $\{y \in C : x > y\}$ is infinite. Suppose the former, and let $x < x_1 < x_2 < \ldots < x_k$ in C. Then $p(x) = x_1$, $p(x_i) = x_{i+1}$ for $1 \leq i < k$ is an isomorphism of k-element substructures, so there is an automorphism f extending it. Then $B_x = \{f^n(x) : n > 0\}$ is an infinite chain above x, and $A_x = \{f^{-n}(x) : n > 0\}$ is an infinite chain below x.

(ii) If $x < y$, by (i) we may choose $x_k > x_{k-1} > \ldots > x_2 > y$. Then the isomorphism $p(x) = x$, $p(x_i) = x_{i-1}$ for $2 < i \leq k$, $p(x_2) = y$ extends to an automorphism, and the image of y lies between x and y. This gives density. For 1-transitivity, let $x, y \in M$ be given. Choose $x_k > x_{k-1} > \ldots > x_2 > x$ and $y_k > y_{k-1} > \ldots > y_2 > y$, and take $\{x, x_2, \ldots, x_k\}$ to $\{y, y_2, \ldots, y_k\}$.

(iii) Since M is proper, M contains a V-shape. By (ii), every point has two incomparable points above it. By (i), given x, there are x_i with $x < x_0 < x_1 < x_2 < \ldots$. Since every point has two incomparable points above it, there is y_i such that $x_i < y_i$ and y_i is incomparable with x_{i+1}. Then $Y_x = \{y_i : i \in \omega\}$ is an infinite antichain above x. Similarly for X_x.

(iv) Since there is a V-shape $a < b, c$ say in M, $a \leq \inf(b, c)$ which is an upward ramification point. Let $x < y$ be arbitrary in M. Pick $b_k > b_{k-1} > \ldots > b_3 > b > a$, and $y_k > y_{k-1} > \ldots > y_3 > y > x$, and take $\{a, b, b_3, \ldots, b_k\}$ to $\{x, y, y_3, \ldots, y_k\}$. Then a goes to x, and b to y, so the image of $\inf(b, c)$ lies between x and y (possibly equal to x—though, since we can easily show by 3-CS-transitivity that $M \cap \mathrm{Ram}(M) = \varnothing$, this cannot actually happen). Similarly for $\downarrow\mathrm{Ram}(M)$. q.e.d.

The classification is given in terms of a maximal chain C of M^+ (it is easy to see that, for every two maximal chains, there is an automorphism taking the first to the second).

Case 1: M is dense in M^+.

- If $\uparrow\mathrm{Ram}(M) = \downarrow\mathrm{Ram}(M)$: $\mathfrak{A}_{\kappa\lambda}$, M and $\mathrm{Ram}(M)$ interdense on C.

- If $\uparrow\mathrm{Ram}(M) \cap \downarrow\mathrm{Ram}(M) = \varnothing$: $\mathfrak{B}_{\kappa\lambda}$, M, $\uparrow\mathrm{Ram}(M)$, $\downarrow\mathrm{Ram}(M)$ all interdense.

Now suppose M is not dense in M^+. Define $\sim$ on $C \cap \mathrm{Ram}(M)$ by: $x \sim y$ for $x \leq y$ if $(x, y) \cap M = \varnothing$. Since M is not dense in M^+, some $\sim$-classes are non-trivial.

Case 2: M is not dense in M^+ and $\uparrow\mathrm{Ram}(M) = \downarrow\mathrm{Ram}(M)$. Then each $\sim$-class is a 1-transitive linear order Z (in Morel's list), and they are all isomorphic (hence non-trivial). These and the points of $M \cap C$ are interdense: $\mathfrak{C}^Z_{\kappa\lambda}$.

Case 3: M not dense in M^+, $\uparrow\mathrm{Ram}(M) \cap \downarrow\mathrm{Ram}(M) = \varnothing$, and all $\sim$-classes are finite. A *pair* consists of $x < y$ where $x \in \uparrow\mathrm{Ram}(M)$, $y \in \downarrow\mathrm{Ram}(M)$, and there is nothing in between. An upward ramification point x which equals $\inf\{y > x : y \in \mathrm{Ram}(M) \cap C\}$ is called an *upper limit ramification point*.

- $\mathfrak{D}_{\kappa\lambda}$: only pairs occur as $\sim$-classes.

- $\mathfrak{E}_\lambda$: only pairs and singleton upward ramification points occur.

- $\mathfrak{F}$: pairs and singleton upward and downward ramification points all occur.

Case 4: M not dense in M^+, $\uparrow\mathrm{Ram}(M) \cap \downarrow\mathrm{Ram}(M) = \varnothing$, and all $\sim$-classes are infinite.

Then it turns out that all $\sim$-classes are isomorphic.

There are many more cases, depending on whether or not each $\sim$-class has a pair, upward or downward limit ramification points, etc.

In summary, the structure of all the CFPOs in this part of the classification is specified by a maximal chain C of M^+, viewed as 'coloured' by finitely many (at most 4) colours which will tell us

(i) if the point lies in M or $\mathrm{Ram}(M)$,

(ii) if in $\mathrm{Ram}(M)$, what 'kind' of ramification point it is, upward/downward, ramification order, upper or lower limit, member of a pair, etc. One can show that $\mathrm{Aut}(M)$ acts transitively on each colour (using k-CS-transitivity on M), so what is involved is classifying certain 1-transitive *coloured* linear orders. The cases of the subdivision correspond to different instances of this classification. So this has strong links with the material presented in §3.

Simplicity.

Finally we consider some simplicity questions about the automorphism groups of these structures [DT$_1$W99]. For chains, Higman showed that $\mathrm{Aut}(\mathbb{Q}, \leq)$ and $\mathrm{Aut}(\mathbb{R}, \leq)$ both have exactly three proper non-trivial normal subgroups, so they are not far from being simple. By contrast, Droste, Holland and Macpherson showed that for T a countable 2-transitive tree with infinite chains, $\mathrm{Aut}(T)$ has the greatest possible number, $2^{2^{\aleph_0}}$, of normal subgroups. In other words, these groups, unlike $\mathrm{Aut}(\mathbb{Q}, \leq)$, are very far from being simple.

It is something of a surprise, therefore, that many of the automorphism groups of the CFPOs in our list are simple groups. This adds to quite a long list of groups of infinite 'homogeneous' structures that are known to have simple automorphism groups. Moreover, there are neat characterizations of which ones are or are not simple, and for which a 'bounded number of conjugates' suffices.

I shall outline the proof of the simplicity of the automorphism groups of one or two of the CFPOs in the classification. In some cases we are able to give the proof in the following strong form: if g and h are non-identity elements of G, then h may be expressed as the product of 12 conjugates of g or g^{-1}. From this one deduces simplicity of G thus. Let N be a non-trivial normal subgroup. Then there is some $g \neq 1$ in N. Let h be any other non-identity element of G. Since h may be written as the product of 12 conjugates of g, g^{-1}, and N is normal, also $h \in N$. It follows that $N = G$.

In other cases, there is no 'bounded' number of conjugates, such as 12. In other words, for given $g, h \neq 1$, there is always a number n such that h is equal to a product of n conjugates of g, g^{-1}, but this n may depend on g and h. This is no disadvantage from the point of view of establishing simplicity, but it does mean that we may not be able to express simplicity by a first order formula. In our case, the point is that there is quite an attractive demarcation between the cases non-simple, simple with ≤ 12 conjugates, and simple with unboundedly many conjugates required, that corresponds to combinatorial properties of the CFPOs in question.

First, as an easy case, let us consider $\mathcal{M}_{\kappa\lambda}$ with $\kappa, \lambda \geq 3$ (see Figure 9), and we show that $G = \text{Aut}(\mathcal{M}_{\kappa\lambda})$ is simple. (Here there is a bound on the number of conjugates.) Here, all maximal chains even of $\mathcal{M}_{\kappa\lambda}^+$ have length 2, all the minimal points have upward ramification order κ, and all maximal points have downward ramification order λ. Suppose that $1 < N \triangleleft G$, and let $g \in N - \{1\}$. Then g moves some point. If g fixes all minimal points, then it must move some maximal point a to b, say. But there is a minimal point c below a not below b, and this would also have to be moved. Hence g moves a minimal point, a say, to ga. Let the path from a to ga be $a = x_0 < x_1 > x_2 < \ldots > x_{2n} = ga$, and let $h \in G$ fix all the points y such that the path from a to y passes through x_1, and permute the other points above a in any given way (there are at least 2 such as $\kappa \geq 3$). Then $hx_i = x_i$ for each i, and $h(ga) = ga$ and $h(gx_1) = gx_1$. Hence $hg^{-1}h^{-1}g$ fixes $\{a, x_1\}$ and permutes the points above a in the same way that h does. As $hg^{-1}h^{-1}g \in N$, N contains an element fixing a and permuting all points above a except x_1 in any given manner. By varying these, and also the point above a that is fixed, we can show that N contains the stabilizer of a. The proof is concluded by showing that any element of G is a product of two elements in stabilizers of points.

For the more complicated cases—skeletal and infinite chain case, we need to isolate a particular class Σ of elements of the group. If M is skeletal and $\text{Aut}(M)$ acts 'sufficiently transitively' on some maximal chain of M^+, we let $g \in \Sigma$ provided that for some $a < b$ in M, $ga = a$ and $gb = b$ and for every $x \in (a, b)^{M^D}$, $x < gx$, with a similar condition in the infinite chain case. Then under the transitivity condition, one can show that any two members of Σ are conjugate. Next one shows that if $1 \neq N \triangleleft G$, then N contains a member of Σ, hence it contains all of Σ, and finally that Σ generates G.

Let us give some examples: $\mathcal{A}_{\kappa\lambda}^{\mathbb{Q}}$ has a simple automorphism group, with a bounded number of conjugates; $\mathcal{C}_{\kappa\lambda}$ has a simple automorphism group, but with no bound.

In contrast, $\mathcal{A}^{\mathbb{Z}}_{\kappa\lambda}$ does not have simple automorphism group. Here, there is a notion of 'level' on the points of M^+, and the permutations which preserve the levels form a proper non-trivial normal subgroup.

References.

[AN98]   S. A. Adeleke and P. M. Neumann, Relations related to betweenness: their structure and automorphisms, Memoirs of the American Mathematical Society vol. 131 (1998), no. 623.

[$C_0$02]   G. Campero-Arena, Transitivity properties of countable coloured orderings, PhD thesis, University of Leeds, 2002.

[$C_0T_1$04a]   G. Campero-Arena and J. K. Truss, Countable, 1-transitive, coloured linear orders I, Journal of Combinatorial Theory, Series A, 105 (2004), 1-13.

[$C_0T_1$04b]   G. Campero-Arena and J. K. Truss, Countable, 1-transitive, coloured linear orders II, Fundamenta Mathematicae 183 (2004), 185-213.

[$C_1$98]   G. Cherlin, The classification of countable homogeneous directed graphs and countable n-tournaments, Memoirs of the American Mathematical Society vol. 131 (1998), 621.

[$C_2$04]   K. M. Chicot, Transitivity properties of countable trees, Ph.D. thesis, University of Leeds, 2004.

[C_3T_1W99]   P. Creed, J. K. Truss, and R. Warren, The structure of k-CS transitive cycle-free partial orders with infinite chains, Mathematical Proceedings of the Cambridge Philosophical Society, 126 (1999), 175-194.

[D85]   M. Droste, Structure of partially ordered sets with transitive automorphism groups, Memoirs of the American Mathematical Society vol. 57 (1985), no. 334.

[DT_1W99]   M. Droste, J. K. Truss, and R. Warren, Simple automorphism groups of cycle-free partial orders, Forum Mathematicum 11 (1999), 279-294.

[$DH_1M_0$89]   M. Droste, W. C. Holland, and H. D. Macpherson, Automorphism groups of infinite semilinear orders, I and II, Proceedings of the London Mathematical Society 58 (1989), 454-478 and 479-494.

[$G_0$76]   A. Gardiner, Homogeneous graphs, Journal of Combinatorial Theory, Series B, 20 (1976), 94-102.

[$G_1$81]   A. M. W. Glass, Ordered permutation groups, London Mathematical Society Lecture Note Series 55 (1981), Cambridge University Press.

[$G_1M_1R_1$93]   A. M. W. Glass, S. H. McCleary and M. Rubin, Automorphism groups of countable highly homogeneous partially ordered sets, Mathematische Zeitschrift 214 (1993), 55-66.

[G_2G_3K96]   M. Goldstern, R. Grossberg and M. Kojman, Infinite homogeneous bipartite graphs with unequal sides, Discrete Mathematics 149 (1996), 69-82.

[$H_0$72]   C. W. Henson, Countable homogeneous relational structures and $\aleph_0$-categorical theories, Journal of Symbolic Logic 37 (1972), 494-500.

[L84]   A. H. Lachlan, Countable homogeneous tournaments, Transactions of the American Mathematical Society 284 (1984), 431-461.

[LW80]   A. H. Lachlan and R. Woodrow, Countable ultrahomogeneous undirected graphs, Transactions of the American Mathematical Society 262 (1980), 51-94.

[M₂65]   A. C. Morel, A class of relation types isomorphic to the ordinals, Michigan Mathematics Journal 12 (1965), 203-215.

[M₃80]   Y. N. Moschovakis, Descriptive set theory, North-Holland, Amsterdam, 1980 [Studies in Logic and the Foundations of Mathematics 100].

[R₀82]   J. G. Rosenstein, Linear Orderings, Academic Press, 1982.

[R₁93]   M. Rubin, The reconstruction of trees from their automorphism groups, Contemporary Mathematics, 151, American Mathematical Society 1993.

[S79]   J. H. Schmerl, Countable homogeneous partially ordered sets, Algebra Universalis 9 (1979) 317-321,

[T₀05]   S. Torrezão de Sousa, Countable homogeneous coloured partial orders, Ph D thesis, University of Leeds, 2005.

[T₁96]   J. K. Truss, Betweenness relations and cycle-free partial orders, Mathematical Proceedings of the Cambridge Philosophical Society 119 (1996), 631-643.

[T₁99]   J. K. Truss, On k-CS-transitive cycle-free partial orders with finite alternating chains, Order 15 (1999), 151-165.

[T₁01]   J. K. Truss, Elementary properties of cycle-free partial orders and their automorphism groups, Order 18 (2001), 359-379.

[W97]   R. Warren, The structure of k-CS-transitive cycle-free partial orders, Memoirs of the American Mathematical Society, vol 129 (1997), 614.

Received: March 17, 2006;
In revised version: March 28, 2006; May 31, 2006;
Accepted by the editors: December 7, 2006.

Theories of Abelian Groups and Modules Preserved under Extensions

ROGER VILLEMAIRE[1]
MICHEL HÉBERT[2]

[1] Département d'Informatique,
Université du Québec à Montréal
C.P. 8888, succ. centre-ville,
Montréal QC, H3C 3P8, Canada

[2] Department of Mathematics,
The American University in Cairo,
113 Sharia Kasr el Aini, P.O.Box 2511,
11511 Cairo, Egypt

E-mail: villemaire.roger@uqam.ca, mhebert@aucegypt.edu

ABSTRACT. A theory T of modules is preserved under extensions if for any submodule A of a module B, B is a model of T as soon as A and B/A are. We give a syntactic characterization of theories of modules preserved under extensions for the case of regular rings and also for the case of complete theory of abelian groups. This answers a question of U. Felgner.

Herrn Prof. Dr. Ulrich Felgner zum 65. Geburtstag gewidmet

1 Introduction

In this paper a *theory* is any consistent set of (first-order, finitary) sentences, a *complete theory* being a theory which contains either φ or $\neg\varphi$ for every sentence φ. Equivalently, it is the set $\text{Th}(M)$ (called the *theory of M*) of all sentences true in some structure M.

We will say that a theory or a sentence is *preserved under* some algebraic operation if its class of models is closed under this operation. Syntactic characterizations of such theories have been intensively studied in model theory, under the name of *preservation theorems*. Perhaps the first preservation theorem was the celebrated G. Birkhoff's 1935 result, stating that a class of algebraic structures is preserved under substructures, direct products and quotients if and only if it can be *defined by equations*, i.e. it is the

This research was partly supported by the National Science and Engineering Council of Canada.

class of all models of a set of sentences of the form $\forall \bar{x}(\phi(\bar{x}))$, where $\phi(\bar{x})$ is a conjunction of atomic formulas.

It turned out to be more difficult to establish preservation theorems for each one of the operations "substructure", "quotient" and (especially) "direct product". In the second half of the 1950's, closure under substructures was characterized by Łoś and Tarski by *universal sentences* [CK90, Theorem 3.2.2], and closure under quotients was shown to correspond to *positive sentences* by R. Lyndon [CK90, Theorem 3.2.4].

The problem of the direct products was much harder, and is more conveniently considered together with the more general case of the so-called *reduced products* (see [CK90]). From the 40's to the 70's, Mostowski, McKinsey, Feferman, Vaught, Horn, Chang, Keisler, Weinstein, Galvin and Shelah have all contributed to the solution of the two problems. [CK90] is a good source to follow the details of this adventure, but we just recall what is needed here.

Definition 1. The set of *Horn formulas* is the smallest set of formulas containing finite disjunctions of negations of atomic formulas with at most one atomic formula, which is closed under conjunction, universal and existential quantifiers.

Horn showed that a theory axiomatized by Horn sentences is preserved under direct products. The converse was proved later for universal-existential theories (which were proved to be precisely the theories preserved under unions of chains of embeddings). However Chang and Morel showed that there are (existential-universal) theories which are not Horn but are nevertheless preserved under direct products. Finally Weinstein [W65] gave a (rather involved) syntactic characterization of theories preserved under direct products, and Keisler proved, under the continuum hypothesis, that the Horn sentences are precisely those preserved under reduced products. Galvin later showed how to get rid of the continuum hypothesis.

In 1976, U. Felgner [F$_0$80] showed that a complete theory of abelian groups is preserved under direct products if and only if it is a Horn theory. This result was further generalized by the first author to every (not necessarily complete) theory of modules [V92].

Note that any theory is preserved under arbitrary products if and only if it is preserved under binary products [CK90, Theorem 6.3.14]. Since direct sums and direct products of modules are elementarily equivalent [P88, Corollary 2.24], a theory of modules is preserved under direct products if and only if it is preserved under binary direct sums.

Now the notion of extension in module theory can be seen as generalizing the concept of binary direct sum: a module B is said to be an *extension of* the module C by the module A if $C \cong B/A$. The direct sum $A \oplus C$ is an extension of C by A since $(A \oplus C)/A \cong C$.

During his doctoral studies at Tübingen in the late 1980's, the first author was asked by Professor Felgner if there could be some natural characterization of the theories of abelian groups preserved under extensions. No satisfactory characterization was found at the time, but the work done then was the starting point of [V92] which was realized during his postdoctoral studies at McGill university.

The main objective of this work is to answer this question. We give such a characterization for a complete theory of abelian groups, in terms of the values of its Szmielew invariants.

The reader must be warned that the concept of preservation under extensions as defined in this paper is not equivalent to the one normally encountered in model theory (such as in [CK90, Exercise 3.2.1]), where an extension of N is just a structure containing N as a substructure. While the usual notion is meaningful for any first-order language, ours is specific to modules.

In Section 2, we review some basic facts about modules and their model theory. We give a simple syntactic characterization of the theories of modules over a regular ring which are preserved under extensions, and indicate possible avenues for more general cases.

Section 3 contains our main result (Theorem 15). After reviewing some basic facts about the model theory of abelian groups, we give a complete proof of our characterization of complete theories preserved under extensions.

2 Modules

We recall some terminology and well-known facts in module theory and its model theory, which we will use throughout the paper. The reader is referred to [P88] for (a lot) more on the subject.

A *(short) exact sequence* is a sequence of homomorphisms

$$0 \to A \xrightarrow{\alpha} B \xrightarrow{\beta} C \to 0 \tag{1}$$

such that α is an embedding, β is surjective, and $\text{im}(\alpha) = \ker(\alpha)$. Note that B is an extension of C by A if only if there exists such a sequence (1).

We will need to combine exact sequences in order to build new ones. Consider two exact sequences

$$0 \to A \xrightarrow{\alpha} B \xrightarrow{\beta} C \to 0 \text{ and}$$

$$0 \to A' \xrightarrow{\alpha'} B' \xrightarrow{\beta'} C' \to 0.$$

The combination of these two exact sequences by component-wise application is the following sequence:

$$0 \to A \oplus A' \xrightarrow{\alpha \oplus \alpha'} B \oplus B' \xrightarrow{\beta \oplus \beta'} C \oplus C' \to 0$$

where $(\alpha \oplus \alpha')(a, a') = (\alpha(a), \alpha'(a'))$, and similarly for $\beta \oplus \beta'$. It is left to the reader to check that this sequence is indeed exact.

If R is a fixed ring with identity, the language of the theory of (right) R-modules is the (first-order, finitary) language containing the neutral element 0, the operation of addition $+$ and, for every $r \in R$, a unary function symbol which we will also denote by r.

The so-called *positive-primitive formulas* will be important in our context. Those are the ones of the form $\exists \bar{y}(\psi(\bar{x}, \bar{y}))$, with ψ a conjunction of atomic formulas. In any R-module, such a sentence is equivalent to one of a simpler form, which we take as our definition here:

Definition 2. A *positive-primitive formula*, for short a *pp-formula*, is a formula which is equivalent to one of the form:

$$\exists \bar{y} \bigwedge_k (\sum_{i=1}^n x_i r_{i,k} + \sum_{j=1}^m y_j s_{j,k} = 0)$$

with $r_{i,k}, s_{j,k} \in R$, and $\bar{y} = y_1...y_m$.

If $A \xrightarrow{\alpha} B$ is an embedding and $\varphi(\bar{x})$ a pp-formula, one has that if $A \models \varphi[\bar{a}]$ then $B \models \varphi[\bar{a}]$. The converse is not necessarily true, but when it is, it gives the following important concept.

Definition 3. An embedding $A \xrightarrow{\alpha} B$ is said to be *pure* if for every pp-formula $\varphi(\bar{x})$, one has that $A \models \varphi[\bar{a}]$ if and only if $B \models \varphi[\bar{a}]$.

In this definition, and in many of the facts about pp-formulas, one can assume that φ has only one free variable.

Definition 4. An exact sequence

$$0 \to A \xrightarrow{\alpha} B \xrightarrow{\beta} C \to 0$$

is said to be *pure-exact* if α is pure.

A theory is *preserved under pure extensions* if for any pure submodule A of B, B is a model of T as soon as A and B/A are. Equivalently, when in every pure-exact sequence as in Definition 4, B is a model of T when A and C are.

If $\varphi(x)$ is a pp-formula and M a module, the set $\varphi(M) = \{m \in M; M \models \varphi[m]\}$ is an abelian subgroup of M.

Definition 5. Let $\varphi(x)$ and $\psi(x)$ be pp-formulas in one variable and M be a module. The *Bauer-Monk invariant* $\mathrm{Inv}(M, \varphi, \psi)$ is the cardinality of the quotient abelian group $\varphi(M)/(\varphi(M) \wedge \psi(M))$ if it is finite, and ∞ otherwise.

The fundamental theorem of the model theory of modules is that it admits *pp-elimination of quantifiers*: every sentence in the language of R-modules is equivalent to a boolean combination of sentences of the form $\mathrm{Inv}(M, \varphi, \psi) < k$, where $\mathrm{Inv}(M, \varphi, \psi)$ is a Bauer-Monk invariant and k a natural number [P88, Corollary 2.15]. We will use mainly the following obvious consequence:

Theorem 6. Two modules are elementarily equivalent if and only if their Bauer-Monk invariants are equal. [P88, Corollary 2.18]

We will see in Section 3 that for abelian groups, one can use still simpler invariants.

Let us now consider theories of modules preserved under extensions. First note that every such theory must be preserved under products (see the Introduction). The syntactic characterization of the theories of structures preserved under products is rather involved, but for modules it takes a particularly simple form, as we will see in the next theorem.

Examples of theories preserved under products but not under extensions are easily found, even for modules over the ring of integers, i.e., the abelian groups (see next section). However, any such theory must be preserved under pure extensions:

Theorem 7. Let T be a theory of R-modules over some ring R. The following conditions are equivalent:

(a) T is preserved under pure extensions;

(b) T is a Horn theory;

(c) T is preserved under reduced products;

(d) T is preserved under products.

If T is complete, then those conditions are equivalent to

(e) every Bauer-Monk invariant of T is either infinite or equal to 1.

Proof. Note that the Bauer-Monk invariants of a *theory* T make sense when T is complete. (d) $\Leftrightarrow$ (c) is the main result of [V92], and (b) $\Leftrightarrow$ (c) is a classical theorem of model theory [CK90, Proposition 6.2.5′]. (a) $\Rightarrow$ (d) follows from the fact that the natural embedding $A \to A \oplus A$ is pure. (d)

⇒ (a) is [P88, Lemma 2.23], stating that if B is a pure extension of C by A, then $B \equiv A \oplus C$. Finally, (d) ⇔ (e) is [P88, Lemma 2.23 & Corollary 2.18].
q.e.d.

Note that the fact that the Bauer-Monk invariants are infinite or equal to 1 can be easily expressed as Horn sentences, so for complete theories the values of the invariants already give an axiomatization in terms of Horn sentences. We deduce immediately:

Corollary 8. Let R be a (von Neumann) regular ring and T be a theory of R-modules. The following conditions are equivalent:

(a) T is preserved under extensions;

(b) T is a Horn theory;

(c) T is preserved under products.

If $T = \text{Th}(M)$ is complete, then those conditions are equivalent to:

(d) for all idempotent r of R, $\text{ann}_M(r)$ is either 0 or infinite.

Proof. (c) ⇒ (a) is clear from Theorem 7 since all extensions are pure when R is regular [P88, Theorem 16.A(iv)]. (c) ⇔ (d) follows from the theorem above and a result of Rothmaler, showing that the Bauer-Monk invariants for modules over a regular ring have the required simple form [P88, Corollary 16.18].
q.e.d.

Corollary 8 suggests a possible approach for the general case. Trying to identify the form of the sentences preserved under extensions, we know already that they are special Horn sentences. However, we know also a condition on the elements of the ring which will make all Horn sentences preserved under extensions, namely that for every element r of the ring, there exists s such that $rsr = r$ (since this is equivalent to being regular by [P88, Theorem 16.a(iii)]. Could we trace down the reason for this connection at the syntactic level? In the case of complete theories, a similar approach could be attempted with the Bauer-Monk invariants instead of the Horn sentences.

We now turn our attention to a special case, namely the complete theories of abelian groups.

3 Abelian groups

In this section we review some basic notions about abelian groups which we will later need. We follow as closely as possible the notation and terminology of the classical reference on abelian groups [F$_2$70-73].

Definition 9. Let p be a prime number. A *p-element* is an element whose order is a power of p.

Definition 10. An abelian group is said to be *p-torsion free* or to have *no p-torsion* if it contains no p-element.

We will need to consider some specific abelian groups which we now describe. The trivial group 0 is the group containing only the neutral element 0. Q will denote as usual the group of rational numbers. Q_p is the subgroup of Q formed of all fractions $\frac{n}{m}$ such that p does not divide m. Z is the group of integers, while mZ is the subgroup of Z formed of all multiples of m. The cyclic group of order m will be denoted by $Z(m)$, and we will see its elements as the cosets $n + mZ$ of Z/mZ. Finally the Prüfer group $Z(p^\infty)$ is the group formed of all p^nth roots of unity for $n \in N$, where N is the set of natural numbers. Note that $Z(p^\infty)$ is also $Q^{(p)}/Z$ where $Q^{(p)}$ is the group of fractions of the form $\frac{m}{p^n}$ (not to be confused with Q_p), where $m \in Z$ and $n \in N$.

Proposition 11. Any extension of two p-divisible abelian groups is p-divisible.

Proof. Let
$$0 \to A \xrightarrow{\alpha} B \xrightarrow{\beta} C \to 0$$
and take $b \in B$. Since $\beta(b)$ is p-divisible it is equal to some $pc \in C$. Take a pre-image $b' \in B$ of c under β. Now b and pb' are both mapped to the same value in C, hence $b - pb' = a$ for some $a \in A$. a being p-divisible we have that $a = pa'$ and therefore $b = p(b' + a')$, showing that b is p-divisible. q.e.d.

Proposition 12. Any extension of two p-torsion-free abelian groups is p-torsion-free.

Proof. Let
$$0 \to A \xrightarrow{\alpha} B \xrightarrow{\beta} C \to 0$$
and take $b \in B$. If $pb = 0$ then $\beta(pb) = p\beta(b) = 0$ and since C has no p-torsion we have that $b \in A$. Again by hypothesis an element of order p of A must be 0 hence $b = 0$ completing the proof. q.e.d.

3.1 Model theory of abelian groups

Szmielew in [S55] showed that the first-order theory of abelian groups is decidable by showing that every formula is equivalent to a boolean combination of pp-formulas with *core sentences*, a concept that we define below.

Szmielew also showed that every abelian group is elementarily equivalent to one of a set of groups of specific forms, which we will call the *Szmielew groups*, following [EF$_1$72]. We now state the results of Szmielew which we will use, again following the presentation of [EF$_1$72].

Let us first introduce the Szmielew invariants. We denote by $\dim_p$ the vector space dimension over the field with p elements if it is finite, and ∞ otherwise. Similarly $|nG|$ is the cardinality of the subgroup $\{ng; g \in G\}$ if it is finite, and ∞ otherwise. Following [F$_2$70-73], $G[p]$ is the subgroup of the abelian group G containing all the elements of order p and $nG[p]$ is a shorthand for $(nG)[p]$. The Szmielew invariants are the following values, where p is a prime number and n a natural number.

$$\begin{aligned}
U(p, n; G) &= \dim_p p^n G[p]/p^{n+1}G[p] \\
\mathrm{Tf}(p, n; G) &= \dim_p p^n G/p^{n+1}G \\
D(p, n; G) &= \dim_p p^n G[p] \\
\mathrm{Exp}(n; G) &= |nG|
\end{aligned}$$

In fact the second and third invariants of [EF$_1$72] are a bit different than ours. They consider instead the values $\mathrm{Tf}(p; G) = \lim_{n \to \infty} \mathrm{Tf}(p, n; G)$ and $D(p; G) = \lim_{n \to \infty} D(p, n; G)$. $\mathrm{Tf}(p; G)$ is well defined since multiplication by p is an epimorphism of $p^n G/p^{n+1}G$ onto $p^{n+1}G/p^{n+2}G$, so $\mathrm{Tf}(p, n; G)$ decreases as n grows. Similarly $D(p; G)$ is well defined since $p^{n+1}G[p] \subseteq p^n G[p]$, so $D(p, n; G)$ decreases as n grows. Our choice of invariants does not change the validity of the results given below, but our version is more convenient for our purpose.

A *core sentence* is just any statement asserting that a Szmielew invariant is smaller than some specific natural number.

As in the case of modules, we have the following fundamental theorem for the model theory of abelian groups:

Theorem 13. Two abelian groups are elementarily equivalent if and only if all their Szmielew invariants are equal. [EF$_1$72, Theorems 2.1 & 2.6]

Finally Szmielew introduced the following kind of groups and showed that every abelian group is elementarily equivalent to one of them [EF$_1$72, Theorem 2.9].

Definition 14. A *Szmielew group* is an abelian group of the following form where $\alpha_{p,n}$, β_p and γ_p are finite or countably infinite, δ is either 0 or 1, and where p ranges over the prime numbers and n ranges over the natural numbers (here $A^{(\alpha)}$ is the direct sum of α many copies of A):

$$\bigoplus_{p,n} Z(p^n)^{(\alpha_{p,n})} \oplus \bigoplus_p Q_p^{(\beta_p)} \oplus \bigoplus_p Z(p^\infty)^{(\gamma_p)} \oplus Q^{(\delta)} \qquad (2)$$

3.2 Complete theories of abelian groups preserved under extensions

In this section we characterize the complete theories of abelian groups preserved under extensions.

As for the modules, if $T = \mathrm{Th}(G)$ is a complete theory, we can write $\mathrm{Inv}(p,n;T)$ instead of $\mathrm{Inv}(p,n;G)$ for any Szmielew invariant.

Theorem 15. A complete theory of abelian groups T is preserved under extensions if and only if all of the following conditions are satisfied:

(a) if $T \neq \mathrm{Th}(0)$, then $\mathrm{Exp}(n;T) = \infty$, for all $n > 0$;

(b) every other Szmielew invariant is either 0 or ∞;

(c) $\mathrm{U}(p,n;T) = 0$ for all primes p and all natural numbers n;

(d) for any given prime p, $\mathrm{Tf}(p,n;T)$ does not depend on n;

(e) for any given prime p, $\mathrm{D}(p,n;T)$ does not depend on n;

(f) for any given prime p, $\mathrm{Tf}(p,n;T)$ and $\mathrm{D}(p,n;T)$ are not both infinite.

In order to prove the theorem, we will need the following lemmas.

Lemma 16. A complete theory of abelian groups T preserved under extensions is either the theory of the trivial abelian group 0 or satisfies $\mathrm{Exp}(n;T) = \infty$, for all $n > 0$.

Proof. First note that the following sequence

$$0 \to Z(p^n) \to Z(p^{2n}) \to Z(p^n) \to 0, \qquad (3)$$

where the embedding sends the coset $x + p^n Z$ to $p^n \cdot x + p^{2n} Z$, is exact.

Secondly, by Theorem 7, $\mathrm{Exp}(n;T)$ is ∞ or 1. In the latter case, a model G of T is n-torsion and, from the structure of such groups (as direct sums of cyclic groups of order bounded by n), it is clear from equation (3) that G must be 0, since otherwise T would not be closed under extensions. q.e.d.

Lemma 17. If T is a complete theory of abelian groups preserved under extensions, then $\mathrm{U}(p,n;T) = 0$ for all primes p and all natural numbers n.

Proof. Let p be a prime number. We will show that there is a model G of T which is a Szmielew group having no cyclic p-group in its decomposition (2). The result will then follow since this implies that $\mathrm{U}(p, n; G) = 0$.

Let G' be a Szmielew group which is a model of T. Suppose G' has a cyclic p-group in the decomposition (2), and let $Z(p^n)$ be such a summand of smallest order. We then have that $\mathrm{U}(p, n-1; G') \neq 0$. We will now show how to build an extension G of G' by G' which is again a Szmielew group but which has no direct summand that is a cyclic p-group of order equal to p^n. Therefore $\mathrm{U}(p, n-1; G) = 0$ and this is a contradiction since T is complete.

In the decomposition of G', regroup all summands of the form $Z(p^n)$ in order to write $G' = \bigoplus Z(p^n) \oplus G''$ where G'' contains no $Z(p^n)$ summand.

By component-wise application of the exact sequence (3), we also have an exact sequence

$$0 \to \bigoplus Z(p^n) \to \bigoplus Z(p^{2n}) \to \bigoplus Z(p^n) \to 0. \quad (4)$$

Finally since $G'' \oplus G''$ is an extension of G'' by itself we get that

$$0 \to G'' \to G'' \oplus G'' \to G'' \to 0 \quad (5)$$

is again exact. Combining the exact sequences (4) and (5) component-wise, we get the exact sequence

$$0 \to \bigoplus Z(p^n) \oplus G'' \to \bigoplus Z(p^{2n}) \oplus G'' \oplus G'' \to \bigoplus Z(p^n) \oplus G'' \to 0. \quad (6)$$

Taking G to be $\bigoplus Z(p^{2n}) \oplus G'' \oplus G''$ completes the proof. q.e.d.

Lemma 18. *Let T be a complete theory of abelian groups. For any prime p, if $\mathrm{U}(p, n; T) = 0$ for every natural number n, then $\mathrm{Tf}(p, n; T) = \mathrm{Tf}(p, 0; T)$ for every natural number n.*

Proof. This follows from Lemma 1.6 of [EF$_1$72], which is proved by showing the exactness of the following sequence:

$$0 \to p^n G[p]/p^{n+1}G[p] \to p^n G/p^{n+1}G \xrightarrow{\times p} p^{n+1}G/p^{n+2}G \to 0.$$

q.e.d.

Lemma 19. *Let T be a complete theory of abelian groups. For any prime p, if $\mathrm{U}(p, n; T) = 0$ for every natural number n, then $\mathrm{D}(p, n; T) = \mathrm{D}(p, 0; T)$ for every natural number n.*

Proof. This follows from [EF$_1$72, Lemma 1.8], which is proved by showing the exactness of the following sequence:

$$0 \to p^{n+1}G[p] \to p^n G[p] \to p^n G[p]/p^{n+1}G[p] \to 0.$$

q.e.d.

In the following lemmas we will use some standard homomorphisms between Q, Q_p, $Z(p^\infty)$, and $Z(p)$, in order to construct specific extensions. Hence it may be useful to recall some well-known related facts:

(i) Q_p is a subgroup of the rationals Q, and the mapping $x \mapsto \frac{x}{n}$ where n is a non-zero integer is an embedding of Q_p into Q.

(ii) $Q_p/pQ_p \cong Z(p)$.

(iii) $Q/Q_p \cong Z(p^\infty)$.

(iv) The (division by p) mapping $(n + pZ) \mapsto (\frac{n}{p})$ is a well defined embedding of $Z(p)$ into $Z(p^\infty)$.

(v) The (division by p) mapping $(\frac{n}{m} + pQ_p) \mapsto (\frac{n}{p \cdot m} + Q_p)$ is a well defined homomorphism from Q_p/pQ_p to Q/Q_p.

Lemma 20. *There is an exact sequence of the form*

$$0 \to Q_p \to Q \oplus Z(p) \to Z(p^\infty) \to 0. \qquad (7)$$

Proof. By the facts above, it is sufficient to show that the following sequence is exact:

$$0 \to Q_p \xrightarrow{\alpha} Q \oplus Q_p/pQ_p \xrightarrow{\beta} Q/Q_p \to 0, \qquad (8)$$

where $\alpha(q) = (\frac{q}{p}, q + pQ_p)$ and $\beta(q, q' + pQ_p) = (q + Q_p) - (\frac{q'}{p} + Q_p)$

The map α is an embedding by the fact (i), and β is an epimorphism because the canonical homomorphism $Q \to Q/Q_p$ is onto.

The fact that $\text{im}(\alpha) \subseteq \ker(\beta)$ is clear from $\beta(\alpha(q)) = \beta(\frac{q}{p}, q + pQ_p) = (\frac{q}{p} + Q_p) - (\frac{q}{p} + Q_p) = 0$.

In order to show that $\ker(\beta) \subseteq \text{im}(\alpha)$, consider $(q, q' + pQ_p)$ such that $\beta(q, q' + pQ_p) = (q + Q_p) - (\frac{q'}{p} + Q_p) = 0$. This means that $q + Q_p = \frac{q'}{p} + Q_p$ and hence $q = \frac{q'}{p} + \frac{n}{m} = \frac{1}{p}(\frac{q' \cdot m + p \cdot n}{m})$, with $\frac{n}{m} \in Q_p$.

Therefore $(q, q' + pQ_p) = (\frac{1}{p}(\frac{q' \cdot m + p \cdot n}{m}), (\frac{q' \cdot m + p \cdot n}{m}) + pQ_p) = \alpha(\frac{q' \cdot m + p \cdot n}{m})$,
as required.
q.e.d.

Lemma 21. If T is a complete theory of abelian groups preserved under extensions, then for every prime p, $\mathrm{Tf}(p, n; T)$ and $\mathrm{D}(p, n; T)$ are not both infinite.

Proof. This proof is similar in spirit to the one of Lemma 17.

Assuming that for some prime p, $\mathrm{Tf}(p, n; T) = \mathrm{D}(p, n; T) = \infty$, we will build an extension G of two models of T such that $\mathrm{U}(p, 0; G) \neq 0$, contradicting Lemma 17.

So take G' to be a Szmielew group which is a model of T satisfying $\mathrm{Tf}(p, n; T) = \mathrm{D}(p, n; T) = \infty$ for some prime p. Since by Lemma 17 we have that $\mathrm{U}(p, n; G') = 0$ for every natural number n, it follows that the decomposition of G' has at least one copy of Q_p and also one copy of $Z(p^\infty)$. Hence $G' = Q_p \oplus G'' = Z(p^\infty) \oplus G'''$ for some G'' and G'''.

Since $G'' \oplus G'''$ is an extension of G'' by G''' we get the exact sequence

$$0 \to G'' \to G'' \oplus G''' \to G''' \to 0. \tag{9}$$

Combining the exact sequences (7) and (9) component-wise, we obtain the exact sequence

$$0 \to Q_p \oplus G'' \to Q \oplus Z(p) \oplus G'' \oplus G''' \to Z(p^\infty) \oplus G''' \to 0. \tag{10}$$

Taking G to be $Q \oplus Z(p) \oplus G'' \oplus G'''$ completes the proof.
q.e.d.

We can now give the first part of the proof of the theorem.

Proof of Theorem 15 (left to right). Condition (a) follows from Lemma 16. Condition (b) follows from the fact that if T is preserved under extensions, then it is preserved under direct products, therefore its invariants (other than $\mathrm{Exp}(n; T)$) are either 0 or infinite.

The others conditions follow from Lemmas 17, 18, 19, and 21 respectively.
q.e.d. (Theorem 15 (left to right))

In order to complete the proof of Theorem 15, we need some more lemmas.

Lemma 22. Let T be a complete theory of abelian groups. If $\mathrm{Tf}(p, n; T) = \infty$, then $\mathrm{Tf}(p, n; B) = \infty$ for any extension B of two models of T.

Proof. Let A and C be two models of T in the exact sequence

$$0 \to A \xrightarrow{\alpha} B \xrightarrow{\beta} C \to 0.$$

By hypothesis we have that $\dim_p p^n C/p^{n+1} C = \infty$. Take infinitely many $c_i \in p^n C$ which are linearly independent modulo $p^{n+1} C$. Each c_i has a pre-image b_i under β which is in $p^n B$ (if $c_i = p^n c$ take p^n times a pre-image of c). These b_i's must be linearly independent modulo $p^{n+1} B$ since the c_i's are independent modulo $p^{n+1} C$. q.e.d.

Lemma 23. Let T be a complete theory of abelian groups. If $D(p, n; T) = \infty$, then $D(p, n; G) = \infty$ for any extension G of two models of T.

Proof. Let A and C be two models of T in the exact sequence

$$0 \to A \xrightarrow{\alpha} B \xrightarrow{\beta} C \to 0.$$

By hypothesis we have that $\dim_p p^n A[p] = \infty$. Now α sends elements of $p^n A[p]$ to element of $p^n B[p]$, proving the claim. q.e.d.

Lemma 24. Let T be a complete theory of abelian groups. If $\mathrm{Exp}(n; T) = \infty$, then $\mathrm{Exp}(n; G) = \infty$ for any extension G of two models of T.

Proof. Let A and C be two models of T in the exact sequence

$$0 \to A \xrightarrow{\alpha} B \xrightarrow{\beta} C \to 0.$$

By hypothesis we have that $|nA| = \infty$. α sends elements of nA to elements of nB, proving the claim. q.e.d.

Proof of Theorem 15 (right to left). Let T be a complete theory of abelian groups satisfying the conditions (a) to (f) of the theorem. We will show that T is preserved under extensions by showing that the value of every invariant is preserved under extensions, i.e., an extension of two models of T has the same invariants as T.

We first have to show that an extension B of two models A and C of T satisfies $U(p, n; B) = 0$ for all prime numbers p and all natural numbers n.

Consider the following exact sequence:

$$0 \to A \xrightarrow{\alpha} B \xrightarrow{\beta} C \to 0. \tag{11}$$

Take $b \in p^n B$. We will show that $b \in p^{n+1}B[p]$, showing that $\mathrm{U}(p,n;B) = 0$.

Now since β is a homomorphism, we have that $\beta(b) \in p^n C[p]$. Since $\mathrm{U}(p,n;C) = 0$ it follows that $\beta(b) \in p^{n+1}C[p]$, which means that there is a $c \in C$ such that $p^{n+1}c = \beta(b)$. Take a pre-image b' of this c under β and let b'' be $p^{n+1}b'$. Since both b and b'' map to the same element under β we have $b = b'' + a$ for some $a \in A$. Now by hypothesis either $\mathrm{Tf}(p,n;T) = 0$ or $\mathrm{D}(p,n;T) = 0$.

In the first case $\dim_p p^m A/p^{m+1}A = 0$ for every natural number m, and therefore every element of A is divisible by p^m for every m. Hence a is divisible by p^{n+1}, and since $b = b'' + a$ and b'' is also divisible by p^{n+1}, we have that b is also divisible by p^{n+1}. Therefore $b \in p^{n+1}B[p]$, proving the claim.

In the case where $\mathrm{D}(p,n;T) = 0$, from $\beta(b) \in p^n C[p]$ it follows that $\beta(b) = 0$, and hence $b \in A$. Now b is an element of order p of A. Since $\mathrm{U}(p,n;A) = 0$ for every natural number n, it follows by definition of $\mathrm{U}(p,n;A)$ that $p^n A[p]/p^{n+1}A[p] = 0$ for all n. Therefore b is divisible in A (and hence also in B) by every power of p. It follows that $b \in p^{n+1}B[p]$ completing the proof of the claim.

Let us now consider the second invariant $\mathrm{Tf}(p,n;T)$. By Lemma 22, if $\mathrm{Tf}(p,n;T)$ is infinite then this also holds for any extension of two models of T. We will now show that if $\mathrm{Tf}(p,n;T) = 0$, then $\mathrm{Tf}(p,n;G) = 0$ for any extension G of two models of T.

By Lemma 18, $\mathrm{Tf}(p,n;T) = \mathrm{Tf}(p,0;T)$, so $\mathrm{Tf}(p,0;T)$ is also equal to 0. Now $p^0 H/pH = H/pH$, so $\mathrm{Tf}(p,0;H) = 1$ is equivalent to $H/pH = 0$. This means that H is p-divisible. Hence every model H of T is p-divisible. Now if G is an extension of two models of T, then by Proposition 11 G is p-divisible, so that $\mathrm{Tf}(p,0;G) = 0$. We have already shown that $\mathrm{U}(p,n;G) = 0$ for all n since it is an extension of two models of T, so by Lemma 18 it follows that $\mathrm{Tf}(p,n;G) = 0$ for all n. This completes the proof of the preservation of $\mathrm{Tf}(p,n;T)$ by extension.

For the third invariant $\mathrm{D}(p,n;T)$, as in the last case it follows from Lemma 23 that if $\mathrm{D}(p,n;T) = \infty$, then this also holds for any extension of two models of T. We now show that if $\mathrm{D}(p,n;T) = 0$, then $\mathrm{D}(p,n;G) = 0$ for any extension G of two models of T.

By Lemma 19, $\mathrm{D}(p,n;T) = \mathrm{D}(p,0;T)$, so $\mathrm{D}(p,0;T)$ is also equal to 0. Now $\mathrm{D}(p,0;H) = \dim_p H[p]$, so $\mathrm{D}(p,0;H) = 0$ is equivalent to $H[p] = 0$, which means that H has no p-torsion. We therefore have that every model H of T has no p-torsion. Consider an extension G of two models of T. By Proposition 12, G has no p-torsion, so $\mathrm{D}(p,0;G) = 0$. We have already shown that $\mathrm{U}(p,n;G) = 0$ for all n since it is an extension of two models of

T, so by Lemma 19 it follows that $D(p,n;G) = 0$ for all n. This completes the proof of the preservation of $D(p,n;T)$ by extension.

The theory of the trivial abelian group 0 is obviously preserved under extensions. If T is not the theory of 0 and fulfills all conditions of the statement of the theorem, then $\text{Exp}(n;T) = \infty$ for all $n > 0$ by hypothesis. Now by Lemma 24 we have that $\text{Exp}(n;G) = \infty$ for any G which is an extension of two models of T. This completes the proof.

q.e.d. (Theorem 15 (right to left))

Theorem 15 characterizes complete theories of abelian groups preserved under extensions in terms of their Szmielew invariants. Alternatively the following result, whose proof consists in computing the Szmielew invariants, characterizes these theories in terms of their Szmielew groups.

Corollary 25. The theory $\text{Th}(G)$ of the abelian group G is preserved by extensions if and only if G is elementarily equivalent to a group of the following form:

$$\bigoplus_{p \in P_1} Q_p^{(\omega)} \oplus \bigoplus_{p \in P_2} Z(p^\infty)^{(\omega)} \oplus Q^{(\delta)}$$

where P_1, P_2 are disjoint sets of prime numbers and δ is either 0 or 1.

4 Conclusion

We have characterized theories preserved under extensions for modules over regular rings, and for complete theories of abelian groups. Different types of characterizations appear in the paper: structural, syntactical, and in terms of the values of the (modules or groups) invariants.

Obvious open problems remain:

1) find a simple syntactically defined family of sentences which characterize theories of R-modules (or abelian groups) preserved under extensions,

2) formulate a more general characterization in terms of the (modules or abelian groups) invariants for the theories preserved under extensions.

A possible approach for 1) was mentioned in Section 2. Note however that there is no guarantee that preservation under extensions is a uniform property: this means that there might be a theory preserved under extensions which is not equivalent to any set of sentences which are themselves (individually) preserved under extensions. Such a possibility was first recognized by M. Rabin, who showed that preservation under intersection (of substructures) is not uniform in this sense [R62]. For more on this see [H91].

As for 2), an idea could be to try to describe the possible invariants of an extension of two modules or abelian groups A and C in terms of the invariants of A and of C. Both suggestions appear to be rather difficult, but this problem surely deserves further study.

On a more personal level, the first author would like again to thank Professor Felgner for supervising him at the doctoral level, sharing his enthusiasm for research and logic. It is a pleasure to have the chance to return after so many years to the field of model theory, particularly by contributing to this longstanding open problem which was initiated by Professor Felgner.

References.

[CK90]   Chen Chung Chang and H. Jerome Keisler. *Model theory, 3rd ed.*, volume 73. North-Holland, 1990.

[$EF_1 72$]   Paul C. Eklof and Edward R. Fisher. The elementary theory of abelian groups. *Annals of Mathematical Logic*, 4(2):115–171, 1972.

[$F_0 80$]   Ulrich Felgner. Horn theories of abelian groups. In *Model theory of algebra and arithmetic*, volume 834 of *Lecture Notes in Mathematics*, pages 169–173. Springer-Verlag, 1980.

[$F_2 70$-73]   Laszlo Fuchs. *Infinite abelian groups*, volume 1-2. Academic Press, 1970-1973.

[H91]   Michel Hébert. Syntactic characterization of closure under connected limits. *Archive for Mathematical Logic*, 31(2):133–143, 1991.

[P88]   Mike Prest. *Model theory and Modules*, volume 130. Cambridge University Press, 1988.

[R62]   Michael O. Rabin. Classes of models and sets of sentences with the intersection property. *Annales scientifiques de l'université de Clermont*, 7:39–53, 1962.

[S55]   Wanda Szmielew. Elementary properties of abelian groups. *Fundamentae Mathematica*, 41:203–71, 1955.

[V92]   Roger Villemaire. Theories of modules closed under direct products. *Journal of Symbolic Logic*, 57(2):515–521, 1992.

[W65]   Joseph Michael Weinstein. *First-order properties preserved by direct product*. PhD thesis, University of Wisconsin, Madison, 1965.

Received: March 19, 2006;
In revised version: November 28, 2006;
Accepted by the editors: December 7, 2006.

A Model Theoretic Approach to Set Theory without the Axiom of Choice

AGATHA C. WALCZAK-TYPKE

Kurt Gödel Research Center for Mathematical Logic
Universität Wien
Währingerstraße 25
1090 Wien, Austria
E-mail: agatha@logic.univie.ac.at

1 Introduction

1.1 Motivation

In the study of set theory without the axiom of choice (AC), non-well-orderable (NWO) structures have been typically constructed in efforts to discern the relative strengths of various weak forms of AC. We believe that such structures are also interesting in their own right.

In this paper, we survey certain model theoretic methods that give general results regarding the types of NWO structures that can be constructed, and the properties certain NWO structures must necessarily have.

We assume that the reader is familiar with Fraenkel-Mostowski permutation and symmetric forcing model constructions. Should this not be the case, a very good introduction to these topics can be found in [J73] or [F71]. We will define here various classes of NWO sets. Many of these classes are only interesting in contexts where the full axiom of choice is not assumed — these definitions will be flagged as such.

We will also assume that the reader is familiar with classical model theory as presented in [$H_2$93, $R_0$00, $B_2$96]. For model theoretic concepts that extend beyond those covered in these standard textbooks, suggested references will be provided as necessary.

The reader will find that many notions from set theory without choice are very similar to notions from model theory. A prime example is the evident similarity between the notion of an amorphous set (Definition 3) and that of a strongly minimal model (Definition 4), though others will be mentioned in this survey. The similarity between amorphousness and strong minimality in fact spurred this line of research. We will point out that historically the development of model theory and set theory without choice has been parallel; one did not generally follow after the other. Rather, the two disciplines demonstrate the same natural methods of generalization from the finite to the infinite.

The subject of this survey represents a confluence of two themes of Ulrich Felgner's research: the study of set theory without choice and the model theory of groups and other algebraic structures, in particular those that are $\aleph_0$-categorical. Set theory without choice was the main theme of Ulrich Felgner's early publications. Many students of set theory without choice have been introduced to the subject via his book [F71]. Later, his interests tended more towards model theory. His earlier papers in this field centered on $\aleph_0$-categorical structures. Certain results therein are used in the work of this survey to obtain structural information about some NWO groups. Here, I would like to add: *Herzlichen Glückwunsch zum Geburtstag, Herr Professor Felgner!*

1.2 Fundamental notions

In the interest of clarity, note that a set X is *finite* if there exists a one-to-one and onto (bijective) mapping from X to some $n \in \omega$. A set is *infinite* if it fails this definition of finite. This definition of finite is not dependent on the axiom of choice, and is absolute for models of set theory. Obviously, finite sets are well ordered. Because we are interested in NWO structures, all structures discussed in this survey are assumed to be infinite.

There are other definitions of "finiteness" that are equivalent to finite only if one assumes AC. In the absence of AC, there exist infinite sets that satisfy these finiteness notions. In these cases, the sets are NWO. The most well known example of such a notion, and the class of sets that most students of mathematics associate most with models of set theory without AC, is perhaps that of Dedekind-finite sets:

Definition 1 ($\neg$AC)**.** A set X is *Dedekind-finite* if X has no proper subset that can be mapped bijectively to X. Equivalently, X is Dedekind-finite if there does not exist a bijection from ω to a subset of X.

Definition 1 was the first formal definition of finiteness or infinity, and was given by Richard Dedekind in the 1870's [D88]. We note that the assumption of the axiom of choice for countable families of sets (AC_ω) is sufficient to show that all Dedekind-finite sets are finite.

Dedekind-finite *sets* are rather well studied, however there are still many open questions as to their possible *structures*. Few general techniques have been developed to discern the sorts of structures that can be carried by a Dedekind-finite set. We believe that the techniques shown in this survey suggest that it is not so much which ordinals can be embedded into a set, but rather *onto* which ordinal a set can or cannot be embedded, that determine the kinds of structures a set can carry. The author of [M$_4$75] showed that for an arbitrary ordinal α, one can construct a model containing a Dedekind-finite set X that does not map onto α. This perhaps explains the enormously wide variety of structures that can be carried by Dedekind-finite sets.

Definition 2 (¬AC)**.** Let α be an infinite ordinal. We call a set X that cannot be mapped surjectively onto α an α^*-*set*. If X is also a NWO set, then we say that X is an α^*-*nwo set*.

In particular, we call an ω^*-set X *weakly Dedekind-finite*. Equivalently (see [T$_1$74]), X is weakly Dedekind-finite if $P(X)$ is Dedekind-finite.

In this survey, we will concentrate on weakly Dedekind-finite sets and on ω_1^*-nwo sets because model theoretic methods seem to lend themselves particularly to these classes of sets. In the latter case, we assume that ω_1 is regular; under this assumption the class of ω_1^*-nwo sets is closed under subsets, unions, and products. The class of weakly Dedekind-finite sets is closed under subsets, unions, and products in any case [W05b, W05a].

We point out that weakly Dedekind-finite sets are Dedekind-finite. On the other hand, ω_1^*-nwo sets can, but need not be, Dedekind-finite. If AC_ω is assumed, all weakly Dedekind-finite sets are finite, but there can exist ω_1^*-nwo sets.

The notion of weak Dedekind-finiteness is an old one, appearing (though under a different name) in [T$_0$24].

We will define other classes of NWO sets as necessary. Figure 1 illustrates the relative strengths of some of the notions mentioned in this survey.

Section 2 will concentrate on techniques from first-order model theory. These will include ideas from geometric stability theory and classical model theory. The effects of model theoretic assumptions on model constructions will also be discussed. Finally, §3 will contain some uses of classical infinitary model theory. Unexplored ideas are given as questions.

A word on notation: in general, infinite NWO structures will be denoted using *Fraktur* letters, while for infinite well-ordered structures, cursive will be used.

2 Methods from first-order logic

2.1 Geometric methods

In this subsection, we focus on a particular class of non-well-orderable sets, namely that of *amorphous* sets. For all of subsection 2.1, we assume ¬AC.

Definition 3 (¬AC)**.** A set X is *amorphous* if it cannot be expressed as the union of two disjoint infinite sets.

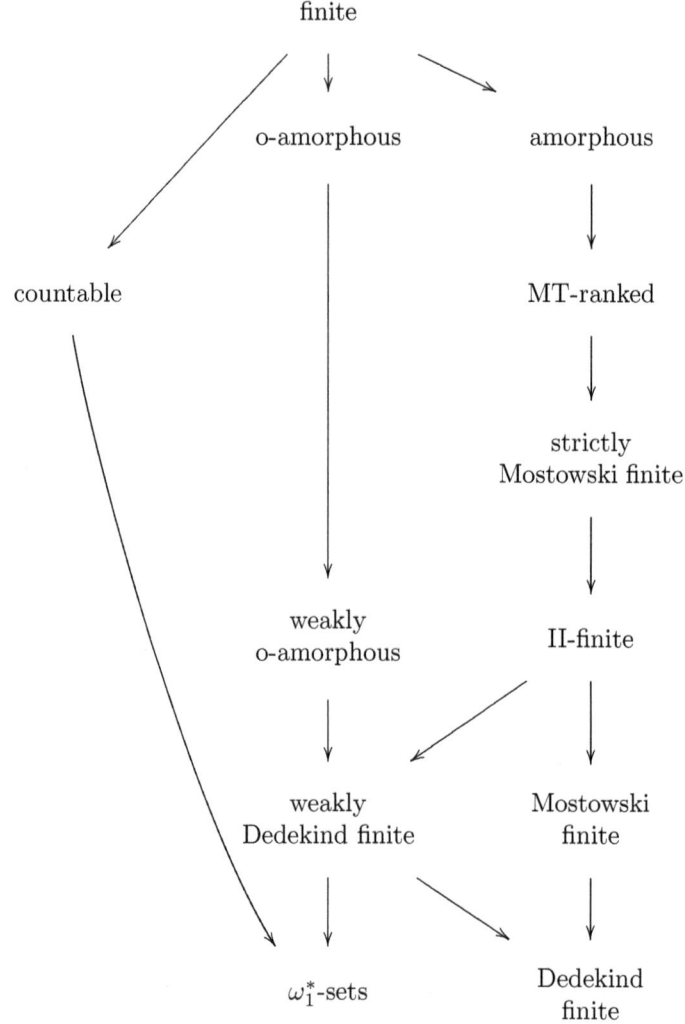

Figure 1. Dependencies between different finiteness notions discussed in this survey. For the purposes of the figure, by "o-amorphous," and "weakly o-amorphous," we mean that the set has an ordering under which it fulfils that definition. See also [$T_1$74] and [$H_3R_1$98, Note 94] and updates on the associated website.

Note that other authors use the term "amorphous" for *infinite* sets that cannot be expressed as the union of two disjoint infinite sets. This notion was introduced as *Ia-finiteness* in [L$_2$58]. As shown in Figure 1, amorphous sets are weakly Dedekind-finite.

Amorphousness is typically considered in the light that it is a finiteness notion, i.e., when AC is assumed, the notion of amorphous as defined here is equivalent to the definition of finiteness.

We mention a fundamental notion from first-order model theory:

Definition 4. A complete theory T is said to be *strongly minimal* if every parametrically definable subset of each of its models is either finite or cofinite.

Strongly minimal theories were introduced by William Marsh in his 1966 Ph.D. thesis, eight years *after* the notion of amorphousness was first defined.

In view of the evident analogy between amorphous sets and strongly minimal theories, as pointed out by Hodges [T$_1$95], Truss attempted to provide a "classification" of amorphous sets [T$_1$95] using Zilber's geometric stability theory methods - an approach fruitful in the study of strongly minimal theories. A good reference for this is [P$_0$96].

We assume all amorphous sets henceforth to be infinite.

In [T$_1$95], Truss begins by finding a broad division of amorphous sets: Denote by $\mathfrak{A}$ an amorphous set, and $\pi(\mathfrak{A})$ a partition of $\mathfrak{A}$ into (infinitely many) finite sets. Because an amorphous set is weakly Dedekind-finite, almost all sets in $\pi(\mathfrak{A})$ will be of a fixed finite cardinality $n_\pi \in \omega$. Call n_π the *gauge* of the partition π. Denote by $\Pi(\mathfrak{A})$ the set of possible partitions $\pi(\mathfrak{A})$ as above, as allowed by the model containing $\mathfrak{A}$. Let $n_\mathfrak{A} = \sup \{n_\pi : \pi \in \Pi(\mathfrak{A})\}$.

An amorphous set $\mathfrak{A}$ is said to be *bounded* if $n_\mathfrak{A}$ is finite. Otherwise, the set is called *unbounded*. Bounded amorphous sets are in some sense structurally straightforward; unbounded amorphous sets exhibit a richer structure, and it is here that Truss applies model theoretic methods.

We continue by giving some definitions basic to geometric stability theory:

Definition 5. A set S together with a certain closure operation $\operatorname{cl}(*)$ from the power set of S, $P(S)$, to $P(S)$ is called a *(combinatorial) pregeometry* if the following hold:

1) $X \subseteq \operatorname{cl}(X)$ for any $X \subseteq S$;

2) $\operatorname{cl}(\operatorname{cl} X) = \operatorname{cl}(X)$ for any $X \subseteq S$;

3) if $a \in \operatorname{cl}(X \cup \{b\})$ then $b \in \operatorname{cl}(X \cup \{a\})$, for any $X \subseteq S$, $a, b \in S$;

4) if $a \in \operatorname{cl}(X)$, then $a \in \operatorname{cl}(Y)$ for some finite subset Y of X.

We say a set X is *closed* if $X = \text{cl}(X)$. The pregeometry (S, cl) is said to be a *geometry* if $\text{cl}(\varnothing)) = \varnothing$ and $\text{cl}(a) = a$ for all $a \in S$, i.e, if the empty set and all points are closed. An *automorphism* of a pregeometry (S, cl) is a $\text{cl}(*)$-preserving permutation of S. The pregeometry (S, cl) is said to be *homogeneous* if for any closed $X \subseteq S$ and $a, b \in S \setminus X$, there is an automorphism of (S, cl) which fixes X pointwise and takes a to b. The pregeometry (S, cl) is said to be *locally homogeneous* if for any finite sets $X \subseteq Y \subseteq S$ and $a, b \in \text{cl}(X) \setminus \text{cl}(Y)$, any automorphism of $(\text{cl}(Y), \text{cl})$ that fixes X pointwise and takes a to b can be extended to an automorphism of $(\text{cl}(Z), \text{cl})$, for any finite set $Z \supseteq Y$. A pregeometry (S, cl) is *degenerate* if for each $X \subseteq S$, $\text{cl}(X) = X$. A pregeometry (S, cl) is *locally finite* if for any finite $X \subseteq S$, $\text{cl}(X)$ is finite.

Note that any pregeometry (S, cl) can be associated with a *canonical geometry* (T, cl'). We define $T = \{\text{cl}(a) : a \in S \setminus \text{cl}(\varnothing)\}$, and for $X \subset S$, $\text{cl}'(\{\text{cl}(a) : a \in X\}) = \{\text{cl}(\{b\}) : b \in \text{cl}(X)\}$. If a pregeometry is (locally) homogeneous, so is its associated canonical geometry.

When working with a strongly minimal set, one takes the closure operation in Definition 5 to be model theoretic algebraic closure.

Because an amorphous set is Dedekind-finite, any geometry defined on it must necessarily be locally finite. Thus, Truss defines the closure operation on an amorphous set $\mathfrak{A}$ to be a function f from the set $e(\mathfrak{A})$ of finite subsets of $\mathfrak{A}$ to $e(\mathfrak{A})$ that satisfies the requirements for the closure operation in a locally homogeneous, locally finite pregeometry as above. If f defines a pregeometry, but not a geometry, we work instead with the associated canonical geometry.

We can now divide the unbounded amorphous sets into two types. If the only geometries that can be defined on an amorphous set A are degenerate, then we call $\mathfrak{A}$ *unbounded, not of projective type*. Otherwise, $\mathfrak{A}$ is *of projective type*.

We now recall a result of [E86]:

Theorem 6. *Let (S, cl) be a non-degenerate, infinite, locally finite, locally homogeneous geometry. Then (S, cl) is either a projective or affine geometry over a finite field.*

We can thus deduce that a geometry defined from an amorphous set of projective type is either affine, or projective.

Assume $\mathfrak{A}$ is amorphous of projective type, and f is a closure operation witnessing this. By remarks in $[\text{C}_0\text{H}_0\text{L}_0 85]$ it is possible to modify f to obtain a projective geometry. This is done in the model theoretic case by naming an element in the language, thereby altering the algebraic closure operation. In the amorphous case, we are free to name finite sets of elements.

Thus, we can disregard the affine case.

Another difference between amorphous sets and strongly minimal theories arises from the fact that we are not bound to one closure operation, algebraic closure. Because we may have several functions that fit the requirements of a closure operation, the same amorphous set may be of projective type over more than one field. It can be shown that any two such fields must have the same characteristic. For some amorphous sets, there may be a bound on the possible cardinality of the underlying field, in others, not.

Thus we come to the conclusion that amorphous sets can be "classified" into the following types:

- Bounded;

- Unbounded, not of projective type;

- Of projective type with a bound on the possible cardinality of the underlying field;

- Of projective type over a field with no bound on the possible cardinality.

Question 7. Unbounded amorphous sets that are not of projective type are largely unexplored, and clearly they are impervious to the methods from first-order model theory outlined in this subsection. Can anything perhaps be said about their structure in terms of the techniques recently developed in the model theory of non-elementary classes?

Question 8. "Classifiers" for the different types of amorphous sets are given in [$T_1$95]. These are ordered tuples of objects that capture some of the important properties of an amorphous set. However, these do not give a real classification because some sets that are different can have the same classifier, and some sets can have more than one classifier. This is particularly evident for amorphous sets of projective type.

Can a genuine classification be given?

2.2 Counting first-order types

In this subsection, we mention the use of more classical first-order model theoretic methods in conjunction with the study of NWO sets. We aim to determine what kinds of first-order structures can be carried by certain classes of sets. Because first-order statements are expressible in arithmetic, and arithmetic is absolute for models of set theory, we can apply any knowledge about the possible first-order definable structures of the theories to the corresponding classes of non-choice sets. In this way, we can apply theorems whose proofs ostensibly rely on the Axiom of Choice to sets in a non-choice setting. Much of the material included in this subsection appeared recently in [W05b]. The remainder can be found in [W05a].

For all of subsection 2.2, we assume $\neg$AC. Furthermore, we will assume that all sets under consideration are infinite.

We begin by mentioning the main result of this subsection.

Theorem 9. *Let $\mathfrak{A}$ be a weakly Dedekind-finite set admitting a structure axiomatizable in a countable (finite or infinite) language. Let $T = \mathrm{Th}(\mathfrak{A})$ be the (complete) set of sentences in the (first-order) language appropriate for the structure true in $\mathfrak{A}$. Then, the set $F_n T$ of formulas in n free variables with respect to T is finite.*

Thus T is $\aleph_0$-categorical.

The key to the proof of Theorem 9 is that weak Dedekind-finiteness limits the number of types that can be realized. This observation then allows us to use part of the usual proof of the Engeler-Ryll-Nardzewski-Svenonius Theorem (see [H$_2$93]) to show that countable models of T are atomic, and that countable models are isomorphic.

Proof. We split the proof into a number of claims.

Claim: $F_n T$ is finite.

Let $n \in \omega$. Because the language is countable, the set of formulas with n free variables is at most countable. For each formula $\varphi(x)$ in n free variables (i.e., $F_n T$, in the usual notation), consider $\{\bar{a} \in \mathfrak{A}^n : \mathfrak{A} \models \varphi(\bar{a})\}$. This is a countable family of subsets of $\mathfrak{A}^n$. However, since the class of weakly Dedekind-finite sets is closed under products, $\mathfrak{A}^n$ is weakly Dedekind-finite. Hence, this family of subsets must be finite. Thus, we map $F_n T$ onto a finite family of sets. Suppose the formulas φ and ψ determine the same set. Then, $\mathfrak{A} \models \forall \bar{x}(\varphi(\bar{x}) \leftrightarrow \psi(\bar{x}))$. Hence, the formula $\forall \bar{x}(\varphi(\bar{x}) \leftrightarrow \psi(\bar{x})) \in T$, and so, $T \vdash \varphi \leftrightarrow \psi$, and ϕ and ψ are identified in $F_n T$. This means that $F_n T$ is finite.

Claim: Every model of T is atomic.

Let $\mathscr{A}$ be a countable model of T, $\bar{a}$ a finite tuple from the set underlying $\mathscr{A}$, and let $\varphi_1, \ldots, \varphi_n$ be a list of the finitely many formulae, up to equivalence in T, which are satisfied by $\bar{a}$ in $\mathscr{A}$. Then, $\varphi_1 \wedge \ldots \wedge \varphi_n = \psi$ is consistent with T, and $T \models \psi \rightarrow \varphi_i$ for each $i \leqslant n$. Thus, $\varphi_1, \ldots, \varphi_n$ are isolated by ψ, and thus $\mathscr{A}$ is atomic.

Claim: Countable atomic models of T are isomorphic.

Let $\mathscr{A}$ and $\mathscr{B}$ be countable atomic models of T. Enumerate both of them with ω. Let a_0 be the first element of $\mathscr{A}$ in the enumeration, and $\varphi(a_0)$ the formula that isolates the finite set of formulae $\psi_{a_0}^1, \ldots, \psi_{a_0}^n$ which are satisfied by a_0 in $\mathscr{A}$. Since $\mathscr{B}$ is also a model of T, there is a first element in the enumeration of $\mathscr{B}$ satisfying $\varphi(x)$ in $\mathscr{B}$. Call this element b_0. Let

b_1 be the first element of $\mathscr{B} \setminus \{b_0\}$, and find the first $a_1 \in \mathscr{A}$ such that the finite set of formulae satisfied by (a_0, a_1) is the same as the finite set of formulae satisfied by (b_0, b_1). Note that the existence of these is guaranteed by the completeness of T. We continue back and forth ω times. This gives enumerations $\{a_i : i < \omega\}$ and $\{b_i : i < \omega\}$ of $\mathscr{A}$ and $\mathscr{B}$ respectively, and the map taking a_i to b_i defines an isomorphism from $\mathscr{A}$ to $\mathscr{B}$.

q.e.d. (Theorem 9)

The results of the rest of this subsection all depend on Theorem 9.

Structures admitting MT-rank.

We now turn to a notion that generalizes amorphousness: MT-rank.

Definition 10 (¬AC)**.** The relation $\mathrm{MT}(X) = \alpha$, for α an ordinal or -1, is defined by the following recursion.

(i) $\mathrm{MT}(X) = -1$ if $X = \varnothing$.

(ii) $\mathrm{MT}(X) = \alpha$ if $\mathrm{MT}(X) \not< \alpha$, and there is some $n \in \omega$ such that if $\{X_i : 0 \leqslant i \leqslant n\}$ are any $n+1$ pairwise disjoint subsets of X, then for some i, $\mathrm{MT}(X_i) < \alpha$.

We refer to this rank as *MT-rank*.

Definition 11 (¬AC)**.** A set X of MT-rank α has *MT-degree* k if X has k pairwise disjoint subsets of MT-rank α, but for any $k+1$ pairwise disjoint subsets, at least one of them has smaller rank.

It follows from Definition 10 that if X has MT-rank α, and k is the least n which witnesses this, then X has MT-degree k. We will write $\mathrm{MT}(X) = (\alpha, k)$ to indicate that X has MT-rank α and MT-degree k. Notice that a set X such that $\mathrm{MT}(X) = (1, 1)$ is amorphous.

MT-rank and degree were first introduced in [M$_3$T$_1$03], and there were referred to as simply "rank" and "degree". The authors of that paper prove basic properties of MT-ranked sets, and give some results about such sets carrying a partial ordering or group structure. In particular, they show that an MT-ranked set is weakly Dedekind-finite.

The definition of MT-rank was originally meant to generalise the notion of an amorphous set in much the same way that Morley rank generalises strong minimality. In addition, the class of amorphous sets is not closed under unions and products, as one would expect of a class of sets satisfying a finiteness notion. On the other hand, the class of MT-ranked sets is closed under unions and products, and thus MT-rank can be considered a closure of amorphousness to a somewhat better-behaved notion of finiteness.

There is more than just a superficial connection between MT-rank and Morley rank.

Theorem 12. Let $\mathfrak{A}$ be a set such that $\mathrm{MT}(\mathfrak{A}) = (\alpha, k)$, and assume $\mathfrak{A}$ admits a first-order structure axiomatizable in a countable (finite or countably infinite) language. Let $\mathscr{A}$ be the unique countable model of $T = \mathrm{Th}(\mathfrak{A})$ guaranteed by Theorem 9. Then, $\mathscr{A}$ has Morley rank, and the value of the Morley rank and degree $\mathrm{MR}(\mathscr{A}) \leqslant \mathrm{MT}(\mathfrak{A}) = (\alpha, k)$, where $\leqslant$ denotes the lexicographic order.

Before we prove Theorem 12, we state an immediate corollary:

Corollary 13. Let $\mathfrak{A}$ be a set admitting MT-rank that carries a structure axiomatizable in a countable (finite or infinite) language. Then $T = \mathrm{Th}(\mathfrak{A})$ is $\aleph_0$-stable.

We turn to the proof of Theorem 12.

Proof. The theory T is $\aleph_0$-categorical by Theorem 9. Thus, it suffices to calculate Morley rank within $\mathscr{A}$.

We proceed with the proof by induction on the MT-rank of $\mathfrak{A}$.

Assume $\mathfrak{A}$ is amorphous, and thus $(n, k) = (1, 1)$. If T is not strongly minimal, then there is a formula φ that definably divides $\mathscr{A}$ into two infinite subsets. Then, both $\exists^{\geqslant n} x \varphi(x) \in T$ and $\exists^{\geqslant n} x \neg \varphi(x) \in T$ for each n. But then, $\{a \in \mathfrak{A} : \mathfrak{A} \models \varphi(a)\}$ violates the amorphousness of $\mathfrak{A}$.

Now, assume that $\mathscr{A}$ does not have Morley rank less than or equal to $(\alpha, k) = \mathrm{MT}(\mathfrak{A})$. Then, there are formulas φ_i that divide $\mathscr{A}$ into $k+1$ subsets not having rank less than α. That is, $\varphi_0[\mathscr{A}], \varphi_1[\mathscr{A}], \ldots, \varphi_k[\mathscr{A}]$ are pairwise disjoint and not of Morley rank $< \alpha$. But then, $\varphi_0[\mathfrak{A}], \varphi_1[\mathfrak{A}], \ldots, \varphi_k[\mathfrak{A}]$ are pairwise disjoint, but by the definition of MT-rank, there is some $\varphi_i[\mathfrak{A}]$ having MT-rank $< \alpha$. By the inductive hypothesis, $\varphi_i[\mathscr{A}]$ has Morley rank $< \alpha$, a contradiction.

q.e.d. (Theorem 12)

Remark 14. Let $\mathfrak{A}, T$ be as in Theorem 12. Then, $\mathrm{MR}(\mathscr{A}) < \omega$.

Proof. $\mathscr{A}$ is $\aleph_0$-categorical and admits Morley rank. From [M$_5$65], $\mathscr{A}$ is $\aleph_0$-stable. By one of the main results of [C$_0$H$_0$L$_0$85], $\mathrm{MR}(\mathscr{A}) < \omega$.

q.e.d. (Remark 14)

We point out a special case of the above theorem.

Corollary 15. Let $\mathfrak{A}$ be an amorphous set that carries a structure axiomatizable in a countable (finite or infinite) language. Then $T = \mathrm{Th}(\mathfrak{A})$ is $\aleph_0$-categorical strongly minimal.

We remark that all of the results cited in the proof of Remark 14 use AC, and so we must justify why we can appeal to these results: $\mathscr{A}$ is itself a (well-ordered) countable structure. So, even if the universe we are working in does not satisfy AC, we can nonetheless carry out all the stability theoretic arguments in a subuniverse containing $\mathscr{A}$ that *does* satisfy AC, for example, $\mathbf{L}[T]$.

All theorems concerning first-order definable substructures of $\aleph_0$-categorical structures with Morley rank apply to structures admitting MT-rank. For example, there can be no infinite MT-ranked fields or Boolean algebras. Likewise, through the work in [F78] and [$B_1C_0M_0$79], we can deduce the following:

Theorem 16. A group $\mathfrak{A}$ whose domain admits MT-rank is abelian-by-finite.

Proof. The statement of Theorem 16 follows from the fact that important group theoretic subgroups of an $\aleph_0$-categorical group are all first-order definable. Thus, these structures are still present in non-choice models, since first-order properties are absolute in this context. In particular, if a group $\mathfrak{A}$ has MT-rank, then $\text{Th}(\mathfrak{A})$ is $\aleph_0$-stable. The unique countable model $\mathscr{A}$ of $\text{Th}(\mathfrak{A})$ is thus abelian-by-finite. An $\aleph_0$-categorical abelian-by-finite group has a 0-definable abelian subgroup of finite index [$B_1C_0M_0$79]. Because the sentence defining the subgroup has no parameters and is first order, it is also satisfied by $\mathfrak{A}$. q.e.d. (Theorem 16)

We can also reaffirm using these results a conclusion of [$T_1$95]: groups with MT-rank 1 and MT-degree 1 are elementary abelian of exponent p, just as $\aleph_0$-categorical strongly minimal groups.

There are cases in which the two ranks are not equal: Let $\mathfrak{A} = \mathfrak{B} \times \mathfrak{B}$ be the direct product of two isomorphic amorphous elementary abelian p-groups. Then, $\text{MT}(\mathfrak{A}) = (2,1)$, but $\text{MR}(\mathscr{A}) = (1,1)$. Thus, the inequality in the statement of the above theorem cannot be improved upon. We can even build a model that contains a group of arbitrarily large MT-rank, but whose theory is strongly minimal. An example of such a construction is demonstrated in Model Construction 17 below.

We construct a model containing a group of MT-rank α, for α an ordinal, using the Fraenkel-Mostowski construction method.

Model Construction 17 (A group having infinite MT-rank). Let $V = \{v_k : k \in \omega\}$ be an $\aleph_0$-dimensional vector space over a finite field of prime order p with an additive group structure (i.e., V is an $\aleph_0$-dimensional elementary abelian p-group). We intend to construct a Fraenkel-Mostowski model that utilizes V in defining an indexing set for the set of atoms.

Indexing the set of atoms: Let V_i, $i \in \alpha$, be isomorphic copies of V, and let $\mathbf{b_i} = (b_{0i}, b_{1i}, \ldots)$ be a fixed basis of V_i. For each $\beta \leqslant \alpha$, let

$$\Upsilon_\beta = \{0_V\} \times \mathrm{Dr}_{i \in \beta} V_i,$$

where $\mathrm{Dr}_{i \in \beta} V_i$ denotes the restricted external direct product of the V_i. Note that each Υ_β has cardinality $\max(\aleph_0, |\beta|)$. Denote by b_{ji}^β the vector

$$b_{ji}^\beta = (0_V, 0_V, \ldots, 0_V, b_{ji}, 0_V, \ldots) \in \Upsilon_\beta,$$

where the b_{ji} is the i-th coordinate. Clearly, each Υ_β has a locally finite structure.

Let U be a set of atoms of cardinality $\max(\aleph_0, |\alpha|)$ indexed by $\bigcup_{\beta \leqslant \alpha} \Upsilon_\beta$. Let

$$U_\beta = \{u_v \in U : v \in \Upsilon_\beta\}.$$

For notational ease, we will refer to atoms by their index when this will not result in confusion.

Defining a permutation group: Let $\mathscr{G}$ be the group of permutations of U with the following properties:

(i) The sets U_β, $\beta \leqslant \alpha$ are fixed.

(ii) For $g \in \mathscr{G}$, the cardinality of the *basic support* S of g is finite:

$$|S| = |\{b_{ji}^\beta : g(b_{ji}^\beta) \neq b_{ji}^\beta, \beta \in \alpha\}| < \omega$$

(iii) The group acts as a linear automorphism on each U_β.

Defining a normal filter: For the purposes of defining a filter, we define the following functions: Let

$$F_\beta : U_\beta \longrightarrow U_{\beta+1}$$

be given by

$$F_\beta(u_v) = u_{v \frown 0},$$

where $v \in \Upsilon_\beta$ and $v \frown 0 \in \Upsilon_\beta \times \{0\} \subset \Upsilon_{\beta+1}$. For limit ordinals γ, let

$$F_{\beta\gamma} : U_\beta \longrightarrow U_\gamma$$

be given by

$$F_{\beta\gamma}(u_v) = u_{(v,\underbrace{0_V, 0_V, 0_V, \ldots}_{\gamma - \beta})}.$$

In the following, for $g \in \mathscr{G}$ and F a function F_β or $F_{\beta\gamma}$, let gF denote the image of F under g, i.e., $\{(gx, gy) : (x, y) \in F\}$.

Let $\mathscr{F}$ be the filter of subgroups generated by

$$\{\mathscr{G}_A : A \subset U, |A| < \omega\} \cup \{\mathscr{G}_{\{g^{-1}F_\beta\}} : \beta \in \alpha, g \in \mathscr{G}\} \cup$$
$$\{\mathscr{G}_{\{g^{-1}F_{\beta\gamma}\}} : \beta \in \alpha, \gamma \in \text{Lim and } \gamma \leqslant \alpha, g \in \mathscr{G}\}. \quad (1)$$

This filter is clearly normal, as $\mathscr{G}_{\{g^{-1}F\}} = g^{-1}\mathscr{G}_F g$, where F is one of F_β or $F_{\beta\gamma}$. Let $\mathscr{N}_{MT_\alpha}$ be the resulting Fraenkel-Mostowski model.

In $\mathscr{N}_{MT_\alpha}$, each U_β is still a vector space over the field of order p, and F_β, $F_{\beta\gamma}$ are linear transformations from U_β to $U_{\beta+1}$, and from U_β to U_γ, respectively. There is, however, no set in the model which gives all these embeddings 'simultaneously'.

Theorem 18. In $\mathscr{N}_{MT_\alpha}$, $\text{MT}(U_\alpha) = (\alpha, 1)$, where U_α is as in the preceding model construction.

Proof. We prove that $\text{MT}(U_\alpha) = (\alpha, 1)$ by induction on $\beta \leqslant \alpha$.

For $\beta = 0$, $|U_0| = 1$.

We show the inductive step in two parts. We first show that the MT-rank of U_β is greater or equal to β, and then show equality.

Assume $\beta \leqslant \alpha$ and that $\text{MT}(U_\delta) = (\delta, 1)$ for all $\delta < \beta$. If $\beta = \delta + 1$ is a successor, then for any n, U_β has n pairwise disjoint subsets of MT-rank δ, namely, $F_\delta[U_\delta] + b_{j(\beta+1)}^{\beta+1}$ for $j < n$. Hence, U_β does not have MT-rank less than β. If β is a limit ordinal, then for any $\delta < \beta$, U_β has a subset of MT-rank greater than δ, namely $F_{(\delta+1)\beta}U_{(\delta+1)}$.

We still must show that U_β has MT-rank exactly β and MT-degree 1: Let $X \subseteq U_\beta$ be any subset in $\mathscr{N}_{MT_\alpha}$. Then, by the definition of the model, $\mathscr{G}_{\{X\}} \in \mathscr{F}$, and thus

$$\mathscr{G}_{\{X\}} \geqslant \mathscr{G}_A \cap \bigcap_{k<l} \mathscr{G}_{\{g_k^{-1}F_{\delta_k}\}} \cap \bigcap_{m<n} \mathscr{G}_{\{g_m^{-1}F_{\delta_m\gamma_m}\}}, \quad (2)$$

with $l, n \in \omega$, and $g_k, g_m \in \mathscr{G}$, and A a finite set. We assume without loss of generality that the set A is closed under the functions $g_k^{-1}F_{\delta_k}$ and $g_k F_{\delta_k}^{-1}$ for $k < l$ and $g_m^{-1}F_{\delta_m\gamma_m}$ and $g_m F_{\delta_m\gamma_m}^{-1}$ for $m < n$.

If $\beta = \delta + 1$ is a successor, let

$$Y = \langle (A \cap U_\beta) \cup \bigcup \{g_k^{-1}F_{\delta_k}U_\delta : \delta_k = \delta\}\rangle,$$

while if $\beta \in \text{Lim}$, let

$$Y = \langle (A \cap U_\beta) \cup \bigcup \{g_m^{-1}F_{\delta_m\gamma_m}U_{\delta_m} : \gamma_m = \beta\}\rangle.$$

Since each g_k has only finite support, Y is generated by the image of U_δ under at most one of the maps together with a finite set. Thus, because the field is finite, Y can be written as a finite union of sets of MT rank at most δ_k, whence $\mathrm{MT}(Y) < \beta$. We show that either $X \subseteq Y$ or $X \supseteq U_\beta \setminus Y$. Thus, we will show that U_β has MT-rank exactly β and degree 1 by showing that for any two disjoint subsets of U_β, one has MT-rank $< \beta$.

Let $u_\upsilon, u_\tau \in U_\beta \setminus Y$. Let $g \in \mathscr{G}$ be such that $g(u_\upsilon) = u_\tau$ and $g(u_\tau) = u_\upsilon$, and $g(F(u_\upsilon)) = F(u_\tau)$ and $F(u_\tau) = F(u_\upsilon)$, where F ranges over all the (finitely many) possible compositions of the $g_k^{-1} F_{\delta_k}$ and $g_m^{-1} F_{\delta_m \gamma_m}$; g fixes all points outside the finitely-generated, and hence finite set, $\langle u_\upsilon, u_\tau, F(u_\upsilon), F(u_\tau) \rangle$. Then,

$$g \in \mathscr{G}_A \cap \bigcap_{k<l} \mathscr{G}_{\{g_k^{-1} F_{\delta_k}\}} \cap \bigcap_{m<n} \mathscr{G}_{\{g_m^{-1} F_{\delta_m \gamma_m}\}},$$

so, by (2), g preserves X, and hence $u_\upsilon \in X \Leftrightarrow u_\tau \in X$, giving the result.

q.e.d. (Theorem 18)

We can apply the Jech-Sochor Embedding theorem to get an analogous result for ZF.

Question 19. What is the relation, if any, between MT-rank and the rank described in [L₁00] in the context of homogeneous model theory/finite diagrams?

Classes with restricted orderings.

Thus far, the correlations between the classes of non-well-ordered structures and traditional model theoretic notions have been surprisingly exact. We give a few further classes of sets where this correlation is not so satisfyingly sharp.

Definition 20 (¬AC). A set X is *Mostowski finite* if every linearly ordered subset of X is finite.

The class of Mostowski finite sets was first discussed in [T₁74] (under the name Δ_3), but arises naturally in models where the ordering principle fails. The following notion of strictly Mostowski finite was introduced in [W05b], and as such, is inspired by the model theoretic strict order property (SOP).

Definition 21 (¬AC). A set X is *strictly Mostowski finite* if any partial ordering on X^n, $n \in \omega$ only has chains of finite bounded length.

Definition 22 (¬AC). A set X is *II-finite* if every family of non-empty subsets of X which can be linearly ordered by inclusion has a maximal element.

II-finiteness was first mentioned in [T$_0$24], while strict Mostowski finiteness makes its first appearance in [W05a]. See Figure 1 for their strengths relative to other classes of sets mentioned in this survey.

Because these subclasses of the class of weakly Dedekind-finite sets are defined according to the relations that they admit, it is natural to examine these sets with the usual model-theoretic notions about definable relations in mind.

We have the following result:

Theorem 23. Let $\mathfrak{A}$ be a strictly Mostowski finite set admitting a structure axiomatizable in a countable (finite or infinite) language. Then $T = \mathrm{Th}(\mathfrak{A})$ is $\aleph_0$-categorical and does not have the strict order property.

Proof. Because strict Mostowski finiteness implies weak Dedekind-finiteness, T is $\aleph_0$-categorical. Let $\mathscr{A}$ be the unique countable model of T.

Assume T has the strict order property. Then T has a formula ϕ which witnesses this. This formula defines a partial ordering on the set of all n-tuples in $\mathscr{A}$ which contains arbitrarily long finite chains. As $\mathfrak{A}$ and $\mathscr{A}$ satisfy the same first-order sentences, $\mathfrak{A}$ admits such a first-order definable partial ordering, contrary to the strict Mostowski finiteness of $\mathfrak{A}$.

q.e.d. (Theorem 23)

A similarly general result for II-finite sets is not known. We will demonstrate in Model Constructions 43 and 45 that II-finite sets can display a variety of stability theoretic properties.

Question 24. Can the class of II-finite sets be categorized according to any model theoretic properties?

Ordered structures.

Techniques from stability theory can be applied to certain unstable theories such as o-minimal and weakly o-minimal theories. Here we discuss such non-well-orderable structures. First, a basic definition:

Definition 25. A subset of a linearly ordered set $(X, \leqslant)$ is called an *interval* if it has the form $(a, b) = \{x : a < x < b\}$, $[a, b]$, $[a, b)$, or $(a, b]$, where $a, b \in X \cup \{\pm\infty\}$. A subset is $Y \subseteq X$ is *convex* if, whenever $a < x < b$ and $a, b \in Y$, then also $x \in Y$.

We recall two definitions that first appeared in [C$_1$T$_1$00]. Both of these were inspired by the model theoretic notions of o-minimality and weak o-minimality (see [H$_1$M$_1$M$_2$NT$_1$00]).

Definition 26 ($\neg$AC)**.** A linearly ordered set $(X, \leqslant)$ is said to be *o-amorphous* if it is infinite and its only subsets (definable or not) are finite unions of intervals and points. We say that a set X is o-amorphous if there exists an ordering $\leqslant$ such that $(X, \leqslant)$ is o-amorphous.

Definition 27 ($\neg$AC). If in Definition 26 we replace the word "intervals" with "convex sets", then X is a *weakly o-amorphous* set.

The similarity between o-amorphous sets and $\aleph_0$-categorical o-minimal theories was studied extensively in [C_1T_100]. There, the authors use precisely the same techinques as those originally used for $\aleph_0$-categorical, o-minimal sets, and find that the structures admitted by o-amorphous sets are precisely those of $\aleph_0$-categorical, o-minimal sets.

The methods from [W05b] presented in this subsection explain why o-amorphous sets and $\aleph_0$-categorical o-minimal sets carry the same structures: o-amorphous sets *are* $\aleph_0$-categorical, o-minimal.

Theorem 28. Let $(\mathfrak{A}, \leqslant)$ be an o-amorphous set, and let $T = \text{Th}(\mathfrak{A})$. Then T is $\aleph_0$-categorical, o-minimal.

Proof. Assume that $(\mathfrak{A}, \leqslant)$ is an o-amorphous set. By [C_1T_100, Lemma 2.5], $\mathfrak{A}$ is weakly Dedekind-finite. Thus the $\aleph_0$-categoricity of T is immediate by Theorem 9.

If T were not o-minimal, then there would be a formula ϕ that would define a subset of $\mathfrak{A}$ that is not a finite union of intervals and points. Then $\{a \in \mathfrak{A} : (\mathfrak{A}, \leqslant) \models \phi(a)\}$ would counter our assumption that $\mathfrak{A}$ is o-amorphous. q.e.d. (Theorem 28)

Thus, we can immediately deduce from [S_2P_086] that, for example, no o-amorphous set can carry a group structure. We can obtain the complete classification of o-amorphous sets in [C_1T_100] from that for $\aleph_0$-categorical o-minimal sets in [S_2P_086].

There is a corresponding relation between weakly o-amorphous sets (as defined in [C_1T_100]) and weakly o-minimal theories (a discussion of which can be found in [$H_1M_1M_2NT_100$]), which may be given by the methods above. For example, $\aleph_0$-categorical weakly o-minimal sets may be given by means of a finite chain of refining equivalence relations on a densely linearly ordered set (which may be formed using a suitable amalgamation class for instance by use of Fraïssé's method). These correspond to one of the classes of weakly o-amorphous sets described in [C_1T_100]. There are more complicated examples also given, but as in the case of amorphous sets, not all can be captured using a first-order language.

2.3 The effect of model theoretic assumptions on model constructions

In the present subsection, we will examine the effects of model theoretic assumptions on the kinds of NWO structures that can be constructed using Fraenkel-Mostowski permutation and symmetric forcing constructions.

We begin by presenting a forcing construction due to Plotkin [$P_1$69]. This construction is analogous to the Fraenkel-Mostowski construction where the atom set is indexed by a countable first order structure and the normal filter employed is that of finite supports. In a sense, this is the minimal non-choice model for a given structure.

Later we discuss the effects of a few model theoretic assumptions on the index set for the atoms. We then present the refinements of the results of [$P_1$69] as given in [W05b].

Plotkin's construction.

The construction in [$P_1$69] is carried out using ramified forcing. We find this obscures and complicates the forcing construction. Hence, we state the construction in terms of unramified forcing. We adopt terminology for forcing as in [$S_1$71].

Let $\mathcal{M}$ be a countable transitive model of $ZF + (\mathbf{V}=\mathbf{L})$. This model will serve as our base model in the forcing construction. Let $\mathcal{A}$ be a relational system in $\mathcal{M}$ such that $\mathcal{A}$ has an infinite domain $A \in \mathcal{M}$ and finitely many relations. For notational simplicity we assume $\mathcal{A}$ has only one binary relation R. We write $\mathcal{A}$ as $\langle A, R \rangle$.

In the following, we assume that $\mathcal{A}$ is a model of a complete first order theory T and equality is modelled in $\mathcal{A}$ by the identity relation on $\mathcal{A}$. The signature of T is denoted by L, $F(L)$ denotes the set of all formulas of L, while $F_n(L)$ denotes the set of L-formulas with at most n free variables. For $\mathcal{B}$, a model of T with domain B, $D_n(B)$ is the set of all n-ary relations on $\mathcal{B}$ which are definable with parameters from the model by $F(L)$, i.e., $s \in D_n(B)$ if

$$s = \{\langle b_1, \ldots, b_n \rangle : \mathcal{B} \models \varphi(b_1, \ldots, b_n, a_1, \ldots, a_m)\}$$

for some $\varphi(y_1, \ldots, y_n, x_1, \ldots, x_m) \in F(L)$ and $a_i \in B$ where $1 \leqslant i \leqslant m$.

All model theory will be done inside $\mathcal{M}$ – so in particular, when we say $\mathcal{A}$ is countable, we mean it is countable in $\mathcal{M}$.

Model Construction 29 ($\mathcal{M}22$ **Plotkin's Model I**). Let C be the set of functions from a finite subset of $A \times \omega \times \omega$ into $\{0,1\}$ partially ordered by reverse extension. We regard C as a forcing notion. Let G be an $\mathcal{M}$-generic set over C, and $\mathcal{M}[G]$ the resulting extension. Then, since $G \in \mathcal{M}[G]$, $F = \bigcup G \in \mathcal{M}[G]$, and F is a function from $A \times \omega \times \omega$ into $\{0,1\}$.

Let $S_{aj} = \{n : F(a,j,n) = 1\} \subseteq \omega$, $\mathfrak{A}_a = \{S_{aj} : j \in \omega\} \subseteq P(\omega)$, $\mathfrak{A} = \{\mathfrak{A}_a : a \in A\} \subseteq PP(\omega)$, and $\mathfrak{R} = \{\langle \mathfrak{A}_a, \mathfrak{A}_b \rangle : \langle a, b \rangle \in R\}$. We use bold symbols for the constants (names) in the forcing language $\mathscr{L}$.

Remark 30. $\mathscr{M}[G] \models \mathbf{S_{aj}} = \mathbf{S_{bk}} \leftrightarrow (a = b \wedge j = k)$. Consequently, $\mathscr{M}[G] \models \mathfrak{A_a} = \mathfrak{A_b} \leftrightarrow a = b$.

Let $\theta(a) = \mathfrak{A}_a$. Note that $\theta \in \mathscr{M}[G]$. Let $\mathscr{N}$ be the submodel of $\mathscr{M}[G]$ consisting of those of its elements which are hereditarily ordinal definable (HOD) over

$$\{S_{aj} : a \in A, j \in \omega\} \cup \{\mathfrak{A}_a : a \in A\} \cup \{\mathfrak{A}\} \cup \{\mathfrak{R}\}.$$

This completes the construction.

Before we continue, we make a remark about the symmetry of the model $\mathscr{N}$: Let $\mathscr{H}_1$ be the group of automorphisms of $\langle A, R \rangle$ which are in the base model $\mathscr{M}$. Let $\mathscr{H}_2$ be the group of permutations on ω whose elements move only finitely many $i \in \omega$ (all of which are in $\mathscr{M}$).

Let $p \in C$. For $\gamma \in \mathscr{H}_1$, define

$$\gamma(p) = \{\langle \gamma(a), j, n, i \rangle : \langle a, j, n, i \rangle \in p\}.$$

For $\pi \in \mathscr{H}_2$, define

$$\pi_a(p) = \{\langle a, \pi j, n, i \rangle : \langle a, j, n, i \rangle \in p\} \cup \{\langle b, j, n, i \rangle : \langle b, j, n, i \rangle \in p \text{ and } b \neq a\}.$$

For formulas $\phi_{\mathfrak{L}}$ of the forcing language $\mathfrak{L}$, we can define $\gamma(\phi_{\mathfrak{L}})$ to be the formula obtained by replacing each occurrence of S_{aj} and $\mathfrak{A}_a$ by $\gamma(S_{aj})$ and $\gamma(\mathfrak{A}_a)$ respectively. The formula $\pi_a(\phi_{\mathfrak{L}})$ is defined analogously. As shown in [P$_1$69, Lemma 2], we then have that

$$p \Vdash \phi_{\mathfrak{L}} \Leftrightarrow \gamma(p) \Vdash \gamma(\phi_{\mathfrak{L}}) \text{ for all } \gamma \in \mathscr{H}_1,$$

and

$$p \Vdash \phi_{\mathfrak{L}} \Leftrightarrow \pi_a(p) \Vdash \pi_a(\phi_{\mathfrak{L}}) \text{ for all } \pi \in \mathscr{H}_2, a \in A.$$

The model $\mathscr{N}$ is thus an (hereditarily) symmetric extension of $\mathscr{M}$ with respect to the group generated by $\mathscr{H}_1$ and $\mathscr{H}_2$.

Finally, we close our exposition of Plotkin's construction with a remark relating it to models built using the Fraenkel-Mostowski method.

Remark 31. Let $\mathscr{U}$ be a model of ZFA+AC (set theory with atoms with the axiom of choice), and let U be its set of atoms, indexed by the domain of a model A of a first-order theory T (as above). Let $\mathscr{M}$ be the kernel of $\mathscr{U}$ (i.e., $P^\infty(\varnothing) \in \mathscr{U}$: a model of ZFC). Let α be an ordinal in $\mathscr{U}$.

Let $\mathscr{V} \subseteq \mathscr{U}$ be the Fraenkel-Mostowski permutation model (a model of ZFA) built using automorphisms of A as permutations, and using the filter of finite supports. Then, there exists a model $\mathscr{N} \supseteq \mathscr{M}$ (a model of ZF) and a set $\mathfrak{A} \in \mathscr{N}$ such that $(P^\alpha(U))^{\mathscr{V}}$ is $\in$-isomorphic to $(P^\alpha(\mathfrak{A}))^{\mathscr{N}}$. Furthermore, the model $\mathscr{N}$ and the set $\mathfrak{A}$ are as defined in the Plotkin model construction.

Proof. The result follows by direct application of the proofs of the Jech-Sochor Embedding Theorem and the Pincus Support Theorem (see [J73, Theorems 6.1 & 6.6]). q.e.d. (Remark 31)

Note that the Plotkin construction predates the Jech-Sochor Embedding Theorem.

The effects of model theoretic assumptions.

We now concentrate on assumptions about the theory T and its model $\mathscr{A}$ in $\mathscr{M}$, and their effect on the structure $\mathfrak{A}$ resulting from the Plotkin construction. We begin with a remark which will be stated without proof.

Remark 32. The correspondence $\theta : a \longrightarrow \mathfrak{A}_a$ is an isomorphism between the relational systems $\langle A, R \rangle$ and $\langle \mathfrak{A}, \mathfrak{R} \rangle$. As noted above, $\theta \in \mathscr{M}[G]$ but $\theta \notin \mathscr{N}$.

Clearly, if $\mathfrak{r}$ is an n-ary relation on $\mathfrak{A}$ in $\mathscr{N}$, then it is HOD over a finite subset of $\mathfrak{A}$ (and also sets of the form S_{aj}, $\mathfrak{A}$, and $\mathfrak{R}$), by the definition of the model $\mathscr{N}$. We wish to show that under certain circumstances, all the n-ary relations on $\mathfrak{A}$ are also L-definable over $\mathfrak{A}$.

We will write formulas in $F(L)$ as small Greek letters, such as φ_L, and the corresponding formulas in the forcing language $\mathfrak{L}$ with quantifiers restricted to $\mathfrak{A}$ as $\varphi_\mathfrak{L}$. Clearly, for $\varphi_L \in F(L)$, we have

$$\mathscr{A} \models \varphi_L(a_1, \ldots, a_n) \leftrightarrow \mathscr{N} \models \varphi_\mathfrak{L}(\mathfrak{A}_{a_1}, \ldots, \mathfrak{A}_{a_n}).$$

It remains to be shown that the relations in $\mathfrak{A}$ can be defined (with appropriate parameters) in L.

Now we define notation that will be in effect for Lemmas 33 and 34 and Corollary 36: Let $\varphi_\mathfrak{L}(\mathfrak{A}_{b_1}, \ldots, \mathfrak{A}_{b_k})$ be a formula in the forcing language that defines a relation $\mathfrak{r}$ on $\mathfrak{A}$ in $\mathscr{N}$. That is,

$$\mathscr{N} \models (\mathfrak{A}_{a_1}, \ldots, \mathfrak{A}_{a_n}) \in \mathfrak{r} \leftrightarrow \varphi_\mathfrak{L}(\mathfrak{A}_{a_1}, \ldots, \mathfrak{A}_{a_n}, \mathfrak{A}_{b_1}, \ldots, \mathfrak{A}_{b_k})$$

where $\mathfrak{A}_{b_1}, \ldots, \mathfrak{A}_{b_k}$ indicates that $\varphi_\mathfrak{L}$ includes mention of $\mathfrak{A}_{b_i}$ or some $S_{b_i b_j}$. We write $\varphi_\mathfrak{L}^{\mathscr{N}}(\mathfrak{A}_{b_1}, \ldots, \mathfrak{A}_{b_k})$ to indicate the interpretation of $\varphi_\mathfrak{L}$ in the model $\mathscr{N}$.

In Lemmas 33 and 34, we fix the $(n+k)$-ary formula $\phi_{\mathfrak{L}}$, hence n and k are both fixed. In the following, $(n+k)$-homogeneous means that tuples of length $n+k$ satisfying the same first-order formulas can be mapped to each other via an automorphism of the structure.

Lemma 33. Assume $\mathscr{A}$ is an $(n+k)$-homogeneous model of a theory T.

Let $\psi_L(u_1, \ldots, u_n, v_1, \ldots, v_k)$ be an atom of the Boolean algebra $F_{n+k}(L)$. Let $\mathfrak{r}_{\psi_{\mathfrak{L}}}(\mathfrak{A}_{b_1}, \ldots, \mathfrak{A}_{b_k})$ be the n-ary relation defined by (here, treating the relation as a set of ordered n-tuples):

$$\mathfrak{r}_{\psi_{\mathfrak{L}}}(\mathfrak{A}_{b_1}, \ldots, \mathfrak{A}_{b_k}) = \{\langle \mathfrak{A}_{a_1}, \ldots, \mathfrak{A}_{a_n}\rangle \in \mathfrak{A}^n : \psi_{\mathfrak{L}}(\mathfrak{A}_{a_1}, \ldots, \mathfrak{A}_{a_n}, \mathfrak{A}_{b_1}, \ldots, \mathfrak{A}_{b_k})\}.$$

If $\mathfrak{r}_{\psi_{\mathfrak{L}}}(\mathfrak{A}_{b_1}, \ldots, \mathfrak{A}_{b_k}) \cap \varphi_{\mathfrak{L}}^{\mathscr{N}}(\mathfrak{A}_{b_1}, \ldots, \mathfrak{A}_{b_k}) \neq \emptyset$, then $\mathfrak{r}_{\psi_{\mathfrak{L}}} \subseteq \varphi_{\mathfrak{L}}^{\mathscr{N}}$.

Proof. By hypothesis, there is some n-tuple

$$\langle \mathfrak{A}_{d_1}, \ldots, \mathfrak{A}_{d_n}\rangle \in \mathfrak{r}_{\psi_{\mathfrak{L}}} \cap \varphi_{\mathfrak{L}}^{\mathscr{N}}.$$

Let $\langle \mathfrak{A}_{c_1}, \ldots, \mathfrak{A}_{c_n}\rangle$ be any other tuple in $\mathfrak{r}_{\psi_{\mathfrak{L}}}$. We assume that

$$\langle \mathfrak{A}_{c_1}, \ldots, \mathfrak{A}_{c_n}\rangle \notin \varphi_{\mathfrak{L}}^{\mathscr{N}}$$

and derive a contradiction.

Since both $\langle \mathfrak{A}_{d_1}, \ldots, \mathfrak{A}_{d_n}\rangle, \langle \mathfrak{A}_{c_1}, \ldots, \mathfrak{A}_{c_n}\rangle \in \mathfrak{r}_{\psi_{\mathfrak{L}}}$, by definition we have

$$\mathscr{N} \models \psi_{\mathfrak{L}}(\mathfrak{A}_{d_1}, \ldots, \mathfrak{A}_{d_n}, \mathfrak{A}_{b_1}, \ldots, \mathfrak{A}_{b_k})$$

and

$$\mathscr{N} \models \psi_{\mathfrak{L}}(\mathfrak{A}_{c_1}, \ldots, \mathfrak{A}_{c_n}, \mathfrak{A}_{b_1}, \ldots, \mathfrak{A}_{b_k}).$$

Because $\theta : a \longrightarrow \mathfrak{A}_a$ is an isomorphism, we have

$$\mathscr{A} \models \psi_L(d_1, \ldots, d_n, b_1, \ldots, b_k)$$

and

$$\mathscr{A} \models \psi_L(c_1, \ldots, c_n, b_1, \ldots, b_k).$$

By assumption, ψ_L is an atom in $F_{n+k}(L)$. Thus, both $n+k$-tuples $\langle d_1, \ldots, d_n, b_1, \ldots, b_k\rangle$ and $\langle c_1, \ldots, c_n, b_1, \ldots, b_k\rangle$ satisfy the same formulas of L. Because we have assumed that $\mathscr{A}$ is $(n+k)$-homogeneous, there is an automorphism γ of $\mathscr{A}$ such that $\gamma(b_i) = (b_i)$ for $1 \leqslant i \leqslant k$, and $\gamma(d_j) = (c_j)$ for $1 \leqslant j \leqslant n$. However, the model $\mathscr{N}$ is symmetric with regards to the automorphism γ, and so we come to a contradiction. q.e.d. (Lemma 33)

Lemma 34. Assume $\mathscr{A}$ is an atomic, $(n+k)$-homogeneous model of T.
Let $\varphi_{\mathfrak{L}}^{\mathscr{N}}$ be the $(n+k)$-ary formula defining an n-ary relation on $\mathfrak{A}$, as above. Then, $\varphi_{\mathfrak{L}}^{\mathscr{N}}$ is a disjoint union of the satisfaction sets of certain formulas which correspond to atoms in $F_{n+k}(L)$.

Proof. We continue with the notation as in the proof of Lemma 33.
Let $\langle \mathfrak{A}_{d_1}, \ldots, \mathfrak{A}_{d_n} \rangle \in \varphi_{\mathfrak{L}}^{\mathscr{N}}(\mathfrak{A}_{b_1}, \ldots, \mathfrak{A}_{b_k})$. Since $\mathscr{A}$ is atomic, there is an atom
$$\psi_L(u_1, \ldots, u_n, v_1, \ldots, v_k)$$
in $F_{n+k}(L)$ such that
$$\mathscr{A} \models \psi_L(d_1, \ldots, d_n, b_1, \ldots, b_k).$$
But then, $\mathscr{N} \models \psi_{\mathfrak{L}}(\mathfrak{A}_{d_1}, \ldots, \mathfrak{A}_{d_n}, \mathfrak{A}_{b_1}, \ldots, \mathfrak{A}_{b_k})$. Therefore,
$$\langle \mathfrak{A}_{d_1}, \ldots, \mathfrak{A}_{d_n} \rangle \in \mathfrak{r}_{\psi_{\mathfrak{L}}}(\mathfrak{A}_{b_1}, \ldots, \mathfrak{A}_{b_k}).$$
By Lemma 33, $\mathfrak{r}_{\psi_{\mathfrak{L}}} \subseteq \varphi_{\mathfrak{L}}^{\mathscr{N}}$. Therefore we have
$$\varphi_{\mathfrak{L}}^{\mathscr{N}} = \bigcup_{i \in I} \mathfrak{r}_{\psi_i \, \mathfrak{L}},$$
where I indexes a set of atoms in $F_{n+k}(L)$. Furthermore,
$$\mathfrak{r}_{\psi_i \, \mathfrak{L}}(\mathfrak{A}_{b_1}, \ldots, \mathfrak{A}_{b_k}) \cap \mathfrak{r}_{\psi_j \, \mathfrak{L}}(\mathfrak{A}_{b_1}, \ldots, \mathfrak{A}_{b_k}) = \emptyset$$
because ψ_i and ψ_j are atoms of $F_{n+k}(L)$. q.e.d. (Lemma 34)

Before we come to an important Corollary, we need a clarifying definition.

Definition 35. A structure S is *strongly $\aleph_0$-homogeneous* if for every pair $\bar{a}, \bar{b}$ of finite tuples of elements of S, if $(S, \bar{a}) \equiv (S, \bar{b})$, then there is an automorphism of S that takes $\bar{a}$ to $\bar{b}$.

Corollary 36.

1) If $\mathscr{A}$ is an atomic strongly $\aleph_0$-homogeneous model of T, every n-ary relation on $\mathfrak{A}$ has the form of a disjoint union $\bigsqcup_{i \in I} \mathfrak{r}_{\psi_i \, \mathfrak{L}}$, where $\psi_{i \, L}$ is an atom of $F_{n+k}(L)$, and I indexes the atoms of $F_{n+k}(L)$.

2) If $\mathscr{A}$ is a prime model of T, every n-ary relation on $\mathfrak{A}$ has the form of a disjoint union $\bigsqcup_{i \in I} \mathfrak{r}_{\psi_i \, \mathfrak{L}}$, where $\psi_{i \, L}$ is as in statement 1).

3) If $\mathscr{A}$ is a countable model of an $\aleph_0$-categorical theory T, then every n-ary relation on $\mathfrak{A}$ has the form of a finite disjoint union $\bigsqcup_{i=1}^{m} \mathfrak{r}_{\psi_i \mathcal{L}}$. Thus, any n-ary relation on $\mathfrak{A}$ is definable with a single first-order L-formula. Thus, in this case, T is the complete first order theory of $\mathfrak{A}$, Th($\mathfrak{A}$).

Proof. The first statement of the corollary is clear from Lemmas 34 and 33. The second statement follows from the fact that a model of a countable first order theory is prime if and only if it is countable and atomic, as well as the fact that a countable and atomic model is (strongly) $\aleph_0$-homogeneous.

The third statement follows easily from the fact that if T is $\aleph_0$-categorical, then for all $n \in \omega$, the set of atoms of $F_n(L)$ is finite. q.e.d. (Corollary 36)

In [P$_1$69], Plotkin concludes that for an $\aleph_0$-categorical theory T, $\mathfrak{A}$ (as defined above) is Dedekind-finite. We now mention the refinements of this result as given in [W05b, W05a].

Theorem 37. Let T be $\aleph_0$-categorical. Then, $\mathfrak{A}$ (as defined above) is weakly Dedekind-finite.

Proof. Assume that $\mathfrak{A}$ is not weakly Dedekind-finite. Then, there are Z_i, $i \in \omega$, pairwise disjoint and non-empty, such that $\mathfrak{A} = \bigcup_{i \in \omega} Z_i$. Since each Z_i is a subset of $\mathfrak{A}$, each can be regarded as a unary relation on $\mathfrak{A}$. Since any relation can be expressed as the finite sum of atoms, each Z_i can be defined using a single formula. Thus we can find $\aleph_0$ many pairwise disjoint formulas, contrary to the $\aleph_0$-categoricity of T. Hence, $\mathfrak{A}$ is weakly Dedekind-finite.

q.e.d. (Theorem 37)

The following theorem gives the exactness of the correlation described in Theorem 23.

Theorem 38. Let T be an $\aleph_0$-categorical theory that does not have the SOP. Then $\mathfrak{A}$ is strictly Mostowski finite.

Proof. Let T be a theory that is $\aleph_0$-categorical and does not have the strict order property. Then $\mathfrak{A}$ is weakly Dedekind-finite. Assume that $\mathfrak{A}$ is not strictly Mostowski finite. Then, for some $n \in \omega$, $\mathfrak{A}^n$ admits a partial order having chains of unbounded length. Using the same arguments as in the previous theorems, this would indicate that T has the strict order property, a contradiction. q.e.d. (Theorem 38)

Theorem 39 gives the exactness of the correlation established in Corollary 13.

Theorem 39. If T is $\aleph_0$-categorical, $\aleph_0$-stable, then $\mathfrak{A}$ admits MT-rank. Furthermore, the Morley rank of T and the MT-rank of $\mathfrak{A}$ are of equal value.

Proof. Assume that T is $\aleph_0$-categorical, $\aleph_0$-stable. Then, T has finite Morley rank n, say, and Morley degree d. Thus, the formula $\phi =$ "$x = x$" has Morley rank n, degree d. Thus, there are L-formulas $\phi_i(\bar{x})$, $i \in \omega$ such that the sets $\phi(\mathscr{A}) \cap \phi_i(\mathscr{A})$ are pairwise disjoint and $MR(\phi \wedge \phi_i) \geqslant n-1$ for each $i \in \omega$. Assume further that $\mathfrak{A}$ does not admit MT-rank. Then for all ordinals α, for each $m \in \omega$, there exist m pairwise disjoint subsets $X_i : 0 \leqslant i \leqslant m$ of $\mathfrak{A}$ such that $MT(X_i) \geqslant \alpha$, for all $i \in m$. However, as subsets of $\mathfrak{A}$ can be regarded as unary relations, each X_i corresponds to a single formula. This clearly contradicts the value of the Morley rank of the theory T.

The values of the ranks must be equal: Assume first that T has Morley rank n and Morley degree d, $\mathfrak{A}$ has MT-rank m, MT-degree c, and that $n < m$. Thus, if we divide $\mathfrak{A}$ into $c+1$ pairwise disjoint sets, one piece must have rank $< m$. However, since each subset of $\mathfrak{A}$ can be defined by a single formula, this would give a contradiction with the Morley rank of T. Now, let us assume that $m < n$. Then, there are first-order formulas ϕ_i in one variable whose satisfactions sets in the unique countable model of T satisfy the terms of the definition of Morley rank. By the construction of the model containing $\mathfrak{A}$, these formulas are then also satisfied in $\mathfrak{A}$, and thus there are subsets of $\mathfrak{A}$ that contradict the assumed MT-rank of $\mathfrak{A}$. q.e.d. (Theorem 39)

Note that the rank values are the same because the model $\mathfrak{A}$ only exhibits the first-order structure of the theory. Later, we will give an example where the ranks are not equal, showing that MT-rank measures more than just first-order structure.

The following Theorems 40, 41, and 42, are proved analogously to those results proved thus far in this subsection. We state them without proof.

Theorem 40. If T is an $\aleph_0$-categorical o-minimal theory, then $\mathfrak{A}$ is o-amorphous.

Theorem 41. If T is an $\aleph_0$-categorical weakly o-minimal theory, then $\mathfrak{A}$ is weakly o-amorphous.

Theorem 42. If T is an $\aleph_0$-categorical strongly minimal theory, then $\mathfrak{A}$ is an amorphous set.

Thus we have shown that all of the correspondences mentioned in Theorems 37, 38, 39, 40, 41, and 42 are exact.

We now return to the classical finiteness notion, II-finiteness, to determine some of the behaviours exhibited by sets of this type. As mentioned, a neat and tidy correlation for sets that are II-finite has not been identified. For example, a model constructed using the method of [P$_1$69] based on the random graph yields a set which is II-finite and has the independence property (IP). Similarly, it is also possible to construct a II-finite set whose theory has the SOP using the generic partial order. Because strictly Mostowski finite sets correspond to those sets that do not have the SOP, a model built using the method of [P$_1$69] based on Hrushovski's stable (not $\aleph_0$-stable) $\aleph_0$-categorical pseudoplane (see [H$_4$88]) is II-finite (because it is strictly Mostowski finite) but has neither the SOP nor the IP. The same can be said about models built from $\aleph_0$-stable, $\aleph_0$-categorical theories. Thus it seems that II-finite sets can exhibit a wide variety of first-order behaviours from a stability theoretic point of view.

We will construct the examples mentioned based on the random graph and the generic partial order. For clarity, we use the Fraenkel-Mostowski method, and implicitly assume application of the Jech-Sochor Theorem. Conclusions about a model based on the Hrushovski pseudoplane can be reached through theorems proved thus far. We omit the pseudoplane's construction since little would be gained from a detailed exposition compared to the effort involved.

Model Construction 43 (A II-finite set with the IP). Let Γ be the random graph. This is a countable, universal, ultrahomogeneous [H$_2$93], $\aleph_0$-categorical graph whose theory has the IP. We will build a Fraenkel-Mostowski model using a finite support structure.

Indexing the set of atoms: Let U be a countable set of atoms. We index using the elements of Γ: Let $U = \{u_\gamma : \gamma \in \Gamma\}$.

Defining a permutation group: Let $\mathscr{G}$ be the group of permutations of U induced by $\mathrm{Aut}(\Gamma)$.

Defining a normal filter: Let $\mathscr{F}$ be the filter generated by $\{\mathscr{G}_A : A \subset \Gamma, |A| \in \omega\}$. This is a filter based on the ideal of finite sets, and is thus normal.

Let $\mathscr{N}_\Gamma$ be the resulting Fraenkel-Mostowski model.

Theorem 44. Let $\mathscr{N}_\Gamma$ and $U \in \mathscr{N}_\Gamma$ be as defined in Model Construction 43. Then, $\mathrm{Th}(U) = \mathrm{Th}(\Gamma)$ and so has the IP, and U is II-finite.

Proof. That $\mathrm{Th}(U)$ has the IP is clear by part 3 of Corollary 36. Assume for a contradiction that U is not II-finite. Let $\{X_i : i \in I\}$ be a partition of U, where I is an infinite ordered set. Since we assume this ordered partition is in the model $\mathscr{N}_\Gamma$, it must be supported by a finite set $S \subset \Gamma$, where $|S| = s$. We write $S = \{\gamma_1, \ldots, \gamma_s\}$. Because Γ is $\aleph_0$-categorical, there are finitely many 1-types over S in $\mathrm{Th}(\Gamma)$. In other words, $\mathrm{Aut}(\Gamma)$ is oligomorphic, and thus has finitely many orbits of $(s+1)$-tuples. Choose distinct members a, b of an infinite orbit over S that lie in different partition pieces, say $a \in X_i, b \in X_j, i \neq j$. Two such elements can be found because the partition is infinite, while the number of orbits is finite. Thanks to the symmetry of the graph relation, $(\gamma_1, \ldots, \gamma_s, a, b)$ and $(\gamma_1, \ldots, \gamma_s, b, a)$ have the same quantifier-free type, so g given by $ga = b$, $gb = a$, and $g\gamma_i = \gamma_i$, for all $\gamma_i \in S$, is a partial automorphism. Because the random graph is ultra-homogeneous, this automorphism extends to an automorphism on the whole of Γ. This automorphism is clearly in the subgroup supported by S, but interchanges elements of the partition, contrary to the antisymmetry of the ordering. Hence we come to a contradiction. q.e.d. (Theorem 44)

Model Construction 45 (A II-finite set with the SOP).

Let $(\Psi, <)$ be the generic partial order. This is a countable, universal, ultrahomogeneous, $\aleph_0$-categorical partial order whose theory has the SOP. We will build a Fraenkel-Mostowski model using a finite support structure.

Indexing the set of atoms: Let U be a countable set of atoms. We index using the elements of Ψ: Let $U = \{u_x : x \in \Psi\}$.

Defining a permutation group: Let $\mathscr{G}$ be the group of permutations of U induced by $\mathrm{Aut}(\Psi)$.

Defining a normal filter: Let $\mathscr{F}$ be the filter generated by $\{\mathscr{G}_A : A \subset \Psi, |A| \in \omega\}$. This is a normal filter because it is generated by the ideal of finite sets.

Let $\mathscr{N}_\Psi$ be the resulting Fraenkel-Mostowski model.

Theorem 46. Let $\mathscr{N}_\Psi$ and $U \in \mathscr{N}_\Psi$ be as defined in Model Construction 45. Then, $\mathrm{Th}(U, <) = \mathrm{Th}(\Psi, <)$ and so has the SOP, and U is II-finite.

Proof. That $\mathrm{Th}(U)$ has the SOP is clear by part 3 of Corollary 36. We note that $(U, <)$ has arbitrarily long finite chains in $\mathscr{N}$, however, in this model, $(U, <)$ does not have *infinite* chains, as we will show.

We note that the generic partial order is characterised by the following property: If $A, B,$ and C are pairwise disjoint finite subsets of Ψ, such that for all $a \in A, b \in B, c \in C$,

$$b < c$$
$$c \not< a$$
$$a \not< b,$$

then there exists $z \in \Psi$ such that

$$b < z < c$$
$$a \not< z \wedge a \not> z,$$

for all $a \in A$, $b \in B$, $c \in C$.

Assume for a contradiction that U is not II-finite. Let $\{X_i : i \in I\}$ be a partition of U, where I is an infinite ordered set. Since we assume this ordered partition is in the model $\mathcal{N}_\Psi$, it must be supported by a finite set $S \subset \Psi$, where $|S| = s$. We write $S = \{a_1, \ldots, a_s\}$. Because Ψ is $\aleph_0$-categorical, there are finitely many 1-types over S in $\mathrm{Th}(\Psi)$. In other words, $\mathrm{Aut}(\Psi)$ is oligomorphic, and thus has finitely many orbits of $(s+1)$-tuples. Thus, there must be two sets from the partition X_i, X_j, $i \neq j$ that each contain an element that has the same 1-type over S. Call these elements $x \in X_i$ and $y \in X_j$. Thus, there is a permutation g such that $ga_i = a_i$ for all $a_i \in S$, and $gx = y$.

We have two possible cases for the relation between x and y:
First, assume $x \not< y \wedge y \not< x$. Then the relation between x and y is symmetric, and g is a partial automorphism that can be extended to an automorphism of the structure. We can argue for the contradiction as in the proof of Theorem 44.

Second, assume that either $x < y$ or $y < x$. Because

$$\{z : tp(z|S) = tp(x|S)\} \cong \Psi,$$

we can choose an element $z \notin S$ such that $z \not< x \wedge z \not> x$ and $z \not< y \wedge y \not> x$. Since x and y are in different partition pieces, z must be in a different partition piece from at least one of them. Thus we have reduced this second case to the first, and we again can use the same argument as in the proof of Theorem 44, giving the desired contradiction. q.e.d. (Theorem 46)

3 Methods outside first-order logic

3.1 Introduction to ω_1^*-nwo sets

Thus far we have surveyed the applications of first-order model theoretic methods on non-well orderable sets, and have learned that these applications are most successful on sets that cannot be mapped onto ω. Now we turn our attention to the wider class of sets that cannot be mapped onto ω_1.

In this section, the model theory goes beyond that of first-order logic. Appropriate reference texts include [H$_2$93, K$_0$71, B$_0$05, K$_0$K$_1$04].

We assume for the rest of this section that $\mathfrak{A}$ is an ω_1^*-nwo having a structure that can be described using a language $L_{\omega_1\omega}$ with a countable vocabulary. Further, we will assume that $2^{\aleph_0}$, and hence the language $L_{\omega_1\omega}$, is well-orderable.

In Subsection 3.2, the methods will be reminiscent of those of Subsection 2.2. We then will return briefly to the Plotkin model construction. In the last Subsection, we will examine a class of ω_1^*-nwo sets that may lend itself to a geometric approach.

3.2 Results employing classical infinitary logic

Much of the material in this subsection can be found in [W05a].

We state a few definitions.

A *partial isomorphism* from $\mathcal{A}$ to $\mathcal{B}$ is a pair of tuples $(\bar{a}, \bar{b})$, of the same finite length, such that $\bar{a} \subset A$, $\bar{b} \subset B$, and $\bar{a}$ and $\bar{b}$ satisfy the same quantifier-free $L_{\omega_1\omega}$-formulas. We note that $(\bar{a}, \bar{b})$ is a partial isomorphism from $\mathcal{A}$ to $\mathcal{B}$ if and only if the empty pair $(\varnothing, \varnothing)$ is a partial isomorphism from $(\mathcal{A}, \bar{a})$ to $(\mathcal{B}, \bar{b})$. A *back-and-forth family* for $\mathcal{A}$ and $\mathcal{B}$ is a set $\mathcal{P}$ of partial isomorphisms from $\mathcal{A}$ to $\mathcal{B}$ such that:

- $\mathcal{P} \neq \varnothing$;

- for each $(\bar{a}, \bar{b}) \in \mathcal{P}$ and $c \in A$, there exists $d \in B$ such that $(\bar{a}c, \bar{b}d) \in \mathcal{P}$;

- for each $(\bar{a}, \bar{b}) \in \mathcal{P}$ and $d \in B$, there exists $c \in A$ such that $(\bar{a}c, \bar{b}d) \in \mathcal{P}$.

We say that two structures $\mathcal{A}$ and $\mathcal{B}$, of arbitrary cardinality, are *potentially isomorphic* if there is a back-and-forth family for $\mathcal{A}$ and $\mathcal{B}$.

We can apply certain classical results of infinitary logic, such as Karp's Theorem, directly to our case:

Remark 47. Let $\mathfrak{A}$ be a structure whose domain A is ω_1^*-nwo, and let $\mathscr{B}$ be a countable structure. Then $\mathfrak{A}$ and $\mathscr{B}$ are potentially isomorphic if and only if $\mathfrak{A} \equiv_{L_{\omega_1\omega}} \mathscr{B}$.

Proof. Assume the structures $\mathfrak{A}$ and $\mathscr{B}$ are potentially isomorphic and let $\mathcal{P}$ be a back-and-forth family witnessing this potential isomorphism. We proceed to show by a standard argument by induction on complexity of formulas ϕ of $L_{\omega_1\omega}$ that if $(\bar{a}, \bar{b}) \in \mathcal{P}$, then $\mathcal{A} \models \phi(\bar{a})$ iff $\mathcal{B} \models \phi(\bar{b})$. For ϕ atomic we have the implication by assumption that $\mathcal{P}$ is a back-and-forth family. Again, the case is clear for negations, and countably many conjunctions and disjunctions of formulas already shown to satisfy the implication. Suppose we have shown the implication for $\phi(x_1, \ldots, x_n)$. Let $(a_1, \ldots, a_{n-1}, b_1, \ldots, b_{n-1}) \in \mathcal{P}$. By the properties of the back-and-forth property, and the inductive hypothesis, the following are equivalent:

$\mathfrak{A} \models \exists x_n \phi[a_1, \ldots, a_{n-1}]$.
$\mathfrak{A} \models \phi[a_1, \ldots, a_{n-1}, a]$ for some $a \in A$.
$\mathfrak{A} \models \phi[a_1, \ldots, a_{n-1}, a]$ for some a with $(a_1, \ldots, a_{n-1}, a, \bar{b}) \in \mathcal{P}$.
$\mathscr{B} \models \phi[b_1, \ldots, b_{n-1}, b]$ for some b with $(a_1, \ldots, a_{n-1}, a, b_1, \ldots, b_{n-1}, b) \in \mathcal{P}$.
$\mathscr{B} \models \phi[b_1, \ldots, b_{n-1}, b]$ for some $b \in B$.
$\mathscr{B} \models \exists x_n \phi[b_1, \ldots, b_{n-1}]$.

Note that $\forall x_n \phi$ is equivalent to $\neg \exists x_n \neg \phi$. Thus we have shown that the two potentially isomorphic structures satisfy the same $L_{\omega_1 \omega}$ sentences.

Note that because neither A nor B can be mapped onto ω_1, a game that corresponds to the back-and-forth family can have length that is at most countable. Thus, the formulas ϕ in the proof above cannot have quantifier rank $\geqslant \omega_1$, and so must be formulas of $L_{\omega_1 \omega}$.

Now, let us assume that $\mathfrak{A}$ and $\mathscr{B}$ satisfy the same sentences of $L_{\omega_1 \omega}$. We define a back-and-forth family $\mathcal{P}$ consisting of the pairs $(\bar{a}, \bar{b})$ such that the $L_{\omega_1 \omega}$-formulas satisfied by $\bar{a}$ in $\mathfrak{A}$ are the same as those satisfied by $\bar{b}$ in $\mathscr{B}$. First of all, we allow $(\bar{a}, \bar{b}) \in \mathcal{P}$ if the pair extends to an isomorphism of a finitely generated structure $\mathfrak{A}_0$ onto a finitely generated structure $\mathscr{B}_0$. Now we can show that $\mathcal{P}$ is a back-and-forth family just as in the proof of Karp's Theorem (see, e.g., [K$_0$K$_1$04]) to complete the proof.

q.e.d. (Remark 47)

Thus, a countable structure and an ω_1^*-nwo structure that are potentially isomorphic are very similar on a deeper level, despite their apparent differences. This is well illustrated by Theorem 48.

Theorem 48. Let $\mathfrak{A}$ be a structure whose domain is ω_1^*-nwo, and let $\mathscr{A}$ be a countable structure that is potentially isomorphic to $\mathfrak{A}$. Then, there exists a generic extension of the set theoretical universe $\mathscr{M}$ in which $\mathfrak{A}$ is countable and isomorphic to $\mathscr{A}$.

Proof. If $\mathfrak{A}$ and $\mathscr{A}$ are potentially isomorphic, then there exists a back-and-forth family $\mathcal{P}$ for the two structures. We can then take $\mathcal{P}$ to be a forcing notion ordered by extension. Let G be an $\mathscr{M}$-generic set, and let $\mathscr{M}[G]$ be the generic extension of the model $\mathscr{M}$. Then $\bigcup G = f$ is a one-to-one function from a subset of $\mathfrak{A}$ into $\mathscr{A}$ and is in the model $\mathscr{M}[G]$. However, for all $a \in \mathfrak{A}$, $\{p \in \mathcal{P} : a \in \text{dom}(p)\}$ is a dense open set. Hence, the domain of f is the whole of $\mathfrak{A}$, and similarly its range is the whole of $\mathscr{A}$, so f is a bijection. Thus, f provides an enumeration of the domain of $\mathfrak{A}$ showing that both $\mathfrak{A}$ and $\mathscr{A}$ are countable in the universe $\mathscr{M}[G]$. Because pairs of tuples in the back-and-forth family satisfy the same quantifier-free formulas, one can show by induction on the complexity of formulas that f is also an isomorphism.

Notice that the proof does not depend on the choice of G.

q.e.d. (Theorem 48)

We have shown that a countable structure and an ω_1^*-nwo structure that are potentially isomorphic are in some sense equivalent, and that the differences depend largely on the set theoretical universe. Our next goal is to find a canonical way of identifying a countable structure that is potentially isomorphic to a given ω_1^*-nwo structure. This is done by first defining a Scott sentence and Scott rank for the structure $\mathfrak{A}$.

We begin by defining language fragments: We have assumed that the vocabulary of the language is at most countable. Let $L(0, \mathfrak{A})$ be the $L_{\omega\omega}$ (first-order) language whose primitive symbols correspond to the relations, functions, and distinguished elements of $\mathfrak{A}$. Note that this is a countable set of formulas. For each limit ordinal λ, let $L(\lambda, \mathfrak{A})$ be

$$\bigcup \{L(\delta, \mathfrak{A}) : \delta < \lambda\}.$$

For each δ, a successor ordinal, let $T(\delta, \mathfrak{A})$ be the complete theory of $\mathfrak{A}$ in the language $L(\delta, \mathfrak{A})$. Let $L(\delta + 1, \mathfrak{A})$ be the least (countable) fragment of $L_{\omega_1\omega}$ that includes $L(\delta, \mathfrak{A})$ and satisfies the following closure conditions:

if $n < \omega$ and $p(x_1, \ldots, x_n)$ is a non-principal n-type of $T(\delta, \mathfrak{A})$ realised in $\mathfrak{A}$, then $\bigwedge \{F(\bar{x}) : F(\bar{x}) \in p\}$ belongs to the fragment.

Please note that all of the fragments defined above are countable.

Lemma 49. Let $\mathfrak{A}$ be such that its domain is ω_1^*-nwo, and assume that ω_1 is regular. Then, there exists a countable ordinal δ such that $L(\gamma, \mathfrak{A}) = L(\delta, \mathfrak{A})$ for all $\gamma > \delta$.

Proof. Let

$$X = \{(\bar{x}, \bar{y}) : \bar{x}, \bar{y} \text{ finite one-to-one sequences of elements of } A, |\bar{x}| = |\bar{y}|\}.$$

Note that X is ω_1^*-nwo since this class is closed under products and subsets.

If $L(\delta+1, \mathfrak{A}) \neq L(\delta+2, \mathfrak{A})$, then there exists a pair of n-tuples of A that are equivalent with respect to all n-ary formulas of $L(\delta, \mathfrak{A})$, but inequivalent with respect to some formula of $L(\delta+1, \mathfrak{A})$. If a pair of tuples is inequivalent at some ordinal δ, it is clearly inequivalent for all $\gamma > \delta$.

Now, we show that the first ordinal at which equality of the language fragments occurs is countable. For each $(\bar{x}, \bar{y}) \in X$ such that $\bar{x}$ and $\bar{y}$ are inequivalent at some stage, we map this pair to the first stage where the inequivalence is exhibited. Thus, we map a subset of X into the class of ordinal numbers. The range of this mapping must be an ordinal, say δ. Since X cannot be mapped onto ω_1, and we have assumed ω_1 to be regular, δ must be countable.   q.e.d. (Lemma 49)

Definition 50. Let $r(\mathfrak{A})$ be the least ordinal δ such that $L(\delta, \mathfrak{A}) = L(\delta+1, \mathfrak{A})$ in the above construction. We call $r(\mathfrak{A})$ the *Scott rank* of $\mathfrak{A}$.

The canonical *Scott sentence* $F_{\mathfrak{A}}$ for $\mathfrak{A}$ is a sentence of $L_{\omega_1 \omega}$ that asserts "I am an atomic model of $T(r(\mathfrak{A}), \mathfrak{A})$".

Thus, it is clear from Remark 47 that the Scott sentence of a ω_1^*-nwo structure $\mathfrak{A}$ is the same as the Scott sentence of any countable structure $\mathscr{A}$ to which $\mathfrak{A}$ is potentially isomorphic. Furthermore, the Scott ranks of $\mathfrak{A}$ and $\mathscr{A}$ are equal, and thus do not depend on the choice of the generic set G in the proof of Theorem 48. It should be noted that not all non-countable structures have a Scott sentence. It is our assumption that $\mathfrak{A}$ cannot be mapped onto ω_1 that guarantees this.

We are, by the above discussion, able to determine the Scott sentence $F_{\mathfrak{A}}$ for a structure $\mathfrak{A}$. It is furthermore possible to show that there exists a countable model that satisfies $F_{\mathfrak{A}}$.

Theorem 51. Let $\mathfrak{A}$ be a ω_1^*-nwo structure, and let $F_{\mathfrak{A}}$ be its Scott sentence. Then there exists a countably infinite model $\mathscr{A}$, unique up to isomorphism that shares the Scott sentence $F = F_{\mathscr{A}} = F_{\mathfrak{A}}$, has the same Scott rank $r = r(\mathscr{A}) = r(\mathfrak{A})$, and is potentially isomorphic to $\mathfrak{A}$.

This theorem can be proved, for example, by using the Makkai Model Existence Theorem, which can be found in [K$_0$71].

Similarly as in subsection 2.2, we have uniquely associated a countable structure with a NWO structure, in this case, a ω_1^*-nwo structure.

By arguments similar to those in the proof of Theorem 9, it is clear that any ω_1^*-nwo structure satisfies only countably many $L_{\omega_1 \omega}$-types. Thus it is a *small* model, and can be associated with the class of atomic models of a first-order theory T [B$_0$05].

Question 52. Does there exist an ω_1^*-nwo structure whose associated countable structure is not strongly $\aleph_0$-homogeneous?

3.3 Plotkin's construction revisited

Using very similar proofs to those found in Section 2.3, we come to the following result:

Theorem 53. Let $\mathscr{A}$ be a countable strongly $\aleph_0$-homogeneous structure (i.e., having Scott rank $r(\mathscr{A}) \leqslant \omega$).

Then $\mathfrak{A}$ (as constructed in 2.3) is an ω_1^*-nwo structure having the same Scott sentence as $\mathscr{A}$.

In the above theorem, it is probably possible to replace the assumption that $\mathscr{A}$ is countable with the assumption that it is small (i.e., satisfies only countably many $L_{\omega_1\omega}$-types).

As mentioned earlier, Plotkin's construction is very similar to a Fraenkel-Mostowski construction with finite supports. Finite support constructions are thus well understood. This leads to a natural question in a relatively unexplored area:

Question 54. What can be said about the influence of model theoretic assumptions on the structure that indexes an atom set in a Fraenkel-Mostowski permutation construction that uses countable supports?

3.4 Quasi-amorphous sets

In this subsection, we mention some results of $[C_1T_101]$.

Definition 55. A set U is *quasi-amorphous* if it is uncountable, is not the union of two uncountable sets, and every uncountable subset contains a countably infinite subset.

The convoluted nature of the above definition is to exclude sets that are the union of an amorphous set and a countable set, sets whose structures are fairly easy to describe with existing techniques.

The notion of quasi-amorphousness was suggested by E. Hrushovski, intended to mirror B. Zilber's notion of quasi-minimality (see $[B_005]$ for a precise definition). The consequence of the definition of quasi-minimality that is most important here is that an uncountable quasi-minimal structure has only countable or co-countable $L_{\omega_1\omega}$-definable subsets. Quasi-minimality has been actively studied very recently, particularly in connection with questions surrounding complex exponentiation. While quasi-minimality is the older notion, much of its structure theory has been published since $[C_1T_101]$ appeared. Thus, the connections between quasi-amorphous and quasi-minimal structures have not been fully explored, and we will have more questions than results in this subsection.

First a remark that justifies the mention of quasi-amorphous sets in this section.

Remark 56. If a set is quasi-amorphous, then it is ω_1^*-nwo.

The analysis of quasi-amorphous sets in $[C_1T_101]$ begins much like the analysis of amorphous sets in $[T_195]$: through examination of possible partitions.

Let U be a quasi-amorphous set, and let π be a partition of U into infinitely many sets, none of which are uncountable. We say that π has a *gauge* n, a natural number of $\aleph_0$, if $\{u \in U : (\exists A \in \pi)(u \in A \wedge |A| = n)\}$ is uncountable. If no form of AC is assumed, a gauge need not exist; under the assumption of AC_ω, every partition of a quasi-amorphous set into countable subsets has a gauge.

The authors of $[C_1T_101]$ continue by determining the possible natures of gauged partitions of quasi-amorphous sets, both assuming AC_ω, and not.

Remark 57.

- If π is a partition of finite gauge of a quasi-amorphous set U, then π is either quasi-amorphous or countably infinite. (AC not assumed)

- If π is a partition of gauge $\aleph_0$ of a quasi-amorphous set U, then π is countably infinite, quasi-amorphous, amorphous, or the union of an amorphous set and a countably infinite set. (AC not assumed)

- If AC_ω is assumed and π is a (necessarily) gauged partition of a quasi-amorphous set, then π is quasi-amorphous.

The authors then build models containing examples of the behaviours described above. Finally, they prove a few results concerning the natures of quasi-amorphous linear orderings and quasi-amorphous groups under the assumption of AC_ω.

Theorem 58. Assuming AC_ω, any linearly ordered quasi-amorphous set $(U, <)$ has a densely ordered partition into countable convex subsets.

Theorem 59. We assume AC_ω. If G is a quasi-amorphous group, then one of the following holds:

(i) G is elementary abelian with prime exponent.

(ii) G is divisible abelian with countable torsion subgroup.

(iii) G is divisible non-abelian with countable center Z, and all non-central elements are conjugate of infinite order.

Question 60. What can be said about quasi-amorphous linear orders and groups if AC_ω is not assumed?

The geometric properties of quasi-amorphous sets have not been explored.

Question 61. What sorts of pregeometries can be carried by a quasi-homogeneous set?

A recent paper [$H_5L_1S_0$05] discusses certain groups interpreted in a large homogeneous model. These groups act on pregeometries obtained from the realisation in the large homogeneous model of quasi-minimal type. The list of possible kinds of groups mentioned in that paper closely resembles the possible quasi-amorphous groups, though is not identical.

Question 62. Is there any relation between either of the frameworks mentioned in [$H_5L_1S_0$05] and quasi-amorphousness?

Question 63. What can be said about the structures on ω_1^*-nwo sets if AC_ω is assumed to fail?

References.

[$B_0$05]   J. Baldwin. Categoricity. March 2005. http://www.math.uic.edu/~jbaldwin.

[$B_1C_0M_0$79]   W. Baur, G. Cherlin, and A. Macintyre. Totally categorical groups and rings. *Journal of Algebra*, 57:407–440, 1979.

[$B_2$96]   S. Buechler. *Essential Stability Theory*. Perspectives in mathematical logic. Springer, Berlin, 1996.

[$C_0H_0L_0$85]   G. Cherlin, L. Harrington, and A. H. Lachlan. $\aleph_0$-categorical, $\aleph_0$-stable structures. *Annals of Pure and Applied Logic*, 28:103–135, 1985.

[$C_1T_1$00]   P. Creed and J. K. Truss. On o-amorphous sets. *Annals of Pure and Applied Logic*, 101:185–226, 2000.

[$C_1T_1$01]   P. Creed and J. K. Truss. On quasi-amorphous sets. *Archive for Mathematical Logic*, 40(8):581–596, 2001.

[D88]   J. W. R. Dedekind. *Was sind und was sollen die Zahlen?* Vieweg, Braunschweig, 1888.

[E86]   David M. Evans. Homogeneous geometries. *Proceedings of the London Mathematical Society. Third Series*, 52:305–327, 1986.

[F71]   U. Felgner. *Models of ZF-Set Theory*, volume 223 of *Lecture Notes in Mathematics*. Springer, Berlin, 1971.

[F78]   U. Felgner. $\aleph_0$-categorical stable groups. *Mathematische Zeitschrift*, 160:27–49, 1978.

[$H_5L_1S_0$05]   T. Hyttinen, O. Lessmann, and S. Shelah. Intepreting groups and fields in some nonelementary classes. *Journal of Mathematical Logic*, 5(1):1–47, 2005.

[$H_1M_1M_2NT_1$00]   B Herwig, H. D. Macpherson, G. Martin, A. Nurtazin, and J. K. Truss. On $\aleph_0$-categorical weakly o-minimal structures. *Annals of Pure and Applied Logic*, 101(1):65–93, 2000.

[$H_2$93]   W. Hodges. *Model Theory*, volume 42 of *Encyclopedia of mathematics and its applications*. Cambridge University Press, Cambridge, 1993.

[H₃R₁98]   P. Howard and J. E. Rubin. *Consequences of the Axiom of Choice*, volume 59 of *Mathematical Surveys and Monographs*. American Mathematical Society, Providence RI, 1998. Updates at http://www.math.purdue.edu/~jer/cgi-bin/conseq.html.

[H₄88]   E. Hrushovski. A stable $\aleph_0$-categorical pseudoplane. Unpublished notes, 1988.

[J73]   T. Jech. *The Axiom of Choice*, volume 75 of *Studies in Logic and the Foundations of Mathematics*. North Holland, Amsterdam, 1973.

[K₀71]   H. J. Keisler. *Model Theory for Infinitary Logic: logic with countable conjunctions and finite quantifiers*, volume 62 of *Studies in Logic and the foundations of mathematics*. North-Holland, Amsterdam, 1971.

[K₀K₁04]   H. J. Keisler and J. F. Knight. Barwise: Infinitary logic and admissible sets. *Bulletin of Symbolic Logic*, 10(1):4–36, 2004.

[L₁00]   O. Lessmann. Ranks and pregeometries in finite diagrams. *Annals of Pure and Applied Logic*, 106(1-3):49–83, 2000.

[L₂58]   A. Lévy. The independence of various definitions of finiteness. *Fundamenta Mathematicae*, 46:1–13, 1958.

[M₃T₁03]   G. S. Mendick and J. K. Truss. A notion of rank in set theory without choice. *Archive for Mathematical Logic*, 42:165–178, 2003.

[M₄75]   G. P. Monro. Independence results concerning Dedekind-finite sets. *Journal of the Australian Mathematical Society*, 19:35–46, 1975.

[M₅65]   M. Morley. Categoricity in power. *Transactions of the American Mathematical Society*, 114:514–538, 1965.

[P₀96]   A. Pillay. *Geometric stability theory*, volume 32 of *Oxford Logic Guides*. Clarendon Press, Oxford, 1996.

[P₁69]   J. M. Plotkin. Generic embeddings. *Journal of Symbolic Logic*, 34:388–394, 1969.

[R₀00]   P. Rothmaler. *Introduction to model theory*, volume 15 of *Algebra, Logic and Application*. Gordon and Breach Science Publishers, 2000. Translated and revised from the 1995 German original by the author.

[S₁71]   J. R. Shoenfield. Unramified forcing. In *Axiomatic Set Theory, Part I*, volume 13 of *Proceedings of Symposia in Pure Mathematics*, pages 357–381, 1971.

[S₂P₀86]   C. Steinhorn and A. Pillay. Definable sets in ordered structures I. *Transactions of the American Mathematical Society*, 295:565–592, 1986.

[T₀24]   A. Tarski. Sur les ensembles finis. *Fundamenta Mathematicae*, 6:45–95, 1924.

[T₁74]   J. K. Truss. Classes of Dedekind finite cardinals. *Fundamenta Mathematicae*, 84:187–208, 1974.

[T₁95]   J. K. Truss. The structure of amorphous sets. *Annals of Pure and Applied Logic*, 73:191–233, 1995.

[W05a]   A. C. Walczak-Typke. *Dedekind-finite structures*. PhD thesis, University of Leeds, 2005.

[W05b] A. C. Walczak-Typke. The first-order structure of weakly Dedekind-finite sets. *Journal of Symbolic Logic*, 70(4):1161–1170, 2005.

Received: March 7, 2006;
In revised version: May 18, 2006;
Accepted by the editors: December 5, 2006.

Quantifier Elimination on Real Closed Fields and Differential Equations

ANDREAS WEBER

Institut für Informatik II
Rheinische Friedrich-Wilhelms-Universität Bonn
Römerstr. 164
53117 Bonn, Germany
E-mail: weber@informatik.uni-bonn.de

ABSTRACT. This paper surveys some recent applications of quantifier elimination on real closed fields in the context of differential equations. Although polynomial vector fields give rise to solutions involving the exponential and other transcendental functions in general, many questions can be settled within the real closed field without referring to the real exponential field.

The technique of quantifier elimination on real closed fields is not only of theoretical interest, but due to recent advances on the algorithmic side including algorithms for the simplification of quantifier-free formulae the method has gained practical applications, e.g., in the context of computing threshold-conditions in epidemic modeling.

1 Introduction

Differential equations are ubiquitous in real world problems modeling. Often one has to determine their trajectories from their initial conditions. Even in the simplest case of one-dimensional linear problems

$$\frac{d}{dt}x(t) = ax(t) \tag{1}$$
$$x(t_0) = x_0 \tag{2}$$

the solutions involve the exponential function:

$$x(t) = x_0 \cdot e^{a(t-t_0)}. \tag{3}$$

So the study of the field of the real numbers with the exponential function as a primitive is important for investigating differential equations. Nevertheless, we will not focus on the remarkable results obtained for the real exponential field, see, e.g., [M$_0$03] for a survey also discussing some of these

Benedikt **Löwe** (*ed.*)
Algebra, Logic, Set Theory.
Festschrift für Ulrich Felgner zum 65. Geburtstag.
College Publications London 2006 [Studies in Logic 4]; p. 291-311

results. Instead, we will show that many questions in the area of differential equations can be settled within the real closed field without referring to the real exponential field.

This possibility is due to the fact that many questions on dynamical systems can be posed by referring to the vector field only, e.g., the equilibrium points can be defined using the vector field only. Consider, e.g., the autonomous vector valued system

$$\frac{d}{dt}x(t) = f(x(t)), \qquad (4)$$

where $f : \mathbb{R}^n \to \mathbb{R}^n$ and $x : \mathbb{R} \to \mathbb{R}^n$. The set of equilibrium points of this system is the set of the zeros of the vector field f, i.e., $\{e \in \mathbb{R}^n \mid f(e) = 0\}$. Very often the vector field f is in the form of a parameterized polynomial vector field, $f(x) = f(u,x) = (f_1, \ldots, f_n)$, where $f_i \in \mathbb{R}[u,x]$ are polynomials of degree $\leq d$, $x = (x_1, \ldots, x_n)$ is a list of variables and $u = (u_1, \ldots, u_k)$ is a list of parameters. Then the set of equilibrium points is algebraic over the parameters u. The question whether there are equilibrium points can be formulated as an existentially quantified first-order formula in the language of ordered fields and in the theory of the ordered field of the reals. A quantifier-free equivalent formula can be found algorithmically, a famous result due to Tarski [T51].

Although the theoretical significance of these results is widely seen, very often there are doubts about their practical feasibility [$S_4$03]:

> So, quantifier elimination is something that is do-able in principle, but not by any computer that you and I are ever likely to see. Well, I'll retract that last statement because it's probably false.

In this paper we do not only want to survey results showing that the technique of quantifier elimination on real closed fields is of theoretical interest for differential equations, but that due to recent advances on the algorithmic side including algorithms for the simplification of quantifier-free formulae the methods have gained practical applications, e.g., in the context of computing threshold-conditions in epidemic modeling.

2 Quantifier Elimination for Real Closed Fields

2.1 A brief history

Tarski's work on a decision method for elementary algebra and geometry [T51] is important for model theory in many aspects. In the survey paper of Macintyre [$M_0$03] it is contrasted to "Tarski's set-theoretic foundational formulations" being the starting point of

> a quite different development, which still flourishes and owes very little to the set-theoretic development.

However, from a purely algorithmic point of view Tarski's method is rather prohibitive, as its complexity cannot be bound by a tower of exponential functions, i.e., is not even elementary recursive. This asymptotic complexity is also the one of the methods described by Seidenberg [$S_0$54] and Cohen [$C_3$69]. The first elementary recursive method was found by Collins [$C_4$75] using the technique of *Cylindrical Algebraic Decomposition* (CAD), whose complexity is doubly exponential, thus reducing the complexity from an unbounded tower of exponentials to one of hight two.

This is a provable lower bound for the general problem of quantifier elimination on real closed field [$D_0H_2$88, $W_3$88]. More precisely, the lower double exponential bound is on the number of changes of quantifiers. For purely existentially or universally quantified problems methods of single exponential complexity have been described, e.g., by Renegar [$R_3$92].

The results on the worst-case asymptotic complexity do only give partial information about the running times for many concrete instances. A major breakthrough for practically working quantifier-elimination methods have been the so called "virtual substitution" methods. Based on ideas of Ferrante and Rackoff for decision problems [$FR_0$79], virtual substitution methods for quantifier elimination date back to a theoretical paper by Weispfenning [$W_3$88]. They have been devised for problems involving polynomials that are linear (or at most quadratic or cubic) in the quantified variables [$W_3$97, $W_3$94, $L_2W_3$93]. Implementations of these methods are available in the REDLOG system mainly developed by A. Dolzman, A. Seidl, and T. Sturm.[1]

Weispfenning [$W_3$99] also showed that the elementary theory of the real numbers in the language having 0,1 as constants, addition and subtraction and integer part as operations, and equality, order and congruences modulo natural number constants as relations admits an effective quantifier elimination procedure and is decidable. He also showed that this so called "mixed real-integer linear quantifier elimination" sample answers for existentially quantified variables. Moreover, it comprises as special cases linear elimination for the reals, and Presburger arithmetic of asymptotically optimal complexity.

There are also sophisticated implementations of the cylindrical algebraic decomposition available in REDLOG and in the QEPCAD library,[2] which contain substantial improvements in many aspects [$A_2C_4M_1$84, $C_4H_4$91, $B_1$98, $S_1S_5$03, $D_2S_1S_5$04].

Another technique for quantifier elimination on real closed fields was published by Weispfenning in 1998 [$W_3$98] (whereas a technical report describ-

---

[1] Available at http://www.fmi.uni-passau.de/~redlog/.
[2] Available at http://www.cs.usna.edu/~qepcad/B/QEPCAD.html.

ing the method had already been published by him in 1993 and the method was implemented as a Diploma thesis by A. Dolzmann in 1994). It is based on real root counting and is now called *Hermitian quantifier elimination* to acknowledge Hermite's work in the area of real root counting.

Although the worst-case asymptotic complexity of this method is not elementary recursive such as Tarski's method, it has been proved to be a powerful tool for particular classes of elimination problems, e.g., problems involving one quantifier block in front of a conjunction containing as many equations as quantifiers and only few other atomic formulae.

None of these methods is superior to another one in general, and some problems could be solved only by their combination, e.g., Dolzmann [$D_2$99] found an automatic solution of a real algebraic implicitization problem of the so called "Enneper surface" by combining all of these three quantifier elimination methods, namely quantifier elimination by virtual substitution, Hermitian quantifier elimination, and quantifier elimination by partial cylindrical algebraic decomposition, as well as the simplification methods described in [$D_2S_5$97].

2.2 Simplifications of Quantifier-Free Formulae

The same semi-algebraic set can be represented by different quantifier-free formulae. On the equivalence classes of quantifier-free formulae describing the same semi-algebraic sets there are different reasonable (partial) orderings giving the notion of one formula being "simpler" than another one, e.g., is $a < 0 \land b = 0 \lor a < 0 \land c = 0$ simpler than $a < 0 \land [b = 0 \lor c = 0]$?

However, many formulae produced by automated systems like the quantifier elimination package REDLOG, or by substitution and specialization of rules like the Routh-Hurwitz criterion, are large and complex, even when the objects they define are quite simple. It is important to consider simplification of formulae, if these have to be used in further computations or made available for human comprehension. Fortunately, the simplification on this level turns out to be on a much coarser level then the one considered above.

There are two algorithmic techniques that we are aware of for simplification of large quantifier-free formulae—which both define in their way a notion of one formula being "simpler" than another one. One technique is described in [$D_2S_5$97] and implemented in REDLOG, and the other is described in [$B_1$98] and implemented in the SLFQ system.[3] The goals of the two methods are very different. The former is intended primarily to combat intermediate expression swell during virtual term substitution. As such it needs to be fast. The latter is intended to reduce the size of the formula, in

---

[3]Available at http://www.cs.usna.edu/~qepcad/SLFQ/Home.html.

terms of number of irreducible polynomials appearing, as much as possible, regardless of time required to do so.

The SLFQ system uses the QEPCAD as a black box to do formula simplification. QEPCAD is able to simplify formulae, but its time and space requirements become prohibitive when input formulae are large. SLFQ basically breaks large input formulae into small pieces, uses QEPCAD to simplify the pieces, and starts a process of combining simplified subformulae and applying QEPCAD to simplify the combined subformulae. Eventually this process produces a simplification of the entire initial formula. QEPCAD takes a formula and constructs an explicit geometric model of the object that formula defines in real Euclidean space—a *Cylindrical Algebraic Decomposition* (CAD). Using the CAD, it is easy to detect when one of those varieties does not actually define a boundary of the geometric object. Once detected it can be easily removed from the CAD, resulting in a simpler CAD representing the same object. This can be repeated until we reach a minimal CAD—i.e., a CAD from which no polynomial can be removed without violating the requirement that the CAD represents the same geometric object. This CAD simplification process is described in [$B_1$98].

The user may also allow SLFQ to produce a simplified formula that disagrees with the input formula, but only on a set of points that is measure zero in the space of all variable assignments—a similar idea as the one behind the so called generic quantifier elimination [D_2SW98, $S_1S_5$03, $D_2G_0$04]. By allowing SLFQ this limited degree of error, the time and space requirements of its computations can be dramatically reduced, and in some cases simpler formulae may be found. For situations in which variables have physical interpretations, allowing this limited error makes sense, since no physical parameter can be controlled precisely enough to be constrained to a measure zero set.

The following examples, which are taken from [$B_1E_2N_2W_1$04], demonstrate how some of these switches affect SLFQ's results. The following calls are with the same input formula.

```
% cat infile
[r - t > 0 /\ r + t > 0] \/ [r - t < 0 /\ r + t > 0]
% ./slfq infile -q
[ r - t /= 0 /\ r + t > 0 ]
% ./slfq infile -q -a "r>0 /\ t>0"
r - t /= 0
% ./slfq infile -q -a "r>0 /\ t>0" -F
TRUE
```

The first call just simplifies the given formula ("-q" tells SLFQ to run in a "quiet" mode). In the second call both variables are assumed to be positive. In the third call both variables are assumed to be positive and SLFQ is allowed to produce a simplified formula that disagrees with the input, but only on a measure zero subset of the space of all variable assignments.

Refer to §3.2 to see SLFQ applied to large, complex input formulae that arise from applying the Routh-Hurwitz criterion to parameterized equilibrium points, and which can be reduced by SLFQ to small and meaningful formulae.

3 Investigating Equilibrium Points

Formally in the following we will only deal with time independent dynamical systems given by polynomial vector fields. This class covers a wide range of practical applications. Moreover, we can in a certain sense define some functions arising as solutions of differential equations by polynomial vector fields of a higher-dimension.

For a given equilibrium point $\underline{x}$ of a C^∞ vector field f the study of the system near this point is classically done by Taylor expanding f near $\underline{x}$ and considering at first the linear system

$$\frac{\mathrm{d}}{\mathrm{d}t}\zeta = D(f)(\underline{x}) \cdot \zeta, \qquad (5)$$

where $D(f)(\underline{x})$ is the Jacobian matrix of f at the point $\underline{x}$. When the matrix $D(f)(\underline{x})$ is *hyperbolic*, i.e., it has no eigenvalue with zero real part, then the stability study of the nonlinear system near the point $\underline{x}$ reduces to the study of the stability of the linear system near the origin 0. In the presence of eigenvalues with zero real part, the linear system gives only partial information about the local dynamics of the nonlinear system near the point $\underline{x}$. In fact, the local behavior near $\underline{x}$ of the nonlinear system depends on the higher order terms of the Taylor expansion of f near the point $\underline{x}$. However, the number of eigenvalues with zero real part of $D(f)(\underline{x})$ remains a fundamental invariant in the study of the topological nature of the local dynamics near the point $\underline{x}$. A systematic way to deal with non-hyperbolic situations is to use center manifold techniques and normal forms theory (see e.g., [$G_3H_3$90, $C_2H_1$96]).

3.1 Existence of equilibria points

Given the parametric nature of the system of differential equations even the question of the existence of equilibria points is a non-trivial question. For many applications this question not only reduces to the one of multi-dimensional equation solving, but to the one of solving the equations with inequality conditions.

Although this is a conceptually simple idea the following example might show some of the power of currently available systems for quantifier-elimination on real closed fields.

An example.
As an example we will take a system arising in epidemic modeling. In this context inequality constraints arise naturally. Consider the SEIRS model [L$_1$v95], which has also been investigated by quantifier-elimination methods in [B$_1$E$_2$N$_2$W$_1$06].

The SEIRS-model for the transmission of infectious diseases is given by the following system of 4 ordinary differential equations:

$$\frac{d}{dt}S = \mu + \gamma R - \mu S - \beta I S \tag{6}$$

$$\frac{d}{dt}E = \beta I S - (\mu + \sigma)E \tag{7}$$

$$\frac{d}{dt}I = \sigma E - (\nu + \mu)I \tag{8}$$

$$\frac{d}{dt}R = \nu I - (\mu + \gamma)R \tag{9}$$

The informal meaning of the variables and parameters is as follows:

S	susceptibles
E	exposed (not yet infectious)
I	infectious
R	recovered (currently immune)
β	transmission parameter
μ	birth rate = mortality rate
σ	rate of change from exposed to infectious
γ	rate of loss of immunity
ν	rate of loss of infectiousness

A point in $SEIR$-space is an equilibrium point if

$$0 = \mu + \gamma R - \mu S - \beta I S \;\land\; 0 = \beta I S - (\mu + \sigma)E \;\land$$
$$0 = \sigma E - (\nu + \mu)I \;\land\; 0 = \nu I - (\mu + \gamma)R \tag{10}$$

and represents an endemic state if

$$S > 0 \;\land\; E > 0 \;\land\; I > 0 \;\land\; R > 0. \tag{11}$$

Therefore, there is an endemic equilibrium for the SEIRS-model if there exist real numbers S, E, I, R such that both formulae (10) and (11) hold.

Note that, as there are several different specialized and general quantifier elimination methods available, it is by no means clear cut as to how this is best done. In situations in which it applies, the method of virtual term substitution is generally much faster than CAD-based quantifier elimination. On the other hand, its output formulae are often extremely large. For this example virtual term substitution does apply so the approach discussed in [$B_1E_2N_2W_1$06] is to perform quantifier elimination by virtual term substitution and simplify the result using SLFQ.

The input to REDLOG is the following:

```
A := mu+gamma*R - mu*S - beta*J*S = 0
    and beta*J*S - (mu+sigma)*F = 0
    and sigma*F - (nu+mu)*J = 0
    and nu*J - (mu+gamma)*R = 0;

F := ex({S,F,J,R}, A and F > 0 and J > 0 and R > 0 and S > 0);

C :={beta > 0, nu > 0, sigma > 0 , gamma > 0, mu > 0};

G := rlqe(F,C);
```

Note that variables E and I, which have special meaning in Reduce, have been replaced with F and J respectively. REDLOG computes a quantifier-free equivalent formula G consisting of 25 atomic formulae. Given the assumptions on the parameters, SLFQ simplifies this to

$$\sigma\beta - \sigma\nu - \mu\nu - \mu\sigma - \mu^2 > 0 \qquad (12)$$

with the following input:

```
slfq -a "beta > 0 /\ nu > 0 /\ sigma > 0 /\
gamma > 0 /\ mu > 0" G
```

Notice that this formula does not contain all parameters; there is no dependency on γ, the rate of loss of immunity. Using the assumption on the parameter—using that σ is positive—this condition is equivalent to

$$\beta > \frac{(\nu + \mu)(\sigma + \mu)}{\sigma} \qquad (13)$$

the threshold condition obtained by "hand computations" in the epidemiological literature.

3.2 Testing stability for equilibrium points

Let $f(u,x) = (f_1, \ldots, f_n)$ be a parameterized vector field, where $f_i \in \mathbb{R}[u,x]$ are polynomials of degree $\leq d$, $x = (x_1, \ldots, x_n)$ is a list of variables and $u = (u_1, \ldots, u_k)$ is a list of parameters. Let us consider the autonomous ordinary differential system

$$\frac{d}{dt}x = f(u,x) \tag{14}$$

and let us denote by $\Phi_t(u,x)$ the flow generated by the vector field f. A good place to start the study of the nonlinear system $\frac{d}{dt}x = f(u,x)$ is to find its equilibrium points, which are given by the equation

$$f(u,x) = 0. \tag{15}$$

If the list of parameters u is given a value $\underline{u} \in \mathbb{R}^k$, and $(\underline{u}, x)$ is an equilibrium point of the specialized nonlinear system $\frac{d}{dt}x = f(\underline{u},x)$, the study of the behavior of the flow $\Phi_t(\underline{u}, x)$ when starting near the equilibrium point $(\underline{u}, \underline{x})$ is classically done using the linear system

$$\frac{d}{dt}\zeta = D(f)(\underline{u}, \underline{x}) \cdot \zeta \tag{16}$$

where $D(f)(\underline{u}, \underline{x})$ is the Jacobian matrix of the vector field $f(\underline{u}, x)$ at the point $\underline{x}$. The flow generated by this linear system is then $e^{tD(f)(\underline{u},\underline{x})} \cdot \zeta = D(\Phi_t)(\underline{u}, \underline{x}) \cdot \zeta$.

A fundamental result due to Hartman and Grobmann (see, e.g., [A$_1$73]) states that in the case of *hyperbolic* equilibrium points, i.e. the matrix $D(f)(\underline{u}, \underline{x})$ has no eigenvalue with zero real part, the nonlinear flow has the same behavior near the equilibrium point $(\underline{u}, \underline{x})$ as the linear flow near the origin 0. In particular, the nonlinear flow $\Phi_t(u,x)$ is asymptotically stable near the equilibrium point $(\underline{u}, \underline{x})$ if and only if all the eigenvalues of the matrix $D(f)(\underline{u}, \underline{x})$ have negative real part.

According to the well known *Routh-Hurwitz criterion*, see e.g. [H$_4$L$_0$S$_3$97] this last condition is equivalent to the signs conjunction

$$\Delta_1(\underline{u}, \underline{x}) > 0 \wedge \cdots \wedge \Delta_n(\underline{u}, \underline{x}) > 0, \tag{17}$$

where the $\Delta_i(\underline{u}, \underline{x})$'s are the Hurwitz determinants associated to the characteristic polynomial of the matrix $D(f)(\underline{u}, \underline{x})$.

As the nonlinear system $\frac{d}{dt}x = f(u,x)$ is parameterized, a natural question is to ask for which values $\underline{u}$ of the parameter u the specialized system $\frac{d}{dt}x = f(\underline{u},x)$ is asymptotically stable near all its equilibrium points. This can be symbolically expressed by the following first-order formula:

$$\forall x \; (f(u,x) = 0 \Rightarrow \Delta_1(u,x) > 0 \wedge \cdots \wedge \Delta_n(u,x) > 0). \tag{18}$$

One can also ask for which values $\underline{u}$ of the parameter u the specialized system $\dot{x} = f(\underline{u}, x)$ is asymptotically stable near at least one of its equilibrium points:

$$\exists x \, (f(u,x) = 0 \wedge \Delta_1(u,x) > 0 \wedge \cdots \wedge \Delta_n(u,x) > 0). \tag{19}$$

These questions, as many others, are thus reduced to quantifier elimination problems for first-order formulae in the language of real closed fields.

Stability of specific parameterized equilibria points.

In some applications there exist specific equilibrium points for all parameter values of interest. However, the stability of these equilibrium points depends on the parameters. Although these problem is not a quantifier elimination problem per se, as applying the Routh-Hurwitz criterion to the Jacobian at the specific equilibrium point already does contain the parameters only and thus there are no quantified variables.

Nevertheless, the formula obtained by the Routh-Hurwitz criterion is huge and beyond human comprehension in general, whereas in many cases arising from applications it describes a rather simple object. The CAD-based simplification techniques realized in the SLFQ program (see §2.2) have proven to be a very useful tool also for this purpose. This statement shall be exemplified by the following example, which is taken from [$B_1E_2N_2W_1$06], where also some subtleties related to the possibility of being a non-hyperbolic equilibrium point are discussed.

Consider the SEIRS-model from above. For all parameter values it has the equilibrium point $S = 1, E = 0, I = 0, R = 0$, which can be interpreted as the "disease free equilibrium". Applying the Routh-Hurwitz criterion to the Jacobian at this point gives a quantifier-free formula containing a large polynomial of degree 9 consisting of 203 terms.

Using the positivity condition on all parameters SLFQ can simplify this large formula to the following one:

$$\sigma\beta - \sigma\nu - \mu\nu - \mu\sigma - \mu^2 < 0. \tag{20}$$

The required computation time has been 0.15 seconds in SLFQ. Notice that this formula coincides with the negation of the formula asking for "endemic equilibria" for the SEIRS model (12) modulo the measure zero set involving equality. The fact that the condition for the local stability of the disease-free equilibrium and the existence of an endemic equilibrium partition the space of valid parameters tells us that there are no parameter values that produce a so called "sub-threshold endemic equilibrium" for these models cf. [$B_1E_2N_2W_1$06, H_0v97, v$W_0$02, K_0V00].

Remark on "hand computations" done in the epidemiological literature. In the epidemiological literature the stability of the "disease free equilibrium" is very often the starting point for computing threshold conditions. However, very often the threshold conditions are formulated using the concept of *basic reproduction ratio* R_0, which denotes the number of secondary infections from each infected individual. This concept, first introduced by Dietz [D$_1$75], is also applicable for stochastic models. If R_0 exceeds one the disease it will reach an endemic stage, in which the disease is always present in the population, if it less than one it will die out.

In general experts in the field have calculated threshold conditions on the basis of R_0 "by hand", either solely by paper and pencil, or in part using computer algebra systems such as Maple or Mathematica as "symbolic calculators". However, in [C$_1$M$_3$W$_1$94] the QEPCAD system for quantifier elimination on real closed field has been used to parametrically investigate R_0 for a model of the epidemic of the AIDS disease.

Remark on vaccination policies. One of the parameters that can be influenced by change of behavior is the transmission parameter β (or its variants). By estimating the transmission parameter from empirical data and using the estimates for the other parameters out of the medical literature one can see how far away from the threshold one is. A symbolic computation of the threshold condition has the major advantage that the influences of changes on the parameters can be estimated much better than would be the case by numerical estimates. This might be one of the reasons why previously a lot of work involving tedious "hand calculations" have be spent to obtain symbolic threshold conditions.

Another possibility to come below the threshold is to reduce the number of susceptibles by vaccinations. If a proportion p of newborns is vaccinated then it can be easily shown by a simple change of variables for most of the models we are considering—such as the SEIRS model—that the effect on the dynamics is the same as if in the original model the transmission parameter β is replaced by $\beta(1-p)$. We refer to [E$_0$R$_5$B$_0$G$_2$00] for the details in the case of the SEIR model. Thus by vaccinating a sufficiently high fraction of susceptibles it is possible to avoid infections also in the group of remaining susceptibles. In the case of RSV epidemics [W$_1$W$_2$M$_2$01, N$_2$W$_1$03] the numerical value of the threshold for β is about 35, when using the disease specific values for average latency period, average duration of infectiousness and the birth-rates for developed countries. The estimates of β for pre-vaccination epidemics are ranging from 70 to 240 for different locations (and variations of the model). Thus the critical percentages of vaccinations are ranging from 50 % to about 85 % for this example.

3.3 Testing Stability by Quantifier Elimination

A wide variety of stability questions for differential equations –ordinary differential equations, ordinary discrete difference equations, initial-boundary value problems for partial differential equations, and semi-discrete equations– is reduced to first-order formulae in the language of the ordered-field of the reals in [$H_4L_0S_397$]. Also the local stability of Runge-Kutta discretizations is investigated.

In [$H_4L_0S_397$] the quantifier elimination is performed by QEPCAD. Simple problems could be solved in a few seconds, and most textbook examples in some minutes or at least a few hours of computation time. However, they found relatively modest problems that were beyond the reach of direct solutions by QEPCAD.

3.4 Bifurcations

As parameters are varied in a given parameterized dynamical system, the phase portrait may undergo qualitative changes. The parameter values where such changes occur and the corresponding changes are called bifurcations. One of the main goals of bifurcations theory is the location of those parameters regions in which a given dynamical system displays the desired behavior.

The simplest bifurcations take place at equilibrium points, and they are called local bifurcations. For a given equilibrium $(\underline{u}, \underline{x})$ a bifurcation may arise when the matrix $D(f)(\underline{u}, \underline{x})$ has some eigenvalues with zero real part. In this case, and for (u, x) close enough to $(\underline{u}, \underline{x})$ radically new dynamical behavior can occur. For example, equilibrium points can be created or destroyed, and even new orbits such as periodic or quasi-periodic ones can be created.

In general, for an n-dimensional autonomous system there are many distinct bifurcating situations depending on the number of eigenvalues with zero real part. A partial classification of local bifurcations is done by using the concept of codimension. For example, codimension one bifurcations are of two kinds: either the Jacobian matrix has a zero eigenvalue or a pair of pure imaginary eigenvalues. In the first case we have a *Saddle-node* bifurcation and the second case corresponds to the so-called *Hopf* bifurcation.

At a Saddle-node bifurcation a pair of equilibrium points coalesce one another. On one side of the bifurcation in the parameter space there are two equilibrium points, and they disappear on the other side. When the system undergoes a Hopf bifurcation at a equilibrium point $(\underline{u}, \underline{x})$, and the parameters u are subjected to small perturbations, the original equilibrium point $(\underline{u}, \underline{x})$ moves analytically in terms of u and no new equilibrium is created in the neighborhood. However, if the imaginary eigenvalues of the linearized system move away from the imaginary axis, one expects the equilibrium point to change its stability type. This change is typically marked by the appearance of a small periodic orbit encircling the equilibrium point as stated by the Poincaré-Andronov-Hopf theorem, see e.g. [C$_2$H$_1$96]. The local dynamics near an equilibrium point with Hopf bifurcation cannot be determined by the linear approximation of the vector field. In fact, depending on the nonlinear terms of f, the equilibrium point can be unstable, stable or even asymptotically stable.

Semi-algebraic characterizations of Hopf bifurcations.
El Kahoui and Weber [E$_2$W$_1$00] showed that Hopf bifurcation fixed points have a semi-algebraic description. The description is carried out by use of the Hurwitz determinants. Applying techniques from the theory of subresultant sequences and of Gröbner bases they could to come up with efficient reductions, which lead to quantifier elimination questions that can often be handled by existing quantifier elimination packages.

The result of the reduction is as follows: For a parameterized vector field $f(u, x)$ and the autonomous ordinary differential system associated with the semi-algebraic description of the set of parameters values for which a Hopf bifurcation (with empty unstable manifold) occurs for the system can be expressed by the following first-order formula:

$$\exists x (f_1(u,x) = 0 \wedge f_2(u,x) = 0 \wedge \cdots \wedge f_n(u,x) = 0$$
$$\wedge\, a_n > 0 \wedge \Delta_{n-1} = 0 \wedge \Delta_{n-2} > 0 \wedge \cdots \wedge \Delta_1 > 0). \qquad (21)$$

In this formula a_n is $(-1)^n$ times the Jacobian determinant of the matrix $Df(u,x)$, and the Δ_i's are the ith Hurwitz determinants of the characteristic polynomial of the same matrix $Df(u,x)$.

It is also possible to give a semi-algebraic description of the set of parameters values for which the system undergoes at most a Hopf bifurcation and all the rest of its eigenvalues are in the left half-plane. In terms of logical formulae this can be expressed as follows:

$$\forall x ((f_1(u,x) = 0 \wedge \cdots \wedge f_n(u,x) = 0) \Rightarrow$$
$$(a_n > 0 \wedge \Delta_{n-1} \geq 0 \wedge \Delta_{n-2} > 0 \wedge \cdots \wedge \Delta_1 > 0)). \qquad (22)$$

Example of computations for Hopf bifurcation fixed points.
The following examples are taken from [E$_2$W$_1$00]. The quantified formula expressing the condition for the Hopf bifurcation fixed point is computed in Maple. Using a software component architecture—which was also used to connect other mathematical services [W$_1$K$_1$E$_1$98, G$_1$K$_1$M$_4$W$_1$99]—the quantifier elimination was then performed by a combination of REDLOG

and QEPCAD for simplifying the results of REDLOG. In [E₂W₁00] and in more detail in [E₂W₁02] it is also shown how to simplify the (partially) quantified formulae by Gröbner basis techniques.

Canonical example for Hopf bifurcation. The following planar system can be viewed as the typical system undergoing a Hopf bifurcation at the origin $(0,0)$:

$$\frac{d}{dt}x(t) = (du + a(x(t)^2 + y(t)^2))x(t) - (w + cu + b(x(t)^2 + y(t)^2))y(t), \quad (23)$$

$$\frac{d}{dt}y(t) = (w + cu + b(x(t)^2 + y(t)^2))x(t) + (du + ax(t)^2 + y(t)^2))y(t). \quad (24)$$

If a given n-dimensional system $\frac{d}{dt}x = f(u,x)$, with a real parameter u undergoes a Hopf bifurcation at the origin when $u = 0$, then using normal forms techniques after projection on the center manifold (see [C₂H₁96] for normal form techniques), one reduces to study a system of the form above with a, b, c, d, w given specified values.

The computation reported in [E₂W₁00] gives the following signs conditions for the system to undergo a Hopf bifurcation with empty unstable manifold:

$$0 < d^2u^2 + w^2 + c^2u^2 + 2wcu \;\wedge\; -2\,du = 0. \quad (25)$$

Notice that in the case $d \neq 0$ the above formula is equivalent to the following simple formula:

$$u = 0. \quad (26)$$

A system arising in epidemiology. The following example is from [L₁v95]. In this research paper the investigation on the existence of Hopf bifurcations is an important part. The differential equations come from epidemiological models with varying population size and dose-dependent latency period.

The following parameterized system of differential equations describes the so called SEIS models of [L₁v95]:

$$\frac{d}{dt}s(t) = b - b\,s(t) + \gamma\,i(t) - (\beta - \alpha)\,s(t)\,i(t), \quad (27)$$

$$\frac{d}{dt}e(t) = -b\,e(t) + \beta\,s(t)\,i(t) + \alpha\,i(t)\,e(t) - \varepsilon\,e(t), \quad (28)$$

$$\frac{d}{dt}i(t) = -(b + \gamma + \alpha)\,i(t) + \alpha\,i(t)^2 + \varepsilon\,e(t). \quad (29)$$

In [L₁v95] it is proved that this system does not have a Hopf bifurcation for any parameter values for the epidemiological relevant cases: all parameters and variables are positive and $s(t) + e(t) + i(t) = 1$.

In [E$_2$W$_1$00] it is reported that the quantifier elimination programs did not succeed for the general system with 3 variables and 5 parameters within one day of computation time. When specializing 4 of the 5 parameters with various values, the combination of REDLOG and QEPCAD returned the correct result, namely `false`, within some seconds of computation time.

Using the refined implementations of the methods in the current version of REDLOG a recently performed quantifier elimination on the formula was successful within some seconds of computation time. However, the computation resulted in a large quantifier free formula in the 5 parameters (consisting of 236 atomic subformulae). This formula should be equivalent to `false`, i.e., no fulfilling instances of the variables should exist. Unfortunately, SLFQ was not able to do the simplification of this formula. Because of the high degree of the polynomials quantifier elimination on the existential quantification of this formula can not use the currently available virtual substitution methods. A CAD based quantifier elimination on the existential quantification of this formula had to go through the entire tree to find that there are no fulfilling instances. This task was also attempted but not successfully finished by REDLOG within two days of computation time and 512 MB of main memory.

Lorenz system. The famous "Lorenz System" [L$_3$63, G$_3$H$_3$90, R$_1$A$_0$87] is given by the following system of ODEs:

$$\frac{d}{dt}x(t) = \alpha\,(y(t) - x(t)) \tag{30}$$

$$\frac{d}{dt}y(t) = r\,x(t) - y(t) - x(t)\,z(t) \tag{31}$$

$$\frac{d}{dt}z(t) = x(t)\,y(t) - \beta\,z(t) \tag{32}$$

It is named after Edward Lorenz at MIT, who first investigated this system as a simple model arising in connection with fluid convection.

After imposing positivity conditions on the parameters, the following answer is reported in [E$_2$W$_1$00] (requiring some seconds of computation time then):

$$\alpha^2 + \alpha\beta - \alpha r + 3\alpha + \beta r + r = 0 \land$$
$$\alpha r - \alpha - \beta^2 - \beta \geq 0 \land$$
$$2\alpha - 1 \geq 0 \land \beta > 0. \tag{33}$$

Thus a simple closed form description involving three free parameters has been found algorithmically by the use of quantifier elimination on real closed fields, which coincides (after some elementary transformation) with the result of a hand computation given in [G$_3$H$_3$90].

4 Testing for Ellipticity of Partial Differential Equations

An important task in the theory of partial differential equations [R$_2$R$_4$93] is their classification into elliptic and hyperbolic systems. The distinction of these two classes is fundamental not only for the theory but also for the numerical analysis, as it decides what kind of conditions (initial or boundary) should be imposed. Furthermore, their solutions behave very differently. Notice that we do not treat parabolic systems separately here. At the coarse level of the discussion here, parabolicity is a degenerate case of hyperbolicity and only appears when finer notions like strict hyperbolicity are introduced.

For elliptic systems boundary value problems are usually well-posed and their solutions show typically a very high regularity. From an application point of view, they model stationary problems. In hyperbolic systems a distinguished direction ("time") exists and one considers initial value problems for them; thus they represent models for evolutionary problems. Even for regular data their solutions may exhibit shocks.

There are several different notions of ellipticity: classical ellipticity, Petrowsky ellipticity and the notion of ellipticity introduced by Douglis and Nirenberg [D$_3$N$_1$55], which will be called DN-ellipticity in the following. DN-ellipticity is the most general of these.

Ellipticity of a system in general—and DN-ellipticity in particular—is defined at a point in the space of independent variables. Thus a given system may be elliptic at some points and not at other. Therefore, the answer to the question whether a system is elliptic will be a description of the region in the space of independent variables in which the system is elliptic.

Seiler and Weber [S$_2$W03] showed that when the coefficients on the given system of PDEs are algebraic or rational functions, the problem of determining DN-ellipticity at a point can be phrased as a decision problem in the first-order theory of the ordered field of the reals. Characterizing the regions in which the system is DN-elliptic is a problem of quantifier elimination on real closed fields. Notice that the quantifier elimination allows finitely many symbolic constants that can also appear in the input system of PDEs.

The definition of DN-ellipticity involves the introduction of a set of integer weights. In the reduction suggested in [S$_2$W03] these weights become part of the quantified formulae that are produced from an input system of PDEs. So this formulations leads to a mixed real-integer quantifier elimination problem of a type that can be solved by the methods of Weispfenning [W$_3$99]. In [S$_2$W03] it is noted that (because of the special form of the problem) an approach in which integer weights are treated as real variables produces the same result as the mixed real-integer quantifier elimination.

In [B₁dN₀05] algorithms for determining regions of DN-ellipticity based on solving real quantifier elimination without the "non-intrinsic" variables are announced. Given the dependence on the complexity of quantifier elimination on the number of variables and quantifier alternation, this is a substantial performance gain.

Tricomi equation. The following example is taken from [S₂W03]. Consider the following variation of Laplace's equation, the *Tricomi equation*:

$$\frac{\partial^2 u}{\partial y^2} - y\frac{\partial^2 u}{\partial x^2} = 0 \tag{34}$$

Obviously, it is only elliptic for $y < 0$. The logical formula for its DN-ellipticity is obtained in MUPAD as follows:

```
LDF := Dom::LinearDifferentialFunction(Vars=[[x,y],[u]],
                                Rest=[Types="Indep"]):
tricomi := LDF(u([y,y])-y*u([x,x])): LDF::ellCond(tricomi)
```

The first input line creates a domain for linear differential functions in the two independent variables x, y and the unknown function u. The coefficients may be arbitrary expressions in x and y. The second line defines the Tricomi equation using an abbreviated syntax for the derivatives. Finally, the method `ellCond` is called which generates the formula using default names for the weights and vectors (with additional optional arguments the names for these variables may be prescribed and a file for the output specified).

The result is the following first-order formula (in REDLOG syntax):

```
ex(s1,
ex(t1,
  (s1 <= 0) and
  (0 <= t1) and
  all(xi1,
  all(xi2,
  all(w1,
    not(
      ((xi1 <> 0) or (xi2 <> 0)) and
      ((w1 <> 0)) and
      ((((s1 + t1 = 2)) impl (-w1*(xi1**2*y - xi2**2) = 0)))
    )
  )))
))
```

The generated formula has one free parameter (the variable y, as we are dealing with a variable coefficient equation), an inner block of three universally quantified variables and an outer block of two existentially quantified variables and consists of 7 atomic subformulae. The corresponding quantifier free formula was found by REDLOG in less than 1 sec of computation time and is exactly the one we expect, namely $y < 0$.

In [S_2W03] successful computations on examples involving 7 existentially and 6 universally quantified variables consisting of 68 atomic subformulae have been reported.

References.

[$A_1$73]   V. I. Arnold. *Ordinary Differential Equations*. M.I.T Press Cambridge, 1973.

[$A_2C_4M_1$84]   D. S. Arnon, G. E. Collins, and S. McCallum. Cylindrical algebraic decomposition I: The basic algorithm. *SIAM Journal on Computing*, 13:865–877, 1984.

[$B_1$98]   C. W. Brown. Simple CAD construction and its applications. *Journal of Symbolic Computation*, 31(5):521–547, 2001.

[$B_1dN_0$05]   C. W. Brown, S. de Vlaming, and G. Nakos. Quantifier elimination and the ellipticity of systems of partial differential equations. In A. Dolzmann, A. Seidl, and T. Sturm, editors, *Algorithmic Algebra and Logic (A3L 2005)*, pages 55–58, Passau, Germany, 2005. Books on Demand.

[$B_1E_2N_2W_1$04]   C. W. Brown, M. El Kahoui, D. Novotni, and A. Weber. Algorithmic methods for computing threshold conditions in epidemic modelling. In V. G. Ganzha, E. W. Mayr, and E. V. Vorozhtsov, editors, *Computer Algebra in Scientific Computing (CASC '04)*, pages 51–60, St. Petersburg, Russia, 2004.

[$B_1E_2N_2W_1$06]   C. W. Brown, M. El Kahoui, D. Novotni, and A. Weber. Algorithmic methods for investigating equilibria in epidemic modeling. *Journal of Symbolic Computation*, 41(11):1157-1173, 2006.

[C_0J98]   B. F. Caviness and J. R. Johnson, editors. *Quantifier Elimination and Cylindrical Algebraic Decomposition*. Springer-Verlag, 1998.

[$C_1M_3W_1$94]   C. Chauvin, M. Müller, and A. Weber. An application of quantifier elimination to mathematical biology. In J. Fleischer, J. Grabmeier, F. W. Hehl, and W. Küchlin, editors, *Computer Algebra in Science and Engineering*, pages 287–296, Bielefeld, Germany, 1994. Zentrum für Interdisziplinäre Forschung, World Scientific.

[$C_2H_1$96]   S. N. Chow and J. K. Hale. *Methods of Bifurcation Theory*, volume 251 of *Grundlehren der Mathematischen Wissenschaften*. Springer-Verlag, second edition, 1996.

[$C_3$69]   P. J. Cohen. Decision procedures for real and p-adic fields. *Communications in Pure and Applied Logic*, 25:213–231, 1969.

[$C_4$75]   G. Collins. Quantifier elimination for real closed fields by cylindrical algebraic decomposition. In *Second GI Conference on Automata*

	Theory and Formal Languages, volume 33 of *Lecture Notes in Computer Science*, pages 134–183. Springer-Verlag, 1975.
[$C_4H_4$91]	G. Collins and H. Hong. Partial cylindrical algebraic decomposition for quantifier elimination. *Journal of Symbolic Computation*, 12(3):299–328, 1991.
[$D_0H_2$88]	J. H. Davenport and J. Heintz. Real quantifier elimination is doubly exponential. *Journal of Symbolic Computation*, 5:29–35, 1988.
[$D_1$75]	K. Dietz. Transmission and control of arbovirus diseases. In D. Ludwig and K. L. Cooke, editors, *Epidemiology*, pages 104–121. SIAM, 1975.
[$D_2$99]	A. Dolzmann. Solving geometric problems with real quantifier elimination. In X. S. Gao, D. Wang, and L. Yang, editors, *Automated Deduction in Geometry*, volume 1669 of *Lecture Notes in Computer Science*, pages 14–29. Springer-Verlag, 1999.
[$D_2G_0$04]	A. Dolzmann and L. A. Gilch. Generic Hermitian quantifier elimination. In B. Buchberger and J. A. Campbell, editors, *Artificial Intelligence and Symbolic Computation, 7th International Conference, AISC 2004*, volume 3249 of *Lecture Notes in Computer Science*, pages 80–93, Linz, Austria, 2004. Springer-Verlag.
[$D_2S_1S_5$04]	A. Dolzmann, A. Seidl, and T. Sturm. Efficient projection orders for CAD. In *ISSAC '04: Proceedings of the 2004 international symposium on Symbolic and algebraic computation*, pages 111–118, New York, NY, USA, 2004. ACM Press.
[$D_2S_5$97]	A. Dolzmann and T. Sturm. Simplification of quantifier-free formulae over ordered fields. *Journal of Symbolic Computation*, 24(2):209–231, 1997. Special Issue on Applications of Quantifier Elimination.
[D_2SW98]	A. Dolzmann, T. Sturm, and V. Weispfenning. A new approach for automatic theorem proving in real geometry. *Journal of Automated Reasoning*, 21(3):357–380, 1998.
[$D_3N_1$55]	A. Douglis and L. Nirenberg. Interior estimates for elliptic systems of partial differential equations. *Communications on Pure and Applied Mathematics*, 8:503–538, 1955.
[$E_0R_5B_0G_2$00]	D. J. Earn, P. Rohani, B. M. Bolker, and B. T. Grenfell. A simple model for complex dynamical transitions in epidemics. *Science*, 287(5453):667–670, 2000.
[$E_2W_1$00]	M. El Kahoui and A. Weber. Deciding Hopf bifurcations by quantifier elimination in a software-component architecture. *Journal of Symbolic Computation*, 30(2):161–179, 2000.
[$E_2W_1$02]	M. El Kahoui and A. Weber. Symbolic equilibrium point analysis in parameterized polynomial vector fields. In V. G. Ganzha, E. W. Mayr, and E. V. Vorozhtsov, editors, *Computer Algebra in Scientific Computing (CASC 2002)*, pages 71–83, Yalta, Ukraine, 2002.
[$FR_0$79]	J. Ferrante and C. W. Rackoff. *The Computational Complexity of Logical Theories*, volume 718 of *Lecture Notes in Computer Science*. Springer-Verlag, 1979.
[$G_1K_1M_4W_1$99]	M. Göbel, W. Küchlin, S. Müller, and A. Weber. Extending a Java based framework for scientific software-components. In V. G. Ganzha, E. W. Mayr, and E. V. Vorozhtsov, editors, *Computer Algebra in Scientific Computing (CASC '99)*, pages 207–222, München, 1999. Springer-Verlag.

[G_3H_390] J. Guckenheimer and P. Holmes. *Nonlinear Oscillations, Dynamical Systems, and Bifurcations of Vector Fields*, volume 42 of *Applied Mathematical Sciences*. Springer-Verlag, 1990.

[H_0v97] K. P. Hadeler and P. van den Driessche. Backward bifurcation in epidemic control. *Mathematical Biosciences*, 146(1):15–35, 1997.

[$H_4L_0S_397$] H. Hong, R. Liska, and S. Steinberg. Testing stability by quantifier elimination. *Journal of Symbolic Computation*, 24(2):161–187, 1997.

[K_0V00] C. M. Kribs-Zaleta and J. X. Velasco-Hernandez. A simple vaccination model with multiple endemic states. *Mathematical Biosciences*, 164:183–201, 2000.

[L_1v95] W.-M. Liu and P. van den Driessche. Epidemiological models with varying population size and dose-dependent latent period. *Mathematical Biosciences*, 128:57–69, 1995.

[L_2W_393] R. Loos and V. Weispfenning. Applying linear quantifier elimination. *The Computer Journal*, 5:450–462, 1993.

[L_363] E. N. Lorenz. Deterministic non-periodic flow. *J. Atmos. Sci.*, 20:130–141, 1963.

[M_003] A. Macintyre. Model theory: Geometrical and set-theoretic aspects and prospects. *The Bulletin of Symbolic Logic*, 9(2):197–212, 2003.

[N_2W_103] D. Novotni and A. Weber. A stochastic method for solving inverse problems in epidemic modelling. In F. Valafar and H. Valafar, editors, *Proceedings of the International Conference on Mathematics and Engineering Techniques in Medicine and Biological Sciences (METMBS '03)*, pages 467–473, Las Vegas, USA, 2003. CSREA Press.

[R_1A_087] R. H. Rand and D. Armbruster. *Perturbation Methods, Bifurcation Theory and Computer Algebra*, volume 65 of *Applied Mathematical Sciences*. Springer-Verlag, 1987.

[R_2R_493] M. Renardy and R. C. Rogers. *An Introduction to Partial Differential Equations*, volume 13 of *Texts in Applied Mathematics*. Springer-Verlag, New York, 1993.

[R_392] J. Renegar. On the computational complexity and geometry of the first-order theory of the reals. *Journal of Symbolic Computation*, 13(3):255–300, 1992.

[S_054] A. Seidenberg. A new decision method for elementary algebra and geometry. *Annals of Mathematics*, 60:365–374, 1954.

[S_1S_503] A. Seidl and T. Sturm. A generic projection operator for partial cylindrical algebraic decomposition. In *ISSAC '03: Proceedings of the 2003 international symposium on Symbolic and algebraic computation*, pages 240–247, Philadelphia, PA, USA, 2003. ACM Press.

[S_2W03] W. M. Seiler and A. Weber. Deciding ellipticity by quantifier elimination. In V. G. Ganzha, E. W. Mayr, and E. V. Vorozhtsov, editors, *Computer Algebra in Scientific Computing (CASC '03)*, pages 347–355, Passau, Germany, 2003.

[S_403] C. Steinhorn. Tame topology and o-minimal structures. Available at http://cowles.econ.yale.edu/conferences/newdirect/lec/beare-steinhorn.pdf, 2003. Lecture notes taken by Brendan Beare.

[T51] A. Tarski. *A Decision Method for Elementary Algebra and Geometry*. University of California Press, Berkeley, second edition, 1951. Reprinted in [C_0J98].

[vW₀02]   P. van den Driessche and J. Watmough. Reproduction numbers and sub-threshold endemic equilibria for compartmental models of disease transmission. *Mathematical Biosciences*, 180:29–48, 2002.

[W₁K₁E₁98]   A. Weber, W. Küchlin, and B. Eggers. Parallel computer algebra software as a Web component. *Concurrency: Practice and Experience*, 10(11–13):1179–1188, 1998.

[W₁W₂M₂01]   A. Weber, M. Weber, and P. Milligan. Modeling epidemics caused by respiratory syncytial virus (RSV). *Mathematical Biosciences*, 172(2):95–113, 2001.

[W₃88]   V. Weispfenning. The complexity of linear problems in fields. *Journal of Symbolic Computation*, 5(1-2):3–27, 1988.

[W₃94]   V. Weispfenning. Quantifier elimination for real algebra—the cubic case. In *Proceedings of the international symposium on Symbolic and algebraic computation*, pages 258–263. ACM Press, 1994.

[W₃97]   V. Weispfenning. Quantifier elimination for real algebra – the quadratic case and beyond. *Applicable Algebra in Engineering, Communication and Computing*, 8(2):85–101, 1997.

[W₃98]   V. Weispfenning. A new approach to quantifier elimination for real algebra. In Caviness and Johnson [C₀J98], pages 376–392.

[W₃99]   V. Weispfenning. Mixed real-integer linear quantifier elimination. In *Proceedings of the 1999 International Symposium on Symbolic and Algebraic Computation*, pages 129–136. ACM Press, 1999.

Received: March 6, 2006;
In revised version: April 18, 2006;
Accepted by the editors: April 18, 2006.

www.ingramcontent.com/pod-product-compliance
Ingram Content Group UK Ltd.
Pitfield, Milton Keynes, MK11 3LW, UK
UKHW021250180426
11946UKWH00003B/60